Discoveries in the Ruins of Nineveh and Babylon

With Travels in Armenia, Kurdistan and the Desert: Being the Result of a Second Expedition Undertaken for the Trustees of the British Museum

VOLUME 1

AUSTEN HENRY LAYARD

CAMBRIDGE
UNIVERSITY PRESS

CAMBRIDGE UNIVERSITY PRESS

Cambridge, New York, Melbourne, Madrid, Cape Town, Singapore,
São Paolo, Delhi, Dubai, Tokyo

Published in the United States of America by Cambridge University Press, New York

www.cambridge.org
Information on this title: www.cambridge.org/9781108016773

© in this compilation Cambridge University Press 2010

This edition first published 1853
This digitally printed version 2010

ISBN 978-1-108-01677-3 Paperback

CAMBRIDGE LIBRARY COLLECTION

Books of enduring scholarly value

Archaeology

The discovery of material remains from the recent or the ancient past has always been a source of fascination, but the development of archaeology as an academic discipline which interpreted such finds is relatively recent. It was the work of Winckelmann at Pompeii in the 1760s which first revealed the potential of systematic excavation to scholars and the wider public. Pioneering figures of the nineteenth century such as Schliemann, Layard and Petrie transformed archaeology from a search for ancient artifacts, by means as crude as using gunpowder to break into a tomb, to a science which drew from a wide range of disciplines - ancient languages and literature, geology, chemistry, social history - to increase our understanding of human life and society in the remote past.

Discoveries in the Ruins of Nineveh and Babylon

Sir Austen Henry Layard (1817–1894) was one of the leading British archaeologists of the nineteenth century. His excavations provided important evidence about ancient Mesopotamia, particularly about the Assyrian civilisation, and his books – part travel writing, part specialised archeological studies – are beautifully evocative. First published in 1853, this two-volume study follows the earlier *Nineveh and its Remains* (1849). It describes Layard's second expedition to the Near East, in 1845, which led to the identification of Kouyunjik as the great Assyrian capital Nineveh. In this richly illustrated book, Layard focuses on the description and interpretation of ruins, as he tells of the discovery of the lost palace of the Assyrian king Sennacherib (eighth century BCE) in northern Iraq. Volume 1 is an account the excavations at Kouyunjik, and also describes a journey along the Khabur river in Syria, where Layard assesses the influence of Assyrian art on the region.

Cambridge University Press has long been a pioneer in the reissuing of out-of-print titles from its own backlist, producing digital reprints of books that are still sought after by scholars and students but could not be reprinted economically using traditional technology. The Cambridge Library Collection extends this activity to a wider range of books which are still of importance to researchers and professionals, either for the source material they contain, or as landmarks in the history of their academic discipline.

Drawing from the world-renowned collections in the Cambridge University Library, and guided by the advice of experts in each subject area, Cambridge University Press is using state-of-the-art scanning machines in its own Printing House to capture the content of each book selected for inclusion. The files are processed to give a consistently clear, crisp image, and the books finished to the high quality standard for which the Press is recognised around the world. The latest print-on-demand technology ensures that the books will remain available indefinitely, and that orders for single or multiple copies can quickly be supplied.

The Cambridge Library Collection will bring back to life books of enduring scholarly value (including out-of-copyright works originally issued by other publishers) across a wide range of disciplines in the humanities and social sciences and in science and technology.

NORTH-EASTERN FACADE AND GRAND ENTRA

Restored from a Sketch

John Murray, Albem

Ford & Wert. Hatton Garden

RANCE OF SENNACHERIB'S PALACE. (KOUYUNJIK)
etch by J. Fergusson, Esqre

Ibemarle   Street, 1852.

DISCOVERIES

IN THE RUINS OF

NINEVEH AND BABYLON;

WITH TRAVELS IN ARMENIA, KURDISTAN AND THE DESERT:

BEING THE RESULT OF A SECOND EXPEDITION

UNDERTAKEN FOR

THE TRUSTEES OF THE BRITISH MUSEUM.

BY AUSTEN H. LAYARD, M.P.

AUTHOR OF "NINEVEH AND ITS REMAINS

" For thou hast made of a city an heap; of a defenced city a ruin: a palace of strangers to be no city; it shall never be built."—ISAIAH, XXV. 2

WITH MAPS, PLANS, AND ILLUSTRATIONS.

IN TWO PARTS. — PART I.

LONDON:

JOHN MURRAY, ALBEMARLE STREET.

1853.

DIRECTION TO BINDER.

When this Work is arranged for Two Volumes, the Second will
commence with Chapter XVI.

TO

THE RIGHT HONORABLE

THE EARL GRANVILLE

This Volume is dedicated,

IN ADMIRATION OF HIS PUBLIC CHARACTER,

AND AS A GRATEFUL ACKNOWLEDGMENT OF MANY ACTS OF

PERSONAL FRIENDSHIP.

PREFACE.

MANY unavoidable delays have prevented the earlier publication of this volume. I can no longer appeal, as in the preface of my former work, to the indulgence of my readers on the score of complete literary inexperience; but I can express heartfelt gratitude for the kind and generous reception given, both by the press and the public, to my first labors. I will merely add, that the following pages were written at different periods, and amidst numerous interruptions but little favorable to literary occupations. This must be my apology, to a certain extent, for the many defects they contain.

Since the publication of my first work on the discoveries at Nineveh much progress has been made in deciphering the cuneiform character, and the contents of many highly interesting and important inscriptions have been given to the public. For these additions to our knowledge we are mainly indebted to the sagacity and learning of two English scholars, Col. Rawlinson and the Rev. Dr. Hincks. In making use of the results of their researches, I have not omitted to own the sources from which my information has been derived. I trust, also, that I have in no in-

stance availed myself of the labors of other writers, or of
the help of friends, without due acknowledgments. I have
endeavored to assign to every one his proper share in
the discoveries recorded in these pages.

I am aware that several distinguished French scholars,
amongst whom I may mention my friends, M. Botta and
M. de Saulcy, have contributed to the successful deciphering of the Assyrian inscriptions. Unfortunately I have
been unable to consult the published results of their inves-
tigations. If, therefore, I should have overlooked in any
instance their claims to prior discovery, I have to express
my regret for an error arising from ignorance, and not
from any unworthy national prejudice.

Doubts appear to be still entertained by many eminent
critics as to the progress actually made in deciphering the
cuneiform writing. These doubts may have been confirmed by too hasty theories and conclusions, which, on
subsequent investigation, their authors have been the first
to withdraw. But the unbiassed inquirer can scarcely
now reject the evidence which can be brought forward to
confirm the general accuracy of the interpretations of the
inscriptions. Had they rested upon a single word, or an
isolated paragraph, their soundness might reasonably have
been questioned; when, however, several independent in-
vestigators have arrived at the same results, and have not
only detected numerous names of persons, nations, and
cities in historical and geographical series, but have found
them mentioned in proper connection with events recorded
by sacred and profane writers, scarcely any stronger

evidence could be desired. The reader, I would fain hope, will come to this conclusion when I treat of the contents of the various records discovered in the Assyrian palaces.

I have endeavored to introduce into these pages as many illustrations from the sculptures as my limits would admit. I have been obliged to include the larger and more elaborate drawings of the bas-reliefs in a folio volume, which will form a second series of the Monuments of Nineveh, and will be published at the same time as the present work.

I trust it may not be inferred from any remark I have been induced to make in the following pages, that I have any grounds of personal complaint against the Trustees of the British Museum. From them I have experienced uniform courtesy and kindness, which I take this opportunity of acknowledging with gratitude; but I cannot, at the same time, forbear expressing a wish, felt in common with myself by many who have the advancement of national education, knowledge, and taste sincerely at heart, that that great establishment, so eminently calculated to promote this important end, should be speedily placed upon a new and more efficient basis.

To Mr. Thomas Ellis, who has enabled me to add to my work translations of inscriptions on Babylonian bowls, now for the first time, through his sagacity, deciphered; to those who have assisted me in my labors, and especially to my friend and companion, Mr. Hormuzd Rassam, to the Rev. Dr. Hincks, to the Rev.

S. C. Malan, who has kindly allowed me the use of his masterly sketches, to Mr. Fergusson, Mr. Scharf, and to Mr. Hawkins, Mr. Birch, Mr. Vaux, and the other officers of the British Museum, I beg to express my grateful thanks and acknowledgments.

London, January, 1853.

Ivory Ornament, from Nimroud.

CONTENTS.

CHAPTER I.

The Trustees of the British Museum resume Excavations at Nineveh. — Departure from Constantinople. — Description of our Party. — Cawal Yusuf. — Roads from Trebizond to Erzeroom. — Description of the Country. — Varzahan and Armenian Churches. — Erzeroom. — Reshid Pasha. — The Dudjook Tribes. — Shahan Bey. — Turkish Reform. — Journey through Armenia. — An Armenian Bishop. — The Lakes of Shailu and Nazik. — The Lake of Wan - - - - - Page 1

CHAP. II.

The Lake of Wan. — Akhlat. — Tatar Tombs. — Ancient Remains. — A Dervish. — A Friend. — The Mudir. — Armenian Remains. — An Armenian Convent and Bishop. — Journey to Bitlis. — Nimroud Dagh. — Bitlis. — Journey to Kherzan. — Yezidi Village - - - 23

CHAP. III.

Reception by the Yezidis. — Village of Guzelder. — Triumphal March to Redwan. — Redwan. — Armenian Church. — Mirza Agha. — The Melek Taous, or Brazen Bird. — Tilleh. — Valley of the Tigris. — Bas-reliefs. — Journey to Dereboun — to Semil. — Abde Agha. — Journey to Mosul. — The Yezidi Chiefs. — Arrival at Mosul. — Xenophon's March from the Zab to the Black Sea - - - - - - 42

CHAP. IV.

State of the Excavations on my Return to Mosul. — Discoveries at Kouyunjik. — Tunnels in the Mound. — Bas-reliefs representing Assyrian Conquests. — A Well. — Siege of a City. — Nature of Sculptures at Kouyunjik. — Arrangements for Renewal of Excavations. — Description of Mound. — Kiamil Pasha. — Visit to Sheikh Adi. — Yezidi Ceremonies. — Sheikh Jindi. — Yezidi Meeting. — Dress of the Women. — Bavian. — Ceremony of the Kaidi. — Sacred Poem of the Yezidi. — Their Doctrines. — Jerraiyah. — Return to Mosul - - - - 66

CHAP. V.

CHAP. VI.

CHAP. VII.

CHAP. VIII.

CHAP. IX.

CHAP. X.

CHAP. XI.

CHAP. XII.

CHAP. XIII.

CHAP. XIV.

CHAP. XV.

CHAP. XVI.

CHAP. XVII.

CHAP. XVIII.

CHAP. XIX.

CHAP. XX.

CHAP. XXI.

CHAP. XXII.

CHAP. XXIII.

CHAP. XXIV.

CHAP. XXV.

CHAP. XXVI.

LIST OF ILLUSTRATIONS.

Maps and Plates.

Wood-cuts.

a

a 2

Ivory Ornament from Nimroud.

NINEVEH AND BABYLON.

Ruined Mosque and Minarets (Erzeroom).

CHAPTER I.

THE TRUSTEES OF THE BRITISH MUSEUM RESUME EXCAVATIONS AT NINEVEH.—
DEPARTURE FROM CONSTANTINOPLE. — DESCRIPTION OF OUR PARTY.—CAWAL
YUSUF. — ROADS FROM TREBIZOND TO ERZEROOM. — DESCRIPTION OF THE
COUNTRY. — VARZAHAN AND ARMENIAN CHURCHES. — ERZEROOM. — RESHID
PASHA.—THE DUDJOOK TRIBES.—SHAHAN BEY.—TURKISH REFORM.—JOURNEY
THROUGH ARMENIA. — AN ARMENIAN BISHOP. — THE LAKES OF SHAILU AND
NAZIK. — THE LAKE OF WAN.

AFTER a few months' residence in England during the year 1848,
to recruit a constitution worn by long exposure to the extremes

B

of an Eastern climate, I received orders to proceed to my post at Her Majesty's Embassy in Turkey. The Trustees of the British Museum did not, at that time, contemplate further excavations on the site of ancient Nineveh. Ill health and limited time had prevented me from placing before the public, previous to my return to the East, the results of my first researches with the illustrations of the monuments and copies of the inscriptions recovered from the ruins of Assyria. They were not published until some time after my departure, and did not consequently receive that careful superintendence and revision necessary to works of this nature. It was at Constantinople that I first learnt the general interest felt in England in the discoveries, and that they had been universally received as fresh illustrations of Scripture and prophecy, as well as of ancient history sacred and profane.

And let me here, at the very outset, gratefully acknowledge that generous spirit of English criticism which overlooks the incapacity and shortcomings of the laborer when his object is worthy of praise, and that object is sought with sincerity and singleness of purpose. The gratitude, which I deeply felt for encouragement rarely equalled, could be best shown by cheerfully consenting, without hesitation, to the request made to me by the Trustees of the British Museum, urged by public opinion, to undertake the superintendence of a second expedition into Assyria. Being asked to furnish a plan of operations, I stated what appeared to me to be the course best calculated to produce interesting and important results, and to enable us to obtain the most accurate information on the ancient history, language, and arts, not only of Assyria, but of its sister kingdom, Babylonia. Perhaps my plan was too vast and general to admit of performance or warrant adoption. I was merely directed to return to the site of Nineveh, and to continue the researches commenced amongst its ruins.

Arrangements were hastily, and of course inadequately, made in England. The assistance of a competent artist was most desirable, to portray with fidelity those monuments which injury and decay had rendered unfit for removal. Mr. F. Cooper was selected by the Trustees of the British Museum to accompany the expedition in this capacity. Mr. Hormuzd Rassam, already well known to many of my readers for the share he had taken in my first discoveries, quitted England with him. They both joined me at Constantinople. Dr. Sandwith, an English physician on a visit to the East, was induced to form one of our party. One Abd-el-Messiah, a Catholic Syrian of Mardin, an active and trustworthy servant during my former residence in Assyria, was fortunately at this time in the capital, and again entered my service: my other attendants were Mohammed Agha, a cawass, and an Armenian named Serkis. The faithful Bairakdar, who had so well served me during my previous journey, had accompanied the English commission for the settlement of the boundaries between Turkey and Persia; with the understanding, however, that he was to meet me at Mosul, in case I should return. Cawal Yusuf, the head of the Preachers of the Yezidis, with four chiefs of the districts in the neighbourhood of Diarbekir, who had been for some months in Constantinople, completed my party.

After my departure from Mosul, in 1847, the military conscription, enforced amongst the Mussulman inhabitants of the Pashalic, was extended to the Yezidis, who, with the Christians, had been previously exempted from its operation on the general law sanctioned by the Koran, and hitherto acted upon by most Mohammedan nations, that none but true believers can serve in the armies of the state. On the ground that being of no recognised infidel sect, they must necessarily be included, like the Druses and Ansyri of Mount Lebanon, amongst Mussulmans, the Government had recently endeavoured to raise recruits for the regular troops amongst the Yezidis. The new regulations had been carried out with great severity, and had given rise to many acts of cruelty and oppression on the part of the local authorities. Besides the feeling common to all Easterns against compulsory service in the army, the Yezidis had other reasons for opposing the orders of the Government. They could not become *nizam*, or disciplined soldiers, without openly violating the rites and observances enjoined by their faith. The bath, to which Turkish soldiers are compelled weekly to resort, is a pollution to them, when taken in common with Mussulmans; the blue color, and certain portions of the Turkish uniform are absolutely prohibited by their law; and they cannot eat several articles of food included in the rations distributed to the troops. The recruiting officers refused to listen to these objections, enforcing their orders with extreme and unnecessary severity. The Yezidis, always ready to suffer for their faith, resisted, and many died under the tortures inflicted upon them. They were, moreover, still exposed to the oppression and illegal exactions of the local governors. Their children were still lawful objects of public sale, and, notwithstanding the introduction of the reformed system of government into the provinces, the parents were subject to persecution, and even to death, on account of their religion. In this state of things, Hussein Bey and Sheikh Nasr, the chiefs of the whole community, hearing that I was at Constantinople, determined to send a deputation to lay their grievances before the Sultan, hoping that through my assistance they could obtain access to some of the Ministers of State. Cawal Yusuf and his companions were selected for the mission ; and money was raised by subscriptions from the sect to meet the expenses of their journey.

After encountering many difficulties and dangers, they reached the capital and found out my abode. I lost no time in presenting them to Sir Stratford Canning, who, ever ready to exert his powerful influence in the cause of humanity, at once brought their wrongs to the notice of the Porte. Through his kindly intercession a firman, or imperial order, was granted to the Yezidis, which freed them from all illegal impositions, forbade the sale of their children as slaves, secured to them the full enjoyment of their religion, and placed them on the same footing as other sects of the empire. It was further promised that arrangements should be made to release them from such military regulations as rendered their service in the army incompatible with the strict observance of their religious duties. So often can influence, well acquired and well directed, be exercised in the great cause of humanity, without distinction of persons or of creeds ! This is but one of the many instances in which Sir Stratford Canning has added to the best renown of the British name.

Cawal Yusuf, having fulfilled his mission, eagerly accepted my proposal to return with me to Mosul. His companions had yet to obtain certain documents from the Porte, and were to remain at Constantinople until their business should be completed. The Cawal still retained the dress of his sect and office. His dark face and regular and expressive features were shaded by a black turban, and a striped aba of coarse texture was thrown loosely over a robe of red silk.

Our arrangements were complete by the 28th of August (1849), and on that day we left the Bosphorus by an English steamer bound for Trebizond. The size of my party and its consequent incumbrances rendering a caravan journey absolutely necessary, I determined to avoid the usual tracks, and to cross eastern Armenia and Kurdistan, both on account of the novelty of part of the country in a geographical point of view, and its political interest as having only recently been brought under the immediate control of the Turkish government.

We disembarked at Trebizond on the 31st, and on the following day commenced our land journey. The country between this port and Erzeroom has been frequently traversed and described. Through it pass the caravan routes connecting Persia with the Black Sea, the great lines of intercourse and commerce between Europe and central Asia. The roads usually frequented are three in number. The summer, or upper, road is the shortest, but is most precipitous, and, crossing very lofty mountains, is closed after the snows commence; it is called *Tchaïrler*, from its fine upland pastures, on which the horses are usually fed when caravans take this route. The middle road has few advantages over the upper, and is rarely followed by merchants, who prefer the lower, although making a considerable detour by Gumish Khaneh, or the Silver Mines. The three unite at the town of Baiburt, midway between the sea and Erzeroom. Although an active and daily increasing trade is carried on by these roads, no means whatever have until recently been taken to improve them. They consist of mere mountain tracks, deep in mud or dust according to the season of the year. The bridges, built when the erection and repair of public works were imposed upon the local governors, and deemed a sacred duty by the semi-independent hereditary families, who ruled in the provinces as Pashas or Dereh-Beys, have been long permitted to fall into decay, and commerce is frequently stopped for days by the swollen torrent or fordless stream. This has been one of the many evil results of the system of centralisation so vigorously commenced by Sultan Mahmoud, and so steadily carried out during the present reign. The local governors, receiving a fixed salary, and rarely permitted to remain above a few months in one office, take no interest whatever in the prosperity of the districts placed under their care. The funds assigned by the Porte for public works, small and totally inadequate, are squandered away or purloined long before any part can be applied to the objects in view.

Since my visit to Trebizond a road for carts has been commenced, which is to lead from that port to the Persian frontiers; but it will, probably, like other undertakings of the kind, be abandoned long before completed, or if ever completed will be permitted at once to fall to ruin from the want of common repair. And yet the Persian trade is one of the chief sources of revenue of the Turkish empire, and unless conveniences are afforded for its prosecution, will speedily pass into other hands. The southern shores of the Black Sea, twelve years ago rarely visited by a foreign vessel, are now coasted by steamers belonging to three companies, which touch nearly weekly at the principal ports; and there is commerce and traffic enough for more. The establishment of steam communication between the ports and the capital has given an activity previously unknown to internal trade, and has brought the inhabitants of distant provinces of the empire into a contact with the capital, highly favorable to the extension of civilisation, and to the enforcement of the legitimate authority of the government. The want of proper harbours is a considerable drawback in the navigation of a sea so unstable and dangerous as the Euxine. Trebizond has a mere roadstead, and from its position is otherwise little calculated for a great commercial port, which, like many other places, it has become rather from its hereditary claims as the representative of a city once famous, than from any local advantages.

The only harbour on the southern coast is that of Batoun, nor is there any retreat for vessels on the Circassian shores. This place is therefore probably destined to become the emporium of trade, both from its safe and spacious port, and from the facility it affords of internal communication with Persia, Georgia, and Armenia. From it the Turkish government might have been induced to construct the road since commenced at Trebizond, had not a political influence always hostile to any real improvement in the Ottoman empire opposed it with that pertinacity which is generally sure to command success.

At the back of Trebizond, as indeed along the whole of this singularly bold and beautiful coast, the mountains rise in lofty peaks, and are wooded with trees of enormous growth and admirable quality, furnishing an unlimited supply of timber for commerce or war. Innumerable streams force their way to the sea through deep and rocky ravines. The more sheltered spots are occupied by villages and hamlets, chiefly inhabited by a hardy and industrious race of Greeks. In spring the choicest flowers perfume the air, and luxuriant creepers clothe the limbs of gigantic trees. In summer the richest pastures enamel the uplands, and the inhabitants of the coasts drive their flocks and herds to the higher regions of the hills. The forests, nourished by the exhalations and rains engendered by a large expanse of water, form a belt, from thirty to fifty miles in breadth, along the Black Sea. Beyond, the dense woods cease, as do also the rugged

ravine and rocky peak. They are succeeded by still higher mountains, mostly rounded in their forms, some topped with eternal
snow, barren of wood and even of vegetation, except during the
summer, when they are covered with Alpine flowers and herbs.
The villages in the valleys are inhabited by Turks, Lazes (Mussulmans), and Armenians; the soil is fertile, and produces much corn.
    Our journey to Erzeroom was performed without incident. A
heavy and uninterrupted rain for two days tried the patience and
temper of those who for the first time encountered the difficulties
and incidents of Eastern travel. The only place of any interest,
passed during our ride, was a small Armenian village, the remains
of a larger, with the ruins of three early Christian churches, or

Ancient Armenian Church at Varzaban

baptisteries. These remarkable buildings, of which many examples exist, belong to an order of architecture peculiar to the

most eastern districts of Asia Minor and to the ruins of ancient Armenian cities *, on the borders of Turkey and Persia. The one, of which I have given a sketch, is an octagon, and may have been a baptistery. The interior walls are still covered with the remains of elaborate frescoes representing scripture events and national saints. The colors are vivid, and the forms, though rude, not inelegant or incorrect, resembling those of the frescoes of the Lower Empire still seen in the celebrated Byzantine church at Trebizond, and in the chapels of the convents of Mount Athos. The knotted capitals of the thin tapering columns grouped together, the peculiar arrangement of the stones over the doorway, supporting each other by a zigzag, and the decorations in general, call to mind the European Gothic of the middle ages. These churches date probably before the twelfth century : but there are no inscriptions, or other clue, to fix their precise epoch, and the various styles and modifications of the architecture have not been hitherto sufficiently studied to enable us to determine with accuracy the time to which any peculiar ornaments or forms may belong. Yet there are many interesting questions connected with this Armenian architecture which well deserve elucidation. From it was probably derived much that passed into the Gothic, whilst the Tatar conquerors of Asia Minor adopted it, as will be hereafter seen, for their mausoleums and places of worship. It is peculiarly elegant both in its decorations, its proportions, and the general arrangement of the masses, and might with advantage be studied by the modern architect. Indeed, Asia Minor contains a mine of similar materials unexplored and almost unknown.

The churches of Varzahan, according to the information I received from an aged inhabitant of the village, had been destroyed some fifty years before by the Lazes. The oldest people of the place remembered the time when divine worship was still performed within their walls.

We reached Erzeroom on the 8th, and were most hospitably received by the British consul, Mr. Brant, a gentleman who has long, well, and honorably sustained our influence in this part of Turkey, and who was the first to open an important field for our commerce in Asia Minor. With him I visited the commander-in-chief of the Turkish forces in Anatolia, who had recently returned

---

* Particularly of Ani. Mons. Texier is, I believe, the only traveller who has attempted to give elaborate plans, elevations, drawings, and restorations of these interesting edifices.

from a successful expedition against the wild mountain tribes of
central Armenia. Reshid Pasha, known as the " *Guzlu*," or " the
Wearer of Spectacles," enjoyed the advantages of an European
education, and had already distinguished himself in the military
career. With a knowledge of the French language he united a
taste for European literature, which, during his numerous expeditions into districts unknown to western travellers, had led him to
examine their geographical features, and to make inquiries into the
manners and religion of their inhabitants. His last exploit had
been the subjugation of the tribes inhabiting the Dudjook Mountains, to the south-west of Erzeroom, long in open rebellion against
the Sultan. The account he gave me of the country and its occupants, much excited a curiosity which the limited time at my command did not enable me to gratify. According to the Pasha, the
tribes are idolatrous, worshipping venerable oaks, great trees,
huge solitary rocks, and other grand features of nature. He was
inclined to attribute to them mysterious and abominable rites.
This calumny, the resource of ignorance and intolerance, from
which even primitive Christianity did not escape, has generally
been spread in the East against those whose tenets are unknown
or carefully concealed, and who, in Turkey, are included under
the general term, indicating their supposed obscene ceremonies, of
Cheragh-sonderan, or " Extinguishers of Lights." They have a
chief priest, who is, at the same time, a kind of political head of
the sect. He had recently been taken prisoner, sent to Constantinople, and from thence exiled to some town on the Danube.
They speak a Kurdish dialect, though the various septs into which
they are divided have Arabic names, apparently showing a southern origin. Of their history and early migrations, however, the
Pasha could learn nothing. The direct road between Trebizond
and Mesopotamia once passed through their districts, and the ruins
of spacious and well-built khans are still seen at regular intervals
on the remains of the old causeway. But from a remote period,
the country had been closed against the strongest caravans, and no
traveller would venture into the power of tribes notorious for their
cruelty and lawlessness. The Pasha spoke of re-opening the road,
rebuilding caravanserais, and restoring trade to its ancient channel—
good intentions, not wanting amongst Turks of his class, and which,
if carried out, might restore a country rich in natural resources to
more than its ancient prosperity. The account he gave me is not
perhaps to be strictly relied on, but a district hitherto inaccessible

may possibly contain the remains of ancient races, monuments of antiquity, and natural productions of sufficient importance to merit the attention of the traveller in Asia Minor.

The city of Erzeroom is rapidly declining in importance, and is almost solely supported by the Persian transit trade. It would be nearly deserted if that traffic were to be thrown into a new channel by the construction of the direct road from Batoun to the Persian frontiers. It contains no buildings of any interest, with the exception of a few ruins of those monuments of early Mussulman domination, the elaborately ornamented portico and minaret faced with glazed tiles of rich yet harmonious coloring, and the conical mausoleum, peculiar to most cities of early date in Asia Minor. The modern Turkish edifices, dignified with the names of palaces and barracks, are meeting the fate of neglected mud. Their crumbling walls can scarcely shelter their inmates in a climate almost unequalled in the habitable globe for the rigor of its winters.

The districts of Armenia and Kurdistan, through which lay our road from Erzeroom to Mosul, are sufficiently unknown and interesting to merit more than a casual mention. The map will show that our route by the lake of Wan, Bitlis, and Jezirah was nearly a direct one. It had been but recently opened to caravans. The haunts of the last of the Kurdish rebels were on the shores of this lake. After the fall of the most powerful of their chiefs, Beder Khan Bey, they had one by one been subdued and carried away into captivity. Only a few months had, however, elapsed since the Beys of Bitlis, who had longest resisted the Turkish arms, had been captured. With them rebellion was extinguished for the time in Kurdistan.

Our caravan consisted of my own party, with the addition of a muleteer and his two assistants, natives of Bitlis, who furnished me with seventeen horses and mules from Erzeroom to Mosul. The first day's ride, as is customary in the East, where friends accompany the traveller far beyond the city gates, and where the preparations for a journey are so numerous that everything cannot well be remembered, scarcely exceeded nine miles. We rested for the night in the village of Guli, whose owner, one Shahan Bey, had been apprised of my intended visit. He had rendered his newly-built house as comfortable as his means would permit for our accommodation, and, after providing us with an excellent supper, passed the evening with me. Descended from an ancient family of Dereh-Beys he had inherited the hospitality and polished manners of a class now almost extinct, and of which a short account may not be uninteresting.

The Turkish conquerors, after the overthrow of the Greek empire, parcelled out their newly acquired dominions into military fiefs. These tenures varied subsequently in size from the vast possessions of the great families, with their hosts of retainers, such as the Kara Osmans of Magnesia, the Pasvan Oglus, and others, to the small *spahiliks* of Turkey in Europe, whose owners were obliged to perform personal military service when called upon by the state. Between them, of middle rank, were the Dereh-Beys, literally the " Lords of the Valley," who resided in their fortified castles, or villages, and scarcely owned more than a nominal allegiance to the Sultan, although generally ready to accompany him in a great national war against the infidels, or in expeditions against too powerful and usurping subjects.   Sultan Mahmoud, a man of undoubted genius and of vast views for the consolidation and centralisation of his empire, aimed not only at the extirpation of all those great families, which, either by hereditary right or by local influence, had assumed a kind of independence; but of all the smaller Dereh-Beys and Spahis.   This gigantic scheme, which changed the whole system of tenure and local administration, whether political or financial, he nearly carried out, partly by force of arms and partly by treachery.   Sultan Abd-ul-Mejid, freed from the difficulties and embarrassments with which an unfortunate war with Russia and successful rebellions in Albania and Egypt had surrounded his father, has completed what Mahmoud commenced. Not only have the few remaining Dereh-Beys been destroyed or removed one by one, but even military tenure has been entirely abolished by arbitrary enactments, which have given no compensation to the owners, and have destroyed the only hereditary nobility in the empire.   Opinions may differ as to the wisdom of the course pursued, and as to its probable results.   Whilst greater personal security has been undoubtedly established throughout the Ottoman dominions, whilst the subjects of the Sultan are, theoretically at least, no longer exposed to the tyranny of local chiefs, but are governed by the more equitable and tolerant laws of the empire ; his throne has lost the support of a race bred to military life, undisciplined it is true, but brave and devoted, always ready to join the holy standard when unfurled against the enemies of the nation and its religion, a race who carried the Turkish arms into the heart of Europe, and were the terror of Christendom.   Whether a regular army, disciplined as far as possible after the fashion of Europe, will supply the place of the old Turkish irregular cavalry and infantry, remains to be seen, and, for reasons which it is scarcely necessary to enter into, may fairly be doubted. With the old system the spirit which supported it is fast dying away, and it may be questioned whether, in Mussulman Turkey, discipline can ever compensate for its loss. The country has certainly not yet recovered from the change. During the former state of things, with all the acts of tyranny and oppression which absolute power engendered, there was more happiness among the people, and more prosperity in the land. The hereditary chiefs looked upon their Christian subjects as so much property to be improved and protected, like the soil itself. They were a source of revenue; consequently heavy taxes which impeded labor, and drove the laborer from the land, were from interest rarely imposed upon them. The Government left the enforcement of order to the local chiefs; all the tribute received from them was so much clear gain to the treasury, because no collectors were needed to raise it, nor troops to enforce its payment. The revenues of the empire were equal to great wars, and there was neither public debt nor embarrassment. Now that the system of centralisation has been fully carried out, the revenues are more than absorbed in the measures necessary to collect them, and the officers of government, having no interest whatever in the districts over which they are placed, neglect all that may tend to the prosperity and well-being of their inhabitants. It may be objected in extenuation that it is scarcely fair to judge of the working of a system so suddenly introduced, and that Turkey is merely in a transition state; the principle it has adopted, whatever its abuse, being fundamentally correct. One thing is certain, that Turkey must, sooner or later, have gone through this change.

It is customary to regard these old Turkish lords as inexorable tyrants — robber chiefs who lived on the plunder of travellers and of their subjects. That there were many who answered to this description cannot be denied; but they were, I believe, exceptions. Amongst them were some rich in virtues and high and noble feeling. It has been frequently my lot to find a representative of this nearly extinct class in some remote and almost unknown spot in Asia Minor or Albania. I have been received with affectionate warmth at the end of a day's journey by a venerable Bey or Agha in his spacious mansion, now fast crumbling to ruin, but still bright with the remains of rich, yet tasteful, oriental decoration; his long beard, white as snow, falling low on his breast; his many-folded turban shadowing his benevolent yet manly countenance, and his limbs enveloped in the noble garments rejected by the new generation; his hall open to all comers, the guest neither asked from whence he came or whither he was going, dipping his hands with him in the same dish; his servants, standing with reverence before him, rather his children than his servants; his revenues spent in raising fountains * on the wayside for the weary traveller, or in building caravanserais on the dreary plain; not only professing but practising all the duties and virtues enjoined by the Koran, which are Christian duties and virtues too; in his manners, his appearance, his hospitality, and his faithfulness a perfect model for a Christian gentleman. The race is fast passing away, and I feel grateful in being able to testify, with a few others, to its existence once, against prejudice, intolerance, and so called reform.

But to return to our host at Guli. Shahan Bey, although not an old man, was a very favorable specimen of the class I have described. He was truly, in the noble and expressive phraseology of the East, an " Ojiak Zadeh," " a child of the hearth," a gentleman born. His family had originally migrated from Daghistan, and his father, a pasha, had distinguished himself in the wars with Russia. He entertained me with animated accounts of feuds between his ancestors and the neighbouring chiefs, when without their armed retainers neither could venture beyond their immediate territories, contrasting, with good sense and a fair knowledge of his subject, the former with the actual state of the country. On the following morning, when I bade him adieu, he would not allow me to reward either himself or his servants, for hospitality extended to so large a company. He rode with me for some distance on my route, with his greyhounds and followers, and then returned to his village.

From Guli we crossed a high range of mountains, running nearly east and west, by a pass called Ali-Baba, or Ala-Baba, enjoying from the summit an extensive view of the plain of Pasvin, once one of the most thickly peopled and best cultivated districts in Armenia. The Christian inhabitants were partly induced by promises of land and protection, and partly compelled by force, to accompany the Russian army into Georgia after the end

---

* The most unobservant and hasty traveller in Turkey would soon become acquainted with this fact, could he read the modest and pious inscription, carved in relief on a small marble tablet of the purest white, adorning almost every half-ruined fountain at which he stops to refresh himself by the wayside.

of the last war with Turkey. By similar means that part of the Pashalic of Erzeroom adjoining the Russian territories was almost stripped of its most industrious Armenian population. To the south of us rose the snow-capped mountains of the Bin-Ghiul, or the " Thousand Lakes," in which the Araxes and several confluents of the Euphrates have their source. We descended from the pass into undulating and barren downs. The villages, thinly scattered over the low hills, were deserted by their inhabitants, who, at this season of the year, pitch their tents and seek pasture for their flocks in the uplands. We encamped for the night near one of these villages, called Gundi-Miran, or, in Turkish, Bey-Kiui, which has the same meaning, " the village of the chief." A man who remained to watch the crops of corn and barley went to the tents, and brought us such provisions as we required. The inhabitants of this district are Kurds, and are still divided into tribes. The owners of Gundi-Miran, and the surrounding villages, are the Ziraklu (the armour-wearers), who came originally from the neighbourhood of Diarbekir. Within a few months of our visit they were in open rebellion against the government, and the country had been closed against travellers and caravans.

Next day we continued our journey amongst undulating hills, abounding in flocks of the great and lesser bustard. Innumerable sheep-walks branched from the beaten path, a sign that villages were near; but, like those we had passed the day before, they had been deserted for the *yilaks*, or summer pastures. These villages are still such as they were when Xenophon traversed Armenia. " Their houses," says he, " were under ground; the mouth resembling that of a well, but spacious below: there was an entrance dug for the cattle, but the inhabitants descended by ladders. In these houses were goats, sheep, cows, and fowls with their young." * The low hovels, mere holes in the hill-side, and the common refuge of man, poultry, and cattle, cannot be seen from any distance, and they are purposely built away from the road to escape the unwelcome visits of travelling government officers and marching troops. It is not uncommon for a traveller to receive the first intimation of his approach to a village by finding his horse's fore feet down a chimney, and himself taking his place unexpectedly in the family circle through the roof. Numerous small streams wind among the valleys, marking by meandering lines of perpetual green their course to the Arras,

---

* Anabasis, lib. iv. c. 5

or Araxes. We crossed that river about midday by a ford not more than three feet deep, but the bed of the stream is wide, and after rains, and during the spring, is completely filled by an impassible torrent. On its southern bank we found a caravan reposing, the horses and mules feeding in the long grass, the travellers sleeping in the shade of their piled up bales of goods. Amongst the merchants we recognised several natives of Mosul who trade with Erzeroom, changing dates and coarse Mosul fabrics for a fine linen made at Riza,— a small place on the Black Sea, near Trebizond,— and much worn by the wealthy and by women.

During the afternoon we crossed the western spur of the Tiektab Mountains, a high and bold range with three well defined peaks, which had been visible from the summit of the Ala-Baba pass. From the crest we had the first view of Subhan, or Sipan, Dagh*, a magnificent conical peak, covered with eternal snow, and rising abruptly from the plain to the north of Lake Wan. It is a conspicuous and beautiful object from every part of the surrounding country. We descended into the wide and fertile plain of Hinnis. The town was just visible in the distance, but we left it to the right, and halted for the night in the large Armenian village of Kosli, after a ride of more than nine hours. I was received at the guest-house† with great hospitality by one Misrab Agha, a Turk,

---

* Sipan is a Kurdish corruption of Subhan, *i. e.* Praise. The mountain is so called, because a tradition asserts that whilst Noah was carried to and fro by the waters of the deluge, the ark struck against its peak, and the patriarch, alarmed by the shock, exclaimed " Subhanu-llah," " Praise be to God !" It has also been conjectured that the name is derived from " Surp," an Armenian word meaning "holy." It has only been ascended once, as far as I am aware, by Europeans. Mr. Brant, the British consul of Erzeroom, accompanied by Lieut. Glascott and Dr. Dickson, reached the summit on the 1st of September, 1838, after experiencing considerable fatigue and inconvenience from some peculiarity in the atmosphere (not, it would appear, the result of any very considerable elevation). They found within the cone a small lake, apparently filling the hollow of a crater ; and scoria and lava, met with in abundance during the ascent, indicated the existence, at some remote period, of a volcano. Unfortunately, the barometers with which the party were provided, were out of order, and Mr. Brant has only been able to estimate the height of the mountain by approximation, at 10,000 feet, which I believe to be under the mark. (See Mr. Brant's highly interesting memoir in the tenth volume of the Journal of the Royal Geographical Society, p. 49.)

† Almost every village in Turkey, not on a high road, and not provided with a caravanserai or khan, contains a house reserved exclusively for the entertainment of guests, in which travellers are not only lodged, but fed, gratuitously. It is maintained by the joint contribution of the villagers, or sometimes by the charitable bequests of individuals, and is under the care either of the chief of

to whom the village formerly belonged as Spahilik or military tenure, and who, deprived of his hereditary rights, had now farmed its revenues. He hurried with a long stick among the low houses, and heaps of dried dung, piled up in every open space for winter fuel, collecting fowls, curds, bread, and barley, abusing at the same time the *tanzimat*, which compelled such exalted travellers as ourselves, he said, " to pay for the provisions we condescended to accept." The inhabitants were not, however, backward in furnishing us with all we wanted, and the flourish of Misrab Agha's stick was only the remains of an old habit. I invited him to supper with me, an invitation he gladly accepted, having himself contributed a tender lamb roasted whole towards our entertainment.

The inhabitants of Kosli could scarcely be distinguished either by their dress or by their general appearance from the Kurds. They seemed prosperous and were on the best terms with the Mussulman farmer of their tithes. This village, with others in the district, had been nearly deserted after the Russian war, the inhabitants migrating into Georgia. Several families had recently returned, but having finished their harvest were desirous of recrossing the frontier, probably a manœuvre to avoid the payment of certain dues and taxes. Of this Misrab Agha was fully aware. " The ill-mannered fellows," exclaimed he, " having filled their bellies with good things, and taken away the fat of the land, want to go back to the Muscovites ; but they deceive themselves, they must now sit where they are." The emigrants did not indeed speak very favourably of the condition of those who had settled in Russia. Many wish to return to their old villages in Turkey, where they can enjoy far greater liberty and independence. This was subsequently confirmed to me by others who had come back to their native settlements. The Russian government, however, by a strict military surveillance along the Georgian frontiers, prevents as far as possible this desertion.

Kosli stands at the foot of the hills forming the southern boundary of the plain of Hinnis, through which flows a branch of the Murad Su, or Lower Euphrates. We forded this river near the ruins of a bridge at Kara Kupri. The plain is generally well cultivated, the principal produce being corn and hemp. The villages, which are thickly scattered over it, have the appearance of

the village, or of a person expressly named for the purpose, and called the Oda-Bashi, the chief of the guest-room. Since the introduction of the *tanzimat* (reformed system), this custom is rapidly falling into disuse in most parts of Turkey frequented by European travellers.

extreme wretchedness, and, with their low houses and heaps of dried manure piled upon the roofs and in the open spaces around, look more like gigantic dunghills than human habitations. The Kurds and Armenian Christians, both hardy and industrious races, are pretty equally divided in numbers, and live sociably in the same filth and misery. The extreme severity of the winter, — the snow lying deep on the ground for some months, — prevents the cultivation of fruit trees, and the complete absence of wood gives the country a desolate aspect. Bustards, cranes, and waterfowl of various kinds abound.

We left the plain of Hinnis by a pass through the mountain range of Zernak. In the valleys we found clusters of black tents belonging to the nomad Kurds, and the hill-sides were covered with their flocks. The summit of a high peak overhanging the road is occupied by the ruins of a castle formerly held by Kurdish chiefs, who levied black-mail on travellers, and carried their depredations into the plains. On reaching the top of the pass we had an uninterrupted view of the Subhan Dagh. From the village of Karagol, where we halted for the night, it rose abruptly before us. This magnificent peak, with the rugged mountains of Kurdistan, the river Euphrates winding through the plain, the peasants driving the oxen over the corn on the threshing-floor, and the groups of Kurdish horsemen with their long spears and flowing garments, formed one of those scenes of Eastern travel which leave an indelible impression on the imagination, and bring back in after years indescribable feelings of pleasure and repose.

The threshing-floor, which added so much to the beauty and interest of the picture at Karagol, had been seen in all the villages we had passed during our day's journey. The abundant harvest had been gathered in, and the corn was now to be threshed and stored for the winter. The process adopted is simple, and nearly such as it was in patriarchal times. The children either drive horses round and round over the heaps, or standing upon a sledge stuck full of sharp flints on the under part, are drawn by oxen over the scattered sheaves. Such were " the threshing-sledges armed with teeth" mentioned by Isaiah. In no instance are the animals muzzled — " thou shalt not muzzle the ox when he treadeth out the corn;" but they linger to pick up a scanty mouthful as they are urged on by the boys and young girls, to whom the duties of the threshing-floor are chiefly assigned. The grain is winnowed by the men and women, who throw the corn and straw

C

together into the air with a wooden shovel, leaving the wind to
carry away the chaff whilst the seed falls to the ground.   The
wheat is then raked into heaps and left on the threshing-floor

Threshing the Corn in Armenia

until the tithe-gatherer has taken his portion.   The straw is stored
for the winter, as provender for the cattle.*

The Kurdish inhabitants of this plain are chiefly of the tribe of
Mamanli, once very powerful, and mustering nearly 2000 horsemen for war, according to the information I received from one of
their petty chiefs who lodged with us for the night in the guest-
house of Karagol.   After the Russian war, part of the tribe was
included in the ceded territory.   Their chief resides at Malaskert.

---

* These processes of threshing and winnowing appear to have been used from
the earliest time in Asia.   Isaiah alludes to it when addressing the Jews
(xxviii. 27, 28.   See Translation by the Rev. John Jones) · —

"The dill is not threshed with *the threshing sledge*,
   Nor is the wheel of the wain made to roll over the cummin.

   .   .   .   .

   Bread corn is threshed :
   But not for ever will he continue thus to thresh it ,
   Though he driveth along the wheels of his wain,
   And his horses, he will not bruise it to dust."

"The oxen and the young asses, that till the ground
   Shall eat clean provender,
   Which hath been winnowed *with the shovel* and with the fan." (xxx. 24.)

"Behold, I have made thee a new sharp threshing wain (sledge) *armed with
   pointed teeth*." (xli. 15.)

"Thou shalt winnow them, and the wind shall carry them away." (xli. 16.)

We crossed the principal branch of the Euphrates soon after leaving Karagol. Although the river is fordable at this time of year, during the spring it is nearly a mile in breadth, overflowing its banks, and converting the entire plain into one great marsh. We had now to pick our way through a swamp, scaring, as we advanced, myriads of wild-fowl. I have rarely seen game in such abundance and such variety in one spot; the water swarmed with geese, duck, and teal, the marshy ground with herons and snipe, and the stubble with bustards and cranes. After the rains the lower road is impassable, and caravans are obliged to make a considerable circuit along the foot of the hills.

We were not sorry to escape the fever-breeding swamp and mud of the plain, and to enter a line of low hills, separating us from the lake of Gula Shailu. I stopped for a few minutes at an Armenian monastery, situated on a small platform overlooking the plain. The bishop was at his breakfast, his fare frugal and episcopal enough, consisting of nothing more than boiled beans and sour milk. He insisted that I should partake of his repast, and I did so, in a small room scarcely large enough to admit the round tray containing the dishes, into which I dipped my hand with him and his chaplain. I found him profoundly ignorant, like the rest of his class, grumbling about taxes, and abusing the Turkish government. All I could learn of the church was that it contained the body of a much venerated saint, who had lived about the time of St. Gregory the Illuminator, and that it was the resort of the afflicted and diseased who trusted to their faith, rather than to medicine, for relief. The whole establishment belongs to the large Armenian village of Kop, which could be faintly distinguished in the plain below. The Kurds had plundered the convent of its books and its finery, but the church remained pretty well as it had been some fifteen centuries ago.

After a pleasant ride of five hours we reached a deep clear lake, embedded in the mountains, two or three pelicans, " swan and shadow double," and myriads of water-fowl, lazily floating on its blue waters. Piron, the village where we halted for the night, stands at the further end of the Gula Shailu, and is inhabited by Kurds of the tribe of Hasananlu, and by Armenians, all living in good fellowship amidst the dirt and wretchedness of their eternal dung-heaps. Ophthalmia had made sad havoc amongst them, and the doctor was soon surrounded by a crowd of the blind and diseased clamoring for relief. The villagers said that a Persian, professing to be a Hakim, had passed through the place some time before, and had offered to cure all bad eyes on payment of a certain sum in advance. These terms being agreed to, he gave his patients a powder which left the sore eyes as they were, and destroyed the good ones. He then went his way : " And with the money in his pocket too," added a ferocious-looking Kurd, whose appearance certainly threw considerable doubt on the assertion ; " but what can one do in these days of accursed Tanzimat (reform)? "

The district we had now entered formerly belonged to Sheriff Bey, the rebellious chief of Moush, but, since his capture last year, had been made *miri*, or government property. Although all the Mohammedan inhabitants of this part of Kurdistan are Kurds, those alone are called so who live in tents ; those who reside in villages are known simply as "Mussulman."

The lake of Shailu is separated from the larger lake of Nazik, by a range of low hills about six miles in breadth. We reached the small village of Khers, built on its western extremity, in about two hours and a half, and found the chief, surrounded by the principal inhabitants, seated on a raised platform near a well-built stone house. He assured me, stroking a beard of spotless white to confirm his words, that he was above ninety years of age, and had never seen an European before the day of my visit. Half blind, he peered at me through his blear eyes until he had fully satisfied his curiosity ; then spoke contemptuously of the Franks, and abused the Tanzimat, which he declared had destroyed all Mussulman spirit, had turned true believers into infidels, and had brought his own tribe to ruin, meaning, of course, that they could no longer prey upon their neighbours. His son, more of a courtier, and probably thinking that something might be gained by praising the present state of things, spoke less unfavourably of reform, though, I doubt not, entertaining equal aversion to it in his heart. The old gentleman, notwithstanding his rough exterior, was hospitable after his fashion, and would not suffer us to depart until we had eaten of every delicacy the village could afford.

Our path lay along the banks of the lake. The people of Khers declare that the Nazik Gul only contains fish during the spring of the year, and then but of the one kind caught in the lake of Wan. I was unable to account for this fact, repeated by the peasants whom we met on our road, until reaching the eastern end of the lake I found that a communication existed between it and that of Wan, by a deep ravine, through which the waters, swollen during the rains and by the melting of the snows in spring, discharge themselves near Akhlat.* At this season there was only water enough in the ravine to show the difference of level. In spring the fish seek the creeks and fresh-water streams to spawn, and at that time alone are captured by the inhabitants of the shores of the lake of Wan. During the rest of the year, they leave the shallows and are secure from the nets of the fishermen.† The only fish known is of the size and appearance of a herring. It is caught during the season in such abundance that it forms, when dried and salted, provision for the rest of the year, and a considerable article of exportation. I was informed, however, by a Christian, that a large fish, probably of the barbel kind, was found in the Nazik Gul, whose waters, unlike those of Wan, are fresh and sweet.

Leaving the Nazik Gul we entered an undulating country traversed by very deep ravines, mere channels cut into the sandstone by mountain torrents. The villages are built at the bottom of these gulleys, amidst fruit trees and gardens, sheltered by perpendicular rocks and watered by running streams. They are undiscovered until the traveller reaches the very edge of the precipice, when a pleasant and cheerful scene opens suddenly beneath his feet. He would have believed the upper country a mere desert had he not spied here and there in the distance a peasant slowly driving his plough through the rich soil. The inhabitants of this district are more industrious and ingenious than their neighbours. They carry the produce of their harvest not on the backs of animals, as in most parts of Asia Minor, but in carts entirely made of wood, no iron being used even in the

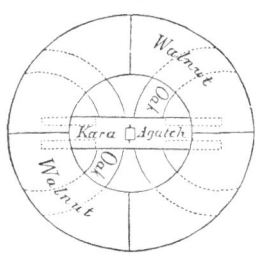

Section of Wheel of Armenian Cart.

wheels, which are ingeniously built of walnut, oak, and kara agatch (literally, black tree —? thorn), the stronger woods being used for rough spokes let into the nave. The plough also differs from that in general use in Asia. To the share are attached two parallel boards, about four feet long and a foot broad, which separate the soil and leave a deep and well defined furrow.

* The Shailu lake has, I was informed, a similar communication with the Murad Su. Both lakes are wrongly placed in the Prussian and other maps, and their outlets unnoticed.

† Yakuti, in his geographical work, the "Moajem el Buldan," mentions this disappearance of the fish, which are only to be seen, he says, during three months of the year. He adds, however, frogs and shellfish.

We rode for two or three hours on these uplands, until, suddenly reaching the edge of a ravine, a beautiful prospect of lake, woodland, and mountain, opened before us.

Armenian Plough, near Akhlat

Early Mussulman Tomb at Akhlat.

CHAP. II.

THE LAKE OF WAN.—AKHLAT.—TATAR TOMBS.—ANCIENT REMAINS.—A DERVISH.
—A FRIEND.—THE MUDIR.—ARMENIAN REMAINS.—AN ARMENIAN CONVENT AND
BISHOP. — JOURNEY TO BITLIS. — NIMROUD DAGH. — BITLIS. — JOURNEY TO
KHERZAN. — YEZIDI VILLAGE.

THE first view the traveller obtains of the Lake of Wan, on descending towards it from the hills above Akhlat, is singularly beautiful. This great inland sea, of the deepest blue, is bounded to the east by ranges of serrated snow-capped mountains, peering one above the other, and springing here and there into the highest peaks of Tiyari and Kurdistan; beneath them lies the sacred island of Akhtamar, just visible in the distance, like a dark shadow on the water. At the further end rises the one sublime cone of the Subhan, and along the lower part of the western shores stretches the Nimroud Dagh, varied in shape, and rich in local traditions.

At our feet, as we drew nigh to the lake, were the gardens of
the ancient city of Akhlat, leaning minarets and pointed mausoleums peeping above the trees. We rode through vast burying-
grounds, a perfect forest of upright stones seven or eight feet high
of the richest red colour, most delicately and tastefully carved
with arabesque ornaments and inscriptions in the massive character
of the early Mussulman age. In the midst of them rose here and
there a conical *turbeh* * of beautiful shape, covered with exquisite
tracery. The monuments of the dead still stand, and have become
the monuments of a city, itself long crumbled into dust. Amidst
orchards and gardens are scattered here and there low houses
rudely built out of the remains of the earlier habitations, and
fragments of cornice and sculpture are piled up into walls around
the cultivated plots.

Leaving the servants to pitch the tents on a lawn near one of the
finest of the old Mussulman tombs, and in a grove of lofty trees,
beneath whose spreading branches we could catch distant views of
the lake, I walked through the ruins. Emerging from the gardens
and crossing a part of the great burying-ground, I came upon a
well-preserved mausoleum of the same deep red stone, now glowing in the rays of the sun; its conical roof rested on columns and
arches, and on a *kubleh*, or place to direct the face in prayer, decorated with all the richness, yet elegance, of Eastern taste. The cornice supporting the roof was formed by many bands of ornament,
each equally graceful though differing one from the other. The
columns stood on a base rising about nine feet from the ground,
the upper part of which was adorned with panels, each varying
in shape, and containing many-angled recesses, decorated with
different patterns, and the lower part projected at an angle with
the rest of the building. In this basement was the chamber; the
mortal remains of its royal occupant had long ago been torn
away and thrown to the dust. Around the turbeh were scattered
richly carved head and foot stones, marking the graves of less
noble men; and the whole was enclosed by a grove of lofty trees,
the dark-blue lake glittering beyond. Whilst the scene was worthy
of the pencil of a Turner, each detail in the building was a study
for an architect. Tradition names the tomb that of Sultan
Baiandour †, one of the chiefs of the great Tatar tribes, who
crossed the frontiers of Persia in the fifteenth century. The

* The small building which sometimes covers a Mohammedan tomb is so
called.

† A sultan of the Ak-Kouyunlu, or White-sheep Tatars, from whom the tribe
derived their name of *Baiandouri*.

F. C. Cooper.

Turbeh, or Tomb, of Sultan Baiandour, at Akhlat.

building still resisting decay is now used as a storehouse for grain and straw by a degenerate race, utterly unmindful of the glories of their ancestors. Near this turbeh were others, less well preserved, but equally remarkable for elegant and varied decoration, their conical roofs fretted with delicate tracery, carved in relief on the red stone. They belong, according to local tradition, to Sultans of the Ak-Kouyunlu and Kara-Kouyunlu Tatars, the well-known tribes of the White and Black Sheep.

Beyond the turbeh of Sultan Baiandour, through a deep ravine such as I have already described, runs a brawling stream, crossed by an old bridge; orchards and gardens make the bottom of the narrow valley, and the cultivated ledges as seen from above, a bed of foliage. The lofty perpendicular rocks rising on both sides are literally honeycombed with entrances to artificial caves,—ancient tombs, or dwelling-places. On a high isolated mass of sandstone stand the walls and towers of a castle, the remains of the ancient city of Khelath, celebrated in Armenian history, and one of the seats of Armenian power. I ascended to the crumbling ruins, and examined the excavations in the rocks. The latter are now used as habitations, and as stables for herds and flocks. The spacious entrances of some are filled up with stones for protection and comfort, a small opening being left for a doorway. Before them, on the ledges overlooking the ravine, stood here and there groups of as noble a race as I have anywhere seen, tall, brawny men, handsome women, and beautiful children. They were Kurds, dressed in the flowing and richly-colored robes of their tribe. I talked with them and found them courteous, intelligent, and communicative.

Many of the tombs are approached by flights of steps, also cut in the rock. An entrance, generally square, unless subsequently widened, and either perfectly plain or decorated with a simple cornice, opens into a spacious chamber, which frequently leads into others on the same level, or by narrow flights of steps into upper rooms. There are no traces of the means by which these entrances were closed: they probably were so by stones, turning on rude hinges, or rolling on rollers.* Excavated in the walls, or some-

---

* Tombs, with entrances closed by stones, ingeniously made to roll back into a groove, still exist in many parts of the East. We learn from both the Old and New Testament, that such tombs were in common use in Palestine, as well as in other countries of Asia. The stone was "rolled away from the sepulchre" in which Christ was laid; which we may gather from the context was a chamber cut into the rock, and intended to receive many bodies, although it had not

times sunk into the floor, are recesses or troughs, in which once lay the bodies of the dead, whilst in small niches, in the sides of the chambers, were placed lamps and sacrificial objects. Tombs in every respect similar are found throughout the mountains of Assyria and Persia, as far south as Shiraz; but I have never met with them in such abundance as at Akhlat. Their contents were long ago the spoil of conquerors, and the ancient chambers of the dead have been for centuries the abodes of the living.

Leaving the valley and winding through a forest of fruit trees, here and there interspersed with a few primitive dwellings, I came to the old Turkish castle, standing on the very edge of the lake. It is a pure Ottoman edifice, less ancient than the turbehs, or the old walls towering above the ravine. Inscriptions over the gateways state that it was partly built by Sultan Selim, and partly by Sultan Suleiman, and over the northern entrance occurs the date of 975 of the Hejira. The walls and towers are still standing, and need but slight repair to be again rendered capable of defence. They inclose a fort, and about 200 houses, with two mosques and baths, fast falling into decay, and only tenanted by a few miserable families, who, too poor or too idle to build anew, linger amongst the ruins. In the fort, separated from the dwelling places by a high thick wall and a ponderous iron-bound gate now hanging half broken away from its rusty hinges, there dwelt, until very recently, a notorious Kurdish freebooter, of the name of Mehemet Bey, who, secure in this stronghold, ravaged the surrounding country, and sorely vexed its Christian inhabitants. He fled on the approach of the Turkish troops, after their successful expedition against Nur-Ullah Bey, and is supposed to be wandering in the mountains of southern Kurdistan.

After the capture of Beder Khan Bey, Osman Pasha, the commander-in-chief of the Turkish army, a man of enterprise and liberal views, formed a plan for restoring to Akhlat its ancient prosperity, by making it the capital of the north-eastern provinces of the Turkish empire. He proposed, by grants of land, to induce the inhabitants of the neighbouring villages to remove to the town, and by peculiar privileges to draw to the new settlement the artizans of Wan, Bitlis, Moush, and even Erzeroom. Its po-

been used before. Such, also, was the tomb of Lazarus. Raphael, who is singularly correct in delineating Eastern habits and costumes in his scriptural pieces, has thus portrayed the tomb of the Saviour in a sketch in the Oxford Collection.

sition on the borders of a vast lake is favourable to traffic, and its
air is considered very salubrious. From its vicinity to the Persian
and Russian frontiers it might become of considerable importance
as a military depôt. Osman Pasha was about to construct a
palace, a bazar, and barracks, and to repair the walls of the old
castle, when death put an end to his schemes. In Turkey a
man in power, from principle, never carries out the plans, or
finishes the buildings of his predecessor; and Akhlat, one of the
most beautiful spots that the imagination can picture, will probably
long remain a heap of ruins. Scarcely a sail flutters on the water.
The only commerce is carried on by a few miserable vessels, which
venture in the finest weather to leave the little harbour of Wan to
search for wood and corn on the southern shores of the lake.

The ancient city of Khelath was the capital of the Armenian
province of Peznouni. It came under the Mohammedan power as
early as the ninth century, but was conquered by the Greeks of
the Lower Empire at the end of the tenth. The Seljuks took it
from them, and it then again became a Mussulman principality.
It was long a place of contention for the early Arab and Tatar
conquerors. Shah Armen * reduced it towards the end of the
twelfth century. It was besieged, without result, by the celebrated Saleh-ed-din, and was finally captured by his nephew, the
son of Melek Adel, in A. D. 1207.

The sun was setting as I returned to the tents. The whole
scene was lighted up with its golden tints, and Claude never
composed a subject more beautiful than was here furnished by
nature herself. I was seated outside my tent gazing listlessly on
the scene, when I was roused by a well-remembered cry, but one
which I had not heard for years. I turned about and saw standing before me a Persian Dervish, clothed in the fawn-colored
gazelle skin, and wearing the conical red cap, edged with fur, and
embroidered in black braid with verses from the Koran and invo-
cations to Ali, the patron of his sect. He was no less surprised
than I had been at his greeting, when I gave him the answer
peculiar to men of his order. He was my devoted friend and servant from that moment, and sent his boy to fetch a dish of pears,
for which he actually refused a present ten times their value. He

* Shah Armen, *i.e.* King of Armenia, was a title assumed by a dynasty
reigning at Akhlat, founded by Sokman Kothby, a slave of the Seljuk prince,
Kothbedin Ismail, who established an independent principality at Akhlat in
A. D. 1100, which lasted eighty years.

declared that I was one of his craft, and was fairly puzzled to make out where I had picked up my knowledge of his mystery and phraseology. But he was not my first Dervish friend; I had had many adventures in company with such as he.

Whilst we were seated chatting in the soft moonlight, Hormuzd was suddenly embraced by a young man resplendent with silk and gold embroidery and armed to the teeth. He was a chief from the district of Mosul and well known to us. Hearing of our arrival he had hastened from his village at some distance to welcome us, and to endeavour to persuade me to move the encampment and partake of his hospitality. Failing, of course, in prevailing upon me to change my quarters for the night, he sent his servant to his wife, who was a lady of Mosul, and formerly a friend of my companion's, for a sheep. We found ourselves thus unexpectedly amongst friends. Our circle was further increased by Christians and Mussulmans of Akhlat, and the night was far spent before we retired to rest.

In the morning, soon after sunrise, I renewed my wanderings amongst the ruins, first calling upon the Mudir, or governor, who received me seated under his own fig-tree. He was an old greybeard, a native of the place, and of a straightforward, honest bearing. I had to listen to the usual complaints of poverty and over-taxation, although, after all, the village, with its extensive gardens, only contributed yearly ten purses, or less than forty-five pounds, to the public revenue. This sum seems small enough, but without trade, and distant from any high road, there was not a para of ready money, according to the Mudir, in the place.

The governor's cottage stood near the northern edge of Akhlat, and a little beyond it the road again emerged into that forest of richly-carved tombs which surrounds the place, like a broad belt—the accumulated remains of successive generations. The triumph of the dead over the living is perhaps only thus seen in the East. In England, where we grudge our dead their last resting places, the habitations of the living encroach on the burial-ground; in the East it is the grave-yard which drives before it the cottage and the mansion. The massive headstones still stand erect long after the dwelling-places of even the descendants of those who placed them there have passed away. Several handsome turbehs, resembling in their general form those I had already visited, though differing from them in their elegant and elaborate details, were scattered amongst the more humble tombs.

From the Mudir's house I rode to the more ancient part of the city and to the rock tombs. The ravine, at no great distance from where it joins the lake, is divided into two branches, each watered by an abundant stream. I followed them both for four or five miles, ascending by the one, then crossing the upland which divides them, and descending by the other. Both afford innumerable pleasant prospects,—the water breaking in frequent cascades over the rocky bottom, beneath thick clusters of gigantic chesnuts and elms, the excavated cliffs forming bold frames to the pictures. I entered many of the rock-tombs, and found all of them to be of the same character, though varying in size. The doors of some have been enlarged, to render the interior more convenient as dwelling-places, and there are but few which have not been blackened by the smoke of the fires of many centuries. The present population of the ravine, small and scanty enough, resides almost entirely in these caves. Amongst the tombs there are galleries and passages in the cliffs without apparent use, and flights of steps, cut out of the rock, which seem to lead nowhere. I searched and inquired in vain for inscriptions and remains of sculpture, and yet the place is of undoubted antiquity, and in the immediate vicinity of cotemporary sites where cunciform inscriptions do exist.

During my wanderings I entered an Armenian church and convent standing on a ledge of rock overhanging the stream, about four miles up the southern ravine. The convent was tenanted by a bishop and two priests. They dwelt in a small low room, scarcely lighted by a hole carefully blocked up with a sheet of oiled paper to shut out the cold; dark, musty, and damp, a very parish clerk in England would have shuddered at the sight of such a residence. Their bed, a carpet worn to threads, spread on the rotten boards; their diet, the coarsest sandy bread and a little sour curds, with beans and mangy meat for a jubilee. A miserable old woman sat in a kind of vault under the staircase preparing their food, and passing her days in pushing to and fro with her skinny hands the goat's skin containing the milk to be shaken into butter. She was the housekeeper and handmaiden of the episcopal establishment. The church was somewhat higher, though even darker than the dwelling-room, and was partly used to store a heap of mouldy corn and some primitive agricultural implements. The whole was well and strongly built, and had the evident marks of antiquity. The bishop showed me a rude cross carved on a rock outside the convent, which, he declared, had been cut by one of the disciples of the Saviour himself. It is, at any rate, considered a relic of very

great sanctity, and is an object of pilgrimage for the surrounding
Christian population. Near the spot are several tombs of former
bishops, the head and foot stones of the same deep mellow red
stone, and as elaborately carved as those of the old Tatar chiefs
near the lake, although differing from them somewhat in the
style of their ornaments; the cross, and the bold, square, ancient
Armenian character being used instead of the flowery scroll-work
and elongated letters of the early Mussulman conquerors. The
bishop, notwithstanding his poverty, was, on the whole, better
informed than others of his order I had met in the provinces. He
had visited the capital, had even studied there, and possessed a
few books, amongst which, fortunately for himself, and I hope for
his congregation, he was not ashamed to include several of the
very useful works issued by the American missionary press, and
by that praiseworthy religious society, the Mekhitarists of Venice.
The older books and MSS. of the church, together with its
little store of plate, its hangings, and its finery, were gone. The
last rummage was made by Mehemet Bey, the Kurdish freebooter of the castle on the lake, who, having been expelled
from his stronghold by the exasperated inhabitants of Akhlat,
took refuge in the Armenian convent, and defended it for nearly
a year against his assailants, living of course, the while, upon the
scanty stores of the priests, and carrying off, when he had no
longer need of the position, the little property he had pulled out
of every nook and corner. The tyranny of this chief had driven
nearly the whole Christian population from Akhlat. About
twenty families only remained, and they were huddled together in the rock tombs, and on the ledges immediately opposite
the convent. They are not allowed to possess the gardens and
orchards near the lake, which are looked upon as the peculiar
property of the ancient Mussulman inhabitants, to be enjoyed by
their orthodox descendants, who employ neither care nor labor in
keeping them up, trusting to a rich soil and a favorable climate
for their annual fruits.

I was again struck during my ride with the beauty of the
children, who assembled round me, issuing, like true Troglodytes,
from their rocky dwelling-places. Near the end of the ravine, on
the edge of a precipice clothed with creepers, is a half-fallen turbeh, of elegant proportions and rich in architectural detail. It
overhangs the transparent stream, which, struggling down its rocky
bed, is crossed by a ruined bridge; a scene calling to mind the
well-known view of Tivoli. Beyond, and nearer to the lake, are

other turbehs, all of which I examined, endeavoring to retain some slight record of their peculiar ornaments. The natives of the place followed me as I wandered about and found names for the ancient chiefs in whose honor the mausoleums had been erected. Amongst them were Iskender, Hassan, and Haroun, the Padishas, or sultans, of the Tatar tribes.*

On my return to our encampment the tents were struck, and the caravan had already began its march. Time would not permit me to delay, and with a deep longing to linger on this favored spot I slowly followed the road leading along the margin of the lake to Bitlis. I have seldom seen a fairer scene, one richer in natural beauties. The artist and the lover of nature may equally find at Akhlat objects of study and delight. The architect, or the traveller, interested in the history of that graceful and highly original branch of art, which attained its full perfection under the Arab rulers of Egypt and Spain, should extend his journey to the remains of ancient Armenian cities, far from high roads and mostly unexplored. He would then trace how that architecture, deriving its name from Byzantium, had taken the same development in the East as it did in the West, and how its subsequent combination with the elaborate decoration, the varied outline, and tasteful coloring of Persia had produced the style termed Saracenic, Arabic, and Moresque. He would discover almost daily, details, ornaments, and forms, recalling to his mind the various orders of architecture, which, at an early period, succeeded to each other in Western Europe and in England†; modifications of style for which we are mainly indebted to the East during its close union with the

---

* Iskender, the son of Kara Yusuf, second sultan of the Tatar dynasty of the Black Sheep, began to reign A.D. 1421, and was murdered by his son, Shah Kobad. Hassan, commonly called Usun, or the Long, the first sultan of the Baiandouri, or White Sheep, Tatars, succeeded to the throne A.D. 1467. Neither of these sultans, however, appear to have died at Akhlat. I have been unable to find the name of Haroun amongst the sultans of these Tatar dynasties. It is possible that the turbehs may be more ancient than the period assigned to them by the inhabitants of Akhlat, and that they may belong to some of the earlier Mussulman conquerors.

† The sketch, not very accurate unfortunately in its details, of the ruined Armenian church at Varzahan (p. 7.), will sufficiently show my meaning, and point out the connection indicated in the text. I would also refer to M. Texier's folio work on Armenia and Persia, for many examples of Armenian churches, illustrating the transition between the Byzantine and what we may undoubtedly term Gothic. It would be of considerable importance to study the remains of churches still scattered over Armenia, and of which no accurate plans or drawings have been published.

West by the bond of Christianity. The Crusaders, too, brought back into Christendom, on their return from Asia, a taste for that rich and harmonious union of color and architecture which had already been so successfully introduced by the Arabs into the countries they had conquered.

This connection between Eastern and Western architecture is one well worthy of study, and cannot be better illustrated than by the early Christian ruins of Armenia, and those of the Arsacian and Sassanian periods still existing in Persia. As yet it has been almost entirely overlooked, nor are there any plans or drawings of even the best known Byzantine, or rather Armenian, remains in Asia Minor, upon which sufficient reliance can be placed to admit of the analogies between the styles being fully proved. The union of early Christian and Persian art and architecture produced a style too little known and studied, yet affording combinations of beauty and grandeur, of extreme delicacy of detail and of boldness of outline, worthy of the highest order of intellect.*

Our road skirted the foot of the Nimroud Dagh, which stretches from Akhlat to the southern extremity of the lake. We crossed several dykes of lava and scoria, and wide mud-torrents now dry, the outpourings of a volcano long since extinct, but the crater of which may probably still be traced in a small lake said to exist on the very summit of the mountain. There are several villages, chiefly inhabited by Christians, built on the water's edge, or in the ravines worn by the streams descending from the hills. Our road gradually led away from the lake. With Cawal Yusuf and my companions I left the caravan far behind. The night came on, and we were shrouded in darkness. We sought in vain for the village which was to afford us a resting-place, and soon lost our uncertain track. The Cawal took the opportunity of relating tales collected during former journeys on this spot, of robber Kurds and murdered travellers, which did not tend to remove the anxiety felt by some of my party. At length, after wandering to

* The Arabs, a wild and uncultivated people, probably derived their first notions of architecture on the conquest of the Persian provinces. The peculiar and highly tasteful style of the Persians, of which traces may still be seen in the remains of the celebrated palace of Chosroes, at Ctesiphon, and in other ruins of southern Persia and Khuzistan, united with the Byzantine churches and palaces of Syria, produced the Saracenic. Already some such modification had, I am convinced, taken place in Armenia by a similar process, the Persian and Imperial power being continually brought into contact in that kingdom. I cannot dwell longer upon this subject, which well merits investigation.

D

and fro for above an hour, we heard the distant jingle of the caravan bells. We rode in the direction of the welcome sound, and soon found ourselves at the Armenian village of Keswak, standing in a small bay, and sheltered by a rocky promontory jutting boldly into the lake.

Next morning we rode along the margin of the lake, still crossing the spurs of the Nimroud Dagh, furrowed by numerous streams of lava and mud. In one of the deep gulleys, opening from the mountain to the water's edge, are a number of isolated masses of sandstone, worn into fantastic shapes by the winter torrents, which sweep down from the hills. The people of the country call them " the Camels of Nimrod." Tradition says that the rebellious patriarch endeavoring to build an inaccessible castle, strong enough to defy both God and man, the Almighty, to punish his arrogance, turned the workmen as they were working into stone. The rocks on the border of the lake are the camels, who with their burdens were petrified into a perpetual memorial of the Divine vengeance. The unfinished walls of the castle are still to be seen on the top of the mountain ; and the surrounding country, the seat of a primæval race, abounds in similar traditions.

We left the southern end of the lake, near the Armenian village of Tadwan, once a place of some importance, and containing a caravanserai, mosques, and baths built by Khosrew Pasha in the sixteenth century. Entering an undulating country we soon gazed for the last time on the deep blue expanse of water, and on the lofty peaks of the Hakkiari mountains. The small trickling streams, now running towards the south, and a gradual descent showed that we had crossed the water-shed of central Asia, and had reached the valleys of Assyria. Here and there the ruins of a fine old khan, its dark recesses, vaulted niches, and spacious stalls, blackened with the smoke of centuries, served to mark one of the great highways, leading in the days of Turkish prosperity from central Armenia to Baghdad. We had crossed this road in the plain of Hinnis. It runs from Erzeroom to Moush and thence to Bitlis, leaving to the east the Nimroud Dagh, which separates it from the lake of Wan. Commerce has deserted it for very many years, and its bridges and caravanserais have long fallen into decay ; when, with the restoration of order and tranquillity to this part of Turkey, trade shall revive, it may become once more an important thoroughfare, uniting the northern and southern provinces of the empire.

We soon entered a rugged ravine worn by the mountain rills, collected into a large stream. This was one of the many

head-waters of the Tigris. It was flowing tumultuously to our
own bourne, and, as we gazed upon the troubled waters, they
seemed to carry us nearer to our journey's end. The ravine was at
first wild and rocky; cultivated spots next appeared, scattered in
the dry bed of the torrent; then a few gigantic trees; gardens and
orchards followed, and at length the narrow valley opened on the
long straggling town of Bitlis.

The governor had provided quarters for us in a large house
belonging to an Armenian, who had been tailor to Beder Khan
Bey. From the terrace before the gate we looked down upon
the bazars built in the bottom of a deep gulley in the centre of
the town. On an isolated rock opposite to us rose a frowning
castle, and, on the top of a lofty barren hill, the fortified dwelling
of Sheriff Bey, the rebel chief, who had for years held Bitlis
and the surrounding country in subjection, defying the authority
and the arms of the Sultan. Here and there on the mountain
sides were little sunny landscapes, gardens, poplar trees, and low
white houses surrounded by trellised vines.

My party was now, for the first time during the journey,
visited with that curse of Eastern travel, fever and ague. The
doctor was prostrate, and having then no experience of the malady,
at once had dreams of typhus and malignant fever. A day's rest
was necessary, and our jaded horses needed it as well as we, for
there were bad mountain roads and long marches before us. I had
a further object in remaining. Three near relations of Cawal
Yusuf returning from their annual visitation to the Yezidi tribes
in Georgia and northern Armenia, had been murdered two years
before, near Bitlis, at the instigation of the Kurdish Bey. The
money collected by the Cawals for the benefit of the sect and
its priesthood, together with their personal effects, had been taken
by Sheriff Bey, and I was desirous of aiding Cawal Yusuf in their
recovery. Reshid Pasha had given me an official order for their
restoration out of the property of the late chief, and it rested
with me to see it enforced. I called early in the morning on the
mudir or governor, one of the household of old Essad Pasha, who
was at that time governor-general of Kurdistan, including Bitlis,
Moush, and the surrounding country, and resided at Diarbekir.
He gave me the assistance I required for the recovery of the
property of the murdered Cawals, and spoke in great contempt of
the Kurds now that they had been subdued, treating like dogs
those who stood humbly before him. The Turks, however, had
but recently dared to assume this haughty tone. Long after the

fall of Beder Khan Bey, the chiefs of Hakkiari, Wan, Moush, and
Bitlis had maintained their independence, and Sheriff Bey had
only been sent that spring to the capital to pass the rest of his
days in exile with the author of the Nestorian massacre.

The governor ordered cawasses to accompany me through the
town. I had been told that ancient inscriptions existed in the
castle, or on the rock, but I searched in vain for them: those
pointed out to me were early Mohammedan. Bitlis contains
many picturesque remains of mosques, baths, and bridges, and was
once a place of considerable size and importance. It is built in
the very bottom of a deep valley, and on the sides of ravines,
worn by small tributaries of the Tigris. The best houses stand
high upon the declivities, and are of stone, ornamented with large
arched windows, trellis work, and porticoes ; many of them being
surrounded by groves of trees. The bazars are in the lowest parts
of the town, and low, ill-built, and dirty. They are generally much
crowded, as in them is carried on the chief trade of this part
of Kurdistan. The export trade is chiefly supplied by the produce of the mountains ; galls, honey, wax, wool, and carpets and
stuffs, woven and dyed in the tents. The dyes of Kurdistan,
and particularly those from the districts around Bitlis, Sert, and
Jezireh, are celebrated for their brilliancy. They are made from
herbs gathered in the mountains, and from indigo, yellow berries,
and other materials, imported into the country. The colors usually
worn by both men and women are a deep dull red and a bright
yellow, mingled with black, a marked taste for these tints, to the
exclusion of almost every other, being a peculiar characteristic of the
Kurdish race from Bayazid to Suleimaniyah. The carpets are of a
rich soft texture, the patterns displaying considerable elegance and
taste : they are much esteemed in Turkey. There was a fair show
of Manchester goods and coarse English cutlery in the shops. The
sale of arms, once extensively carried on, had been prohibited. The
trade is chiefly in the hands of merchants from Mosul and Erzeroom, who come to Bitlis for galls, at present almost the only article
of export from Kurdistan to the European markets. This produce
of the oak was formerly monopolised by Beder Khan Bey, and
other powerful Kurdish chiefs, but the inhabitants are now permitted to gather them without restriction, each village having its
share in the woods. The wool of the mountains is coarse, and
scarcely fit for export to Europe; and the " teftik," a fine under-
hair of the goat, although useful and valuable, is not collected in
sufficient quantity for commerce. There is a race of sheep in Kur-

distan producing a long silken wool, like that of Angora, but it is not common, and the fleeces being much prized as saddle and other ornaments by the natives, are expensive. There are, no doubt, many productions of the mountains, besides valuable minerals, which appear to abound, that would become lucrative objects of commerce were tranquillity fully restored, and trade encouraged. The slaughter-houses, the resort of crowds of mangy dogs, are near the bazars, on the banks of the stream, and the effluvia arising from them is most offensive.

Having examined the town I visited the Armenian bishop, who dwells in a large convent in one of the ravines branching off from the main valley. On my way I passed several hot springs, some gurgling up in the very bed of the torrent. The bishop was maudlin, old, and decrepit ; he cried over his own personal woes, and over those of his community, abused the Turks, and the American missionaries, whispering confidentially in my ear as if the Kurds were at his door. He insisted in the most endearing terms, and occasionally throwing his arms round my neck, that I should drink a couple of glasses of fiery raki, although it was still early morning, pledging me himself in each glass. He showed me his church, an ancient building, well hung with miserable daubs of saints and miracles. On the whole, whatever may have been their condition under the Kurdish chiefs, the Christians of Bitlis at the time of my visit had no very great grounds of complaint. I found them well inclined and exceedingly courteous, those who had shops in the bazar rising as I passed. The town contains about seven hundred Armenian and forty Jacobite families (the former have four churches), but no Nestorians, although formerly a part of the Christian population was of that sect.

There are three roads from Bitlis to Jezireh ; two over the mountains through Sert, generally frequented by caravans, but very difficult and precipitous ; a third more circuitous, and winding through the valleys of the eastern branch of the Tigris. I chose the last, as it enabled me to visit the Yezidi villages of the district of Kherzan. We left Bitlis on the 20th. Soon issuing from the gardens of the town we found ourselves amidst a forest of oaks of various descriptions.* It was one of those deep, narrow, and rocky valleys abounding in Kurdistan ; the

---

* In the appendix will be found a note, with which I have been kindly favored by Dr. Lindley, upon the new and remarkable oaks found in these mountains, and now for the first time grown in this country from acorns sent home by me.

foaming torrent dashing through it, to be crossed and re-crossed, to the great discomfort of the laden mules, almost at every hundred yards, and from the want of bridges generally impassable during the spring and after rains.   In autumn and winter the declivities are covered with the black tents of the Kochers, or wandering Kurds, who move in summer to the higher pastures.   The tribes inhabiting the valley are the Selokeen, the Hamki, and the Babosi, by whom the relatives of Cawal Yusuf were murdered. There are no villages near the road-side.   They stand in deep ravines branching out from the main valley, either perched on precipitous and almost inaccessible ledges of rock, or hid in the recesses of the forest.   Several bridges and spacious khans, whose ruins still attest the ancient commerce and intercourse carried on through these mountains, are attributed, like all other public works in the country, to Sultan Murad during his memorable expedition against Baghdad (A.D. 1638).

About five miles from Bitlis the road is carried by a tunnel, about twenty feet in length, through a mass of calcareous rock, projecting like a huge rib from the mountain's side.   The mineral stream, which in the lapse of ages has formed this deposit, is still at work, projecting great stalactites from its sides, and threatening to close ere long the tunnel itself.   There is no inscription to record by whom and at what period this passage was cut.   It is, of course, assigned to Sultan Murad, but is probably of a far earlier period.   There are many such in the mountains[*]; and the remains of a causeway, evidently of great antiquity, in many places cut out of the solid rock, are traceable in the valley. We pitched our tents for the night near a ruined and deserted khan.

We continued during the following day in the same ravine, crossing by ancient bridges the stream which was gradually gathering strength as it advanced towards the low country.   About noon we passed a large Kurdish village called Goeena, belonging to Sheikh Kassim, one of those religious fanatics who are the curse of Kurdistan.   He was notorious for his hatred of the Yezidis, on whose districts he had committed numerous depredations, murdering those who came within his reach.   His last expedition had not proved successful; he was repulsed with the loss of many of his followers.   We encamped in the afternoon on the

_____

[*] See Col. Sheil's Memoir in the Journal of the Royal Geographical Society, vol. viii. p. 81.

bank of the torrent, near a cluster of Kurdish tents, concealed from view by the brushwood and high reeds. The owners were poor but hospitable, bringing us a lamb, yahgourt, and milk. Late in the evening a party of horsemen rode to our encampment. They were a young Kurdish chief, with his retainers, carrying off a girl with whom he had fallen in love,—not an uncommon occurrence in Kurdistan. They dismounted, eat bread, and then hastened on their journey to escape pursuit.

Starting next morning soon after dawn we rode for two hours along the banks of the stream, and then, turning from the valley, entered a country of low undulating hills. Here we left the Bitlis stream, which is joined about six hours beyond, near a village named Kitchki, by the river of Sert, another great feeder of the Tigris. This district abounds in saline springs and wells, whose waters, led into pans and allowed to evaporate, deposit much salt, which is collected and forms a considerable article of export even to the neighbourhood of Mosul.

We halted for a few minutes in the village of Omais-el-Koran, belonging to one of the innumerable saints of the Kurdish mountains. The Sheikh himself was on his terrace superintending the repair of his house, gratuitously undertaken by the neighbouring villagers, who came eagerly to engage in a good and pious work. Whilst the chief enjoys the full advantages of a holy character the place itself is a Ziorah, or place of pilgrimage, and a visit to it is considered by the ignorant Kurds almost as meritorious as a journey to Mecca; such pilgrimages being usually accompanied by an offering in money, or in kind, are not discouraged by the Sheikh.

Leaving a small plain, we ascended a low range of hills by a precipitous pathway, and halted on the summit at a Kurdish village named Khokhi. It was filled with Bashi-Bozuks, or irregular troops, collecting the revenue, and there was such a general confusion, quarrelling of men and screaming of women, that we could scarcely get bread to eat. Yet the officer assured me that the whole sum to be raised amounted to no more than seventy piastres (about thirteen shillings). The poverty of the village must indeed have been extreme, or the bad will of the inhabitants outrageous.

It was evening before we descended into the plain country of the district of Kherzan. The Yezidi village of Hamki had been visible for some time from the heights, and we turned towards it. As the sun was fast sinking, the peasants were leaving the threshing-floor, and gathering together their implements of husbandry.

They saw the large company of horsemen drawing nigh, and took us for irregular troops, — the terror of an Eastern village. Cawal Yusuf, concealing all but his eyes with the Arab kefieh, which he then wore, rode into the midst of them, and demanded in a preremptory voice provisions and quarters for the night. The poor creatures huddled together, unwilling to grant, yet fearing to refuse. The Cawal having enjoyed their alarm for a moment, threw his kerchief from his face, exclaiming, " O evil ones ! will you refuse bread to your priest, and turn him hungry from your door ? " There was surely then no unwillingness to receive us. Casting aside their shovels and forks, the men threw themselves upon the Cawal, each struggling to kiss his hand. A boy ran to the village to spread the news, and from it soon issued women, children, and old men, to welcome us. A few words sufficed to explain from whence we came, and what we required. Every one was our servant. Horses were unloaded, tents pitched, lambs brought, before we had time to look around. There was a general rejoicing, and the poor Yezidis seemed scarcely able to satiate themselves with looking on their priest; for a report had gone abroad, and had been industriously encouraged by the Mussulmans, who had heard of the departure of the deputation for Constantinople, that Yusuf and his companions had been put to death by the Sultan, and that not only the petition of the Yezidis had been rejected, but that fresh torments were in store for them. For eight months they had received no news of the Cawal, and this long silence had confirmed their fears ; but " he was dead and is alive again, he was lost and is found ; " and they made merry with all that the village could afford.

Yusuf was soon seated in the midst of a circle of the elders. He told his whole history, with such details and illustrations as an Eastern alone can introduce, to bring every fact vividly before his listeners. Nothing was omitted: his arrival at Constantinople, his reception by me, his introduction to the ambassador, his interview with the great ministers of state, the firman of future protection for the Yezidis, prospects of peace and happiness for the tribe, our departure from the capital, the nature of steam-boats, the tossing of the waves, the pains of sea-sickness, and our journey to Kherzan. Not the smallest particular was forgotten ; every person and event were described with equal minuteness ; almost the very number of pipes he had smoked and coffees he had drunk was given. He was continually interrupted by exclamations of

gratitude and wonder; and, when he had finished, it was my turn
to be the object of unbounded welcomes and salutations.

As the Cawal sat on the ground, with his noble features and
flowing robes, surrounded by the elders of the village, eager
listeners to every word which dropped from their priest, and looking towards him with looks of profound veneration, the picture
brought vividly to my mind many scenes described in the sacred
volumes. Let the painter who would throw off the convention-
alities of the age, who would feel as well as portray the incidents
of Holy Writ, wander in the East, and mix, not as the ordinary
traveller, but as a student of men and of nature, with its people.
He will daily meet with customs which he will otherwise be at
a loss to understand, and be brought face to face with those who
have retained with little change the manners, language, and dress
of a patriarchal race.

Yezidi Women.

Kurdish Women at a Spring

CHAP. III.

RECEPTION BY THE YEZIDIS. — VILLAGE OF GUZELDER. — TRIUMPHAL MARCH
TO REDWAN. — REDWAN.—ARMENIAN CHURCH. — MIRZA AGHA. — THE MELEK
TAOUS, OR BRAZEN BIRD.—TILLEH.—VALLEY OF THE TIGRIS.—BAS RELIEFS.—
JOURNEY TO DEREBOUN — TO SEMIL. — ABDE AGHA. — JOURNEY TO MOSUL. —
THE YEZIDI CHIEFS. — ARRIVAL AT MOSUL. — XENOPHON'S MARCH FROM THE
ZAB TO THE BLACK SEA.

I WAS awoke on the following morning by the tread of horses and
the noise of many voices. The good people of Hamki having sent
messengers in the night to the surrounding villages to spread the
news of our arrival, a large body of Yezidis on horse and on foot
had already assembled, although it was not yet dawn, to greet us
and to escort us on our journey. They were dressed in their gayest
garments, and had adorned their turbans with flowers and green
leaves. Their chief was Akko, a warrior well known in the
Yezidi wars, still active and daring, although his beard had long
turned grey. The head of the village of Guzelder, with the principal inhabitants, had come to invite me to eat bread in his house,
and we followed him. As we rode along we were joined by parties

of horsemen and footmen, each man kissing my hand as he arrived, the horsemen alighting for that purpose. Before we reached Guzelder the procession had swollen to many hundreds. The men had assembled at some distance from the village, the women and children, dressed in their holiday attire, and carrying boughs of trees, congregated on the housetops. As I approached sheep were brought into the road and slain before my horse's feet, and as we entered the yard of Akko's house, the women and men joined in the loud and piercing " tahlel." The chief's family were assembled at his door, and his wife and mother insisted upon helping me to dismount. We entered a spacious room completely open to the air on one side, and distinguished by that extreme neatness and cleanliness peculiar to the Yezidis. Many-colored carpets were spread over the floor, and the principal elders took their seats with me.

Soon after our arrival several Fakirs*, in their dark coarse dresses and red and black turbans, came to us from the neighbouring villages. One of them wore round his neck a chain, as a sign that he had renounced the vanities of the world, and had devoted himself to the service of God and his fellow-creatures. Other chiefs and horsemen also flocked in, and were invited to join in the feast, which was not, however, served up until Cawal Yusuf had related his whole history once more, without omitting a single detail. After we had eaten of stuffed lambs, pillaws, and savory dishes and most luscious grapes, the produce of the district, our entertainer placed a present of home-made carpets at my feet, and we rose to depart. The horsemen, the Fakirs, and the principal inhabitants of Guzelder on foot accompanied me. At a short distance from the village we were met by another large body of Yezidis, and by many Jacobites, headed by one Namo, who, by the variety of his arms, the richness of his dress, a figured Indian silk robe, with a cloak of precious fur, and his tastefully decorated Arab mare, might rather have been taken for a Kurdish bey than the head of a Christian village. A bishop and several priests were with him. Two hours' ride, with this great company, the horsemen galloping to and fro, the footmen discharging their firearms, brought us to the large village of Koshana. The whole of the population, mostly dressed in pure white, and wearing leaves and flowers in their turbans, had turned out to meet us; women stood on the road-side with jars of fresh water and bowls of sour milk, whilst others with the children were assembled on the housetops

---

* The lowest order of the Yezidi priesthood.

making the *tahlel.* Resisting an invitation to alight and eat bread, and having merely stopped to exchange salutations with those assembled, I continued on the road to Redwan, our party swollen by a fresh accession of followers from the village. Ere long we were met by three Cawals on their periodical visitation to the district. They were nearly related to Cawal Yusuf, and old friends of my own. With them, amongst others, were several young Mussulmans, who appeared to be on the best terms with their Yezidi friends, but had probably ridden out with them to show their gay dresses and admirable horsemanship. As we passed through the defile leading into the plain of Redwan, we had the appearance of a triumphal procession, but as we approached the small town a still more enthusiastic reception awaited us. First came a large body of horsemen, collected from the place itself, and the neighbouring villages. They were followed by Yezidis on foot, carrying flowers and branches of trees, and preceded by musicians playing on the tubbul and zernai.* Next were the Armenian community headed by their clergy, and then the Jacobite and other Christian sects, also with their respective priests; the women and children lined the entrance to the place and thronged the housetops. I alighted amidst the din of music and the "tahlel" at the house of Nazi, the chief of the whole Yezidi district, two sheep being slain before me as I took my feet from the stirrups.

Nazi's house was soon filled with the chiefs, the principal visitors, and the inhabitants of Redwan. Again had Cawal Yusuf to describe all that had occurred at Constantinople, and to confirm the good tidings of an imperial firman giving the Yezidis equal rights with Mussulmans, a complete toleration of their religion, and relief from the much dreaded laws of the conscription. At length breakfast was brought and devoured. It was then agreed that Nazi's house was likely to be too crowded during the day to permit me to enjoy comfort or quiet, and with a due regard to the duties of hospitality, it was suggested that I should take up my quarters in the Armenian church, dining in the evening with the chiefs to witness the festivities.

The change was indeed grateful to me, and I found at length a little repose and leisure to reflect upon the gratifying scene to which I had that day been witness. I have, perhaps, been too minute in the account of my reception at Redwan, but I record

* A large drum beaten at both ends, and a kind of oboe or pipe.

with pleasure this instance of a sincere and spontaneous display of gratitude on the part of a much maligned and oppressed race. To those, unfortunately too many, who believe that Easterns can only be managed by violence and swayed by fear, let this record be a proof that there are high and generous feelings which may not only be relied and acted upon without interfering with their authority, or compromising their dignity, but with every hope of laying the foundation of real attachment and mutual esteem.

The church stands on the slope of a mound, on the summit of which are the ruins of a castle belonging to the former chiefs of Redwan. It was built expressly for the Christians of the Armenian sect by Mirza Agha, the last semi-independent Yezidi chief, a pleasing example of toleration and liberality well worthy of imitation by more civilised men. The building is peculiar and primitive in its construction; one side of the courtyard is occupied by stables for the cattle of the priests; above them is a low room with a dead wall on three sides and a row of arches on the fourth. On the opposite side of the court is an iwan, or large vaulted chamber, completely open on one side to the air; in its centre, supported on four columns, is a gaudily painted box containing a picture of the Virgin; a few miserable daubs of saints are pasted on the walls. This is the church, when in summer the heat prevents the use of a closed room. It can only be divided from the yard by a curtain of figured cotton print, drawn across when unbelievers enter the building; a low doorway to the left leads into a dark inner church, in which pictures of the Virgin and saints can faintly be distinguished by the light of a few propitiatory lamps struggling with the gloom. Service was performed in the open iwan during the afternoon, the congregation kneeling uncovered in the yard.

The priests of the different communities called upon me as soon as I was ready to receive their visits. The most intelligent amongst them was a Roman Catholic Chaldæan, a good-humoured, tolerant fellow, who with a very small congregation of his own did not bear any ill will to his neighbours. With the principal Yezidi chiefs, too, I had a long and interesting conversation on the state of their people and on their prospects. Nazi is descended from the ancient hereditary lords of Redwan. The last of them was Mirza Agha, his uncle, whose history and end were those of many of the former independent chieftains of Turkey. When the celebrated Reshid Pasha had subdued northern Kurdistan and was marching to the south, Mirza Agha, dreading the approach of the army, submitted to the Sultan, and agreed to receive a Turkish governor in his castle. The officer chosen for the post was one Emin Agha. He had not been long in Redwan before he carried away by force the beautiful wife of the Yezidi chief. Mirza Agha, instead of appealing to arms, went to Reshid Pasha, and feigning that the woman was a slave and not his wife, protested that Emin Agha might come back without fear to his government. The Turk did return, but he and his followers were no sooner in the power of the chief than they fell victims to his revenge. Reshid Pasha then marched against Redwan, but being called away against the rebel Bey of Rahwanduz, was unable to subdue the district. After the successful termination of the expedition against the Kurdish bey, Mirza Agha again made an unqualified submission, was received into favour, and appointed governor over his own people. On the death of Reshid Pasha he was invited to the quarters of the new Turkish commander, and treacherously murdered during his visit. His former wife, who, according to the laws of the sect, could not be received again into the community, had been placed in the harem of the murderer; she died on hearing the fate of her Yezidi husband. The body of Mirza Agha was brought by some faithful attendants to his native place, and lies under a neat turbeh on the banks of the stream to the west of the town. Nazi, his nephew, was his successor, but long oppression has reduced him to poverty; the old castle has been deserted, and is fast falling to ruin, whilst its owner occupies a mud hovel like the meanest of his followers.

Redwan is called a town, because it has a bazar, and is the chief place of a considerable district. It may contain about eight hundred rudely-built huts, and stands on a large stream, which joins the Diarbekir branch of the Tigris, about five or six miles below. The inhabitants are Yezidis, with the exception of about one hundred Armenian, and forty or fifty Jacobite and Chaldæan families. A Turkish Mudir, or petty governor, generally resides in the place, but was absent at the time of my visit.

The sounds of rejoicing had been heard during the whole afternoon; raki had circulated freely, and there were few houses which had not slain a lamb to celebrate the day. After we had dined, the dances commenced in the courtyard of Nazi's house, and were kept up during the greater part of the night, the moon shedding its pale light on the white robes of the Yezidi dancers.

But as the sun was setting we were visited by one of those sudden storms or whirlwinds which frequently riot over the plains of Mesopotamia and through the valleys of Assyria. Although it lasted scarcely more than half an hour, it tore down in its fury tents and more solid dwellings, and swept from the housetops the beds and carpets already spread for the night's repose. After its passage, the air seemed even more calm than it had been before, and those who had been driven to take shelter from its violence within the walls resumed their occupations and their dances.

We slept in the long room opening on the courtyard, and were awoke long before daybreak by the jingling of small bells and the mumbling of priests. It was Sunday, and the Armenians commence their church services betimes. I gazed half dozing, and without rising from my bed, upon the ceremonies, the bowing, raising of crosses, and shaking of bells, which continued for above three hours, until priests and congregation must have been well nigh exhausted. The people, as during the previous afternoon's service, stood and knelt uncovered in the courtyard.

The Cawals, who are sent yearly by Hussein Bey and Sheikh Nasr to instruct the Yezidis in their faith, and to collect the contributions forming the revenues of the great chief, and of the tomb of Sheikh Adi, were now in Redwan. The same Cawals do not take the same rounds every year. The Yezidis are parcelled out into four divisions for the purpose of these annual visitations, those of the Sinjar, of Kherzan, of the pashalic of Aleppo, and of the villages in northern Armenia, and within the Russian frontiers. The Yezidis of the Mosul districts have the Cawals always amongst them. I was aware that on the occasion of these journeys the priests carry with them the celebrated Melek Taous, or brazen peacock, as a warrant for their mission. A favourable opportunity now offered itself to see this mysterious figure, and I asked Cawal Yusuf to gratify my curiosity. He at once acceded to my request, and the Cawals and elders offering no objection, I was conducted early in the morning into a dark inner room in Nazi's house. It was some time before my eyes had become sufficiently accustomed to the dim light to distinguish an object, from which a large red coverlet had been raised on my entry. The Cawals drew near with every sign of respect, bowing and kissing the corner of the cloth on which it was placed. A stand of bright copper or brass, in shape like the candlesticks generally used in Mosul and Baghdad, was surmounted by the rude image of a bird in the same metal, and more like an Indian or Mexican

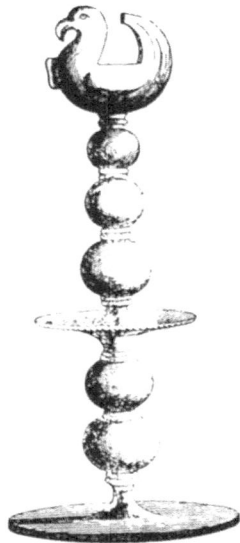

idol than a cock or peacock. Its peculiar workmanship indicated some antiquity, but I could see no traces of inscription upon it. Before it stood a copper bowl to receive contributions, and a bag to contain the bird and stand, which takes to pieces when carried from place to place. There are four such images, one for each district visited by the Cawals. The Yezidis declare that, notwithstanding the frequent wars and massacres to which the sect has been exposed, and the plunder and murder of the priests during their journeys, no Melek Taous has ever fallen into the hands of the Mussulmans. Cawal Yusuf, once crossing the desert on a mission to the Sinjar, and seeing a body of Bedouin horsemen in the distance, buried the Melek Taous. Having been robbed and then left by the Arabs, he dug it up and carried it in safety to its destination.

The Melek Taous, or Copper Bird of the Yezidis.

Mr. Hormuzd Rassam was alone permitted to visit the image with me. As I have elsewhere observed*, it is not looked upon as an idol, but as a symbol or banner, as Sheikh Nasr termed it, of the house of Hussein Bey.

Having breakfasted at Nazi's house we left Redwan, followed by a large company of Yezidis, whom I had great difficulty in persuading to turn back about three or four miles from the town. My party was increased by a very handsome black and tan greyhound with long silky hair, a present from old Akko, the Yezidi chief, who declared that he loved him as his child. The affection was amply returned. No delicacies or caresses would induce Touar, for such was the dog's name, to leave his master. He laid himself down and allowed one of the servants to drag him by a rope over the rough ground, philosophically giving tongue to his complaints in a low howl. This greyhound, a fine specimen of a noble breed, much prized by the Kurds and Persians, became, from his highly original character and complete independence, a great favourite with us. He soon forgot his old masters, and formed an equal attachment for his new. Another dog, a shepherd cur, had

* Nineveh and its Remains, vol. i. p. 298.

accompanied our caravan the whole way from Trebizond. He joined us without invitation, and probably finding the living to his taste, and the exercise conducive to health, remained with us, acknowledging the hospitality shown him by keeping watch over the horses by night.

Cawal Yusuf, and the Yezidi chiefs, had sent messengers even to Hussein Bey to apprise him of our coming. As they travelled along they scattered the news through the country, and I was received outside every village by its inhabitants. At Kunduk, two hours from Redwan, we found a second breakfast prepared for us, and were obliged to alight. Below this place the Redwan stream joins the Diarbekir branch of the Tigris, the two forming a broad river. Near are the remains of Husn Kaifa, and of other ancient cities, which I was unable to visit.

We had scarcely left Kunduk when we were met by a party of Christians, with the Kiayah of the village of Aoudi at their head. I was again obliged to stop, eat bread, and receive an offering of home-made carpets, of which we had now well nigh received a mule-load as presents. The inhabitants of the district were suffering much from oppression and illegal taxation.

The Kiayah, with some horsemen, accompanied us to Tilleh, where the united waters of Bitlis, Sert, and the upper districts of Bohtan, join the western branch of the Tigris. The two streams are about equal in size, and at this time of the year both fordable in certain places. We crossed the lower, or eastern, which we found wide and exceedingly rapid, the water, however, not reaching above the saddle-girths. The villagers raised the luggage, and supported the horses against the current, which rushing over loose and slippery stones, affording an uncertain footing, threatened to sweep the animals down the stream. Our travelling companion, the dog from Trebizond, having made several vain attempts to brave the rapids, quietly retired, thinking our company not worth any further risk. Touar, more fortunate, was carried over in the arms of a servant.

The spot at which we crossed was one of peculiar interest. It was here that the Ten Thousand in their memorable retreat forded this river, called, by Xenophon, the Centritis. The Greeks having fought their way over the lofty mountains of the Carduchians, found their further progress towards Armenia arrested by a rapid stream. The ford was deep, and its passage disputed by a formidable force of Armenians, Mygdonians, and Chaldæans, drawn up on an eminence 300 or 400 feet from the river. In this strait

E

Xenophon dreamt that he was in chains, and that suddenly his fetters burst asunder of their own accord. His dream was fulfilled when two youths casually found a more practicable ford, by which the army, after a skilful stratagem on the part of their commander, safely reached the opposite bank.*

The village of Tilleh belongs to Hassan Agha, a Kurdish chief, who lives in a small mud fort. He maintained, during the time of Beder Khan Bey, a sort of independence, sorely oppressing Christians and Yezidis. Unfortunately the Turks, with their usual want of foresight and justice, had enabled him to continue in his evil ways by selling him the revenues and tithes of the district, and naming him its governor. He came out and invited me into his castle, pressing me to pass the night with him, and regaling us with pipes and coffee. It was near Tilleh that the Sultan's troops, assisted by the Yezidis, completely defeated Khan Mahmoud, who was marching with the tribes of Wan and Hakkiari to the help of Beder Khan Bey.

The sun had set before our baggage had been crossed, and we sought, by the light of the moon, the difficult track along the Tigris, where the river forces its way to the low country of Assyria, through a long, narrow, and deep gorge. Huge rocks rose perpendicularly on either side, broken into many fantastic shapes, and throwing their dark shadows over the water. In some places they scarcely left room for the river to pursue its course; and then a footpath, hardly wide enough to admit the loaded mules, was carried along a mere ledge overhanging the gurgling stream. The gradual deepening of this outlet during countless centuries is strikingly shown by the ledges which jutt out like a succession of cornices from the sides of the cliffs. The last ledge left by the retiring waters formed our pathway. The geological history of the Tigris, and, consequently, of the low country, at its entry into the plain, is strikingly illustrated by this rocky ravine. In winter this drainer of the springs and snows of the highlands of Armenia and Kurdistan is swollen into a most impetuous torrent, whose level is often full thirty feet above the summer average of the river.

We found no village until we reached Chellek. The place had been deserted by its inhabitants for the Yilaks, or mountain pastures. On the opposite side of the river (in the district of Asheeti) danced the lights of a second village, also called Chellek, but dis-

* Anab. book iv. c. 3.

tinguished from the one on the eastern bank by the addition of
" Ali Rummo," the name of a petty Kurdish chief, who owns a
mud fort there.

After some search we found a solitary Kurd, who had been left
to watch the small patches of cultivation belonging to the villagers.
Taking us for Turkish soldiers, he had hidden himself on our arrival.   He offered to walk to the tents, and returned after midnight
with provisions for ourselves and barley for our horses.

For three hours during the following morning we followed the
bold and majestic ravine of the Tigris, scenes rivalling each other
in grandeur and beauty opening at every turn.   Leaving the
river, where it makes a sudden bend to the northward, we commenced a steep ascent, and in an hour and a half reached the
Christian village of Khouara.   We rested during the heat of the
day under the grateful shade of a grove of trees, and in the
afternoon continued our journey, ascending again as soon as we
had left the village, towards the crest of a mountain, from whence,
according to Cawal Yusuf, we were to behold all the world ;
and certainly, when we reached the summit, there was about as
much of the world before us as could well be taken in at one ken.
We stood on the brink of the great platform of Central Asia.
Beneath us were the vast plains of Mesopotamia, lost in the
hazy distance, the undulating land between them and the Taurus confounded, from so great a height, with the plains themselves ; the hills of the Sinjar and of Zakko, like ridges on an
embossed map ; the Tigris and the Khabour, winding through the
low country to their place of junction at Dereboun ; to the right,
facing the setting sun, and catching its last rays, the high cone
of Mardin ; behind, a confused mass of peaks, some snow-capped,
all rugged and broken, of the lofty mountains of Bohtan and
Malataiyah ; between them and the northern range of Taurus,
the deep ravine of the river and the valley of Redwan.   I
watched the shadows as they lengthened over the plain, melting
one by one into the general gloom, and then descended to the
large Kurdish village of Funduk, whose inhabitants, during the
rule of Beder Khan Bey, were notorious amongst even the savage
tribes of Bohtan for their hatred and insolence to Christians.

Although we had now nothing to fear, I preferred seeking
another spot for our night's halt, and we passed through the narrow streets as the families were settling themselves on the housetops for their night's rest.   We had ridden about half a mile when
we heard a confused murmur in the village, and saw several Kurds

running towards us at the top of their speed. Mr. C., had been
fairly frightened into a state of despair by the youngest of our
party, who entered with mischievous minuteness into the details of
the innumerable robberies and murders, authentic and otherwise,
committed by the people of Funduk. He now made up his mind that
his last hour was come, but gallantly prepared his double-barrelled
pistols. Neither Cawal Yusuf nor myself could exactly make
out what was in store for us, until the foremost of the runners,
seizing my bridle, declared that the Kiayah, or chief, would not
allow me to proceed without partaking of his hospitality ; that it
was worse than an insult to pass his house without eating bread
and sleeping under his roof. Other Kurds soon came up with
us, using friendly violence to turn my horse, and swearing that the
chief, although suffering from severe illness, would come out himself unless I consented to retrace my steps. It was useless to per-
sist in a refusal after such a display of hospitality, and notwithstanding the protests of my companion, who believed that we were
rushing into the jaws of destruction, I rode back to the village.

Resoul Kiayah, although laboring under a fit of ague, was
standing at his door to receive me, surrounded by as ferocious
a set of friends as one could well desire to be in company
with. " He had entertained," he exclaimed, as he saluted me,
" Osman Pasha and Ali Pasha, and it would be a disgrace upon
his house if the Bey passed without eating bread in it." In
the meanwhile a sheep had been slain, and comfortable carpets
and cushions spread on the housetop. His greeting of Yusuf,
although he knew him to be a Yezidi, was so warm and evidently
sincere, that I was at a loss to account for it, until the Cawal
explained to me that when Khan Mahmoud and Beder Khan Bey's
troops were defeated near Tilleh, the Kiayah of Funduk fell into
the hands of the men of Redwan, who were about to inflict summary justice upon him by pitching him into the river. He was
rescued by our friend Akko, who concealed him in his house until
he could return to Kurdistan in safety. To show his gratitude he
has since condescended to bestow on the Yezidi chief the title of
father, and to receive with a hearty welcome such travellers of the
sect as may pass through his village. The Kurds of Funduk wear
the Bohtan dress in its full perfection, a turban nearly three feet
in diameter, shalwars or trowsers of enormous width, loose embroidered jackets, and shirt sleeves sweeping the ground ; all being
striped deep dull red and black, except the under-linen and one kerchief tied diagonally across the turban, which is generally of bright

yellow. They are armed, too, to the teeth, and as they crouched round the fires on the housetops, their savage countenances peering through the gloom, my London companion, unused to such scenes, might well have fancied himself in a den of thieves. The Kiayah, notwithstanding his bad reputation, was exact in all the duties of hospitality; the supper was abundant, the coffee flowed perpetually, and he satisfied my curiosity upon many points of revenue, internal administration, tribe-history, and local curiosities.

We passed the night on the roof without any adventure, and resumed our journey before dawn on the following morning, to the great relief of Mr. C., who rejoiced to feel himself well out of the hands of such dangerous hosts. Crossing a mountain wooded with dwarf oaks, by a very difficult pathway, carried along and over rocks containing many excavated tombs, we descended to Fynyk, a village on the Tigris supposed to occupy the site of an ancient town (Phœnica).* We rested during the heat of the day in one of the pleasant gardens with which the village is surrounded. At its entrance was a group of girls and an old Kurd baking bread in a hole in the ground, plastered with clay. " Have you any bread?" we asked. — " No, by the Prophet!" " Any buttermilk?"—" No, by my faith!" " Any fruit?"—" No, by Allah!" — the trees were groaning under the weight of figs, pomegranates, pears, and grapes. He then asked a string of questions in his turn: " Whence do you come?"—" From afar!" " What is your business?"—" What God commands!" " Whither are you going?"—" As God wills!" The old gentleman, having thus satisfied himself as to our character and intentions, although our answers were undoubtedly vague enough, and might have been elsewhere considered evasive, left us without saying a word more, but soon after came back bearing a large bowl of curds, and a basket filled with the finest fruit. Placing these dainties before me, he ordered the girls to bake bread, which they speedily did, bringing us the hot cakes as they drew them from their primitive oven.

---

* It was at the foot of this steep descent that Xenophon was compelled to turn off, as caravans still are, from the river, and to brave the difficulties of a mountain pass, defended by the warlike Carduchi or Kurds. The Rhodian, who offered to construct a bridge with the inflated skins of sheep, goats, oxen, and asses, anchoring them with stones, and covering them with fascines and earth, had perhaps taken his idea from the rafts which were then used for the navigation of the Tigris, as they are to this day. As there was a large body of the enemy on the opposite side, ready to dispute the passage, the Greeks were unable to avail themselves of his ingenious suggestion.

After we had breakfasted, some Kurds who had gathered round us, offered to take me to a rock, sculptured, they said, with unknown Frank figures. We rode up a narrow and shady ravine, through which leapt a brawling torrent, watering fruit trees and melon beds. The rocks on both sides were

honeycombed with tombs. The bas-relief is somewhat above the line of cultivation, and is surrounded by excavated chambers. It consists of two figures, dressed in loose vests and trowsers, one apparently resting his hand on the shoulder of the other. There are the remains of an inscription, but too much weather-worn to be copied with any accuracy. The costume of the figures, and the forms of the characters, as far as they can be distinguished, prove that the tablet belongs to the Parthian period. It closely resembles monuments of the same epoch existing in the mountains of Persia.* Most of the

Sculptured Tablet at Fynyk

surrounding tombs, like those of Akhlat, contain three troughs or niches for the dead, one on each side, and a third facing the entrance.

We quitted Fynyk in the afternoon. Accompanied by Cawal Yusuf and Mr. C., I left the caravan to examine some rock-sculptures, in a valley leading from Jezireh to Derghileh, the former stronghold of Beder Khan Bey. The sculptures are about two miles from the high road, near a small fort built by Mir Saif-ed-din †, and now occupied by a garrison of Arnaouts. There are two tablets, one above the other; the upper contains a warrior on horseback, the lower a single figure. Although no traces of in-

* Particularly those which I discovered near Shimbor, in the mountains of Susiana. (Journal of Geog. Soc. vol. xvi. p. 84.)

† Mir Saif-ed-din was the hereditary chief of Bohtan, in whose name Beder Khan Bey exercised his authority. His son, Asdenshir (a corruption of Ardeshir) Bey, is now under *surveillance* amongst the Turks. So well aware was Beder Khan Bey of the necessity of keeping up the idea amongst the Kurds, that his power was delegated to him by the Mir, that he signed most of his public documents with that chief's seal, although he confined him a close prisoner until his death.

scription remain, the bas-reliefs may confidently be assigned to the
same period as that at Fynyk. Beneath them is a long cutting,
and tunnel in the rock, probably an ancient watercourse for irrigation, to record the construction of which the tablets may have been
sculptured. On our return we passed a solitary Turkish officer,
followed by his servant, winding up the gorge on his way to Derghileh, where one Ali Pasha was stationed with a detachment of
troops; a proof of the change which had taken place in the country

Rock Sculpture near Jezireh.   Rock Sculpture near Jezireh.

since my last visit, when Beder Khan Bey was still powerful, and
no Turk would have ventured into that wild valley.

We found the caravan at Mansouriyah, where they had established themselves for the night. This is one of the very few
Nestorian Chaldæan villages of the plains which has not gone over
to the Roman Catholic faith. It contains a church, and supports
a priest. The inhabitants complained much of oppression, and,
unfortunately, chiefly from brother Christians formerly of their
own creed. I was much struck with the intelligence and beauty of
the children ; one boy, scarcely twelve years of age, was already a

E 4

shamasha, or deacon, and could read with ease the Scriptures and the commentaries.

We left Mansouriyah at four in the morning, passing Jezireh about dawn, its towers and walls just visible through the haze on the opposite bank of the Tigris. Shortly after we were unexpectedly met by a number of Yezidi horsemen, who, having heard of our approach from the messengers sent to Hussein Bey, had ridden through the night from Dereboun to escort us. They were mounted on strong, well-bred Arab mares, and armed with long lances tipped with ostrich-feathers. We learnt from them that the country was in a very disturbed state, on account of the incursions of the Desert Arabs; but as a strong party was waiting to accompany us to Semil, I determined upon taking the shorter, though more dangerous and less frequented, road by Dereboun. This road, impracticable to caravans except when the river Khabour is fordable, winds round the spur of the Zakko hills, and thus avoids a difficult and precipitous pass. We stopped to breakfast at the large Catholic Chaldæan village of Tiekhtan, one of the many settlements of the same sect scattered over the singularly fertile plain of Zakko. The Yezidi Kochers, or Nomades, had begun to descend from the mountain pastures, and their black tents and huts of boughs and dried grass were scattered amongst the villages. We forded the Khabour, where it is divided into several branches, and not far from its junction with the Tigris. The water in no part reached much above the horses' bellies, and the stream was far less rapid than that of the eastern Tigris, at Tilleh. Dereboun is a large Yezidi village standing on the western spur of the Zakko range. Numerous springs burst from the surrounding rocks, and irrigate extensive rice-grounds. Below is the large Christian village of Feshapoor, where there is a ferry across the Tigris. We were most hospitably entertained by the Yezidi chief, one of the horsemen who had met us near Jezireh.

We mounted our horses as the moon rose, and resumed our journey, accompanied by a strong escort, which left us when we were within five or six miles of Semil. It was late in the forenoon before we reached our halting-place, after a dreary and fatiguing ride. We were now fairly in the Assyrian plains; the heat was intense — that heavy heat, which seems to paralyse all nature, causing the very air itself to vibrate. The high artificial mound of the Yezidi village, crowned by a modern mud-built castle, had been visible in the distance long before we reached it, miraged into double its real size, and into an imposing group of towers and

fortifications. Almost overcome with weariness, we toiled up to
it, and found its owner, Abde Agha, the Yezidi chieftain, seated
in the gate, a vaulted entrance with deep recesses on both sides,
used as places of assembly for business during the day*, and as
places of rest for guests during the night. He was of a tall,
commanding figure, with the deepest and most powerful voice
I ever heard. We arrived earlier than he had expected, our
forced march from Dereboun having saved us some hours, and he
apologised for not having ridden out to meet us. His reception
was most hospitable; the lamb was slain and the feast prepared.
But, in the midst of our greetings, a man appeared breathless
before him. The Bedouins had attacked the neighbouring district
and village of Pashai, belonging to Abde Agha's tribe. No time
was lost in idle preparations. The messenger had scarcely delivered
his message, and answered a few necessary inquiries, before the
high bred mare was led out ready saddled from the harem; her
owner leapt on her back, and followed by a small body of horsemen, his immediate dependants, galloped off in the direction of the
Tigris. Wearied by my long night's march I retreated to a cool
dark chamber in the castle, unmindful of the bloody business on
which its owner had sallied forth.

Abde Agha did not return that day, but his wife well performed
all the duties of hospitality in his stead. Messengers occasionally
came running from the scene of the fight with the latest news,
mostly, as in such cases, greatly exaggerated, to the alarm of those
who remained in the castle. But the chief himself did not appear
until near dawn the following morning, as we were preparing to
renew our journey. He had not been idle during his absence, and
his adherents concurred in stating that he had killed five Arabs
with his own hand. His brother, however, had received a dangerous wound, and one of his relations had been slain. He advised
us to make the best of our way to Tel Eskoff, before the Arabs
were either repulsed, or had succeeded in taking Pashai. He could
not furnish us with an escort, as every man capable of bearing

* The custom of assembling and transacting business in the gate is continually referred to in the Bible. See 2 Sam. xix. 8., where king David is
represented as sitting in the gate; comp. 2 Chron. xviii. 9., and Dan. ii. 49.
The gates of Jewish houses were probably similar to that described in the
text. Such entrances are also found in Persia. Frequently in the gates of
cities, as at Mosul, these recesses are used as shops for the sale of wheat and
barley, bread and grocery. Elisha prophesies that a measure of fine flour shall
be sold for a shekel, and two measures of barley for a shekel, *in the gate* of
Samaria. 2 Kings, vii. 1. and 18.

arms was wanted to defend the district against the Bedouins, who were now swarming over the river to support their companions. Taking a hasty leave of us, and changing his tired mare, he rushed again to the fight. We rode off in the direction of the hills, taking an upper road, less likely to be occupied by the Arabs.

About three miles from Semil we saw a horseman closely pursued by a Bedouin, who was fast coming up with him, but on observing us turned back, and soon disappeared in the distance. The fugitive was a Mosuleean Spahi, with broken spear, and speechless with terror. When he had sufficiently recovered himself to speak, he declared that the Bedouins had defeated the Yezidis, and were spreading over the country. Although not putting much faith in the information, I urged on the caravan, and took such precautions as were necessary. Suddenly a large body of horsemen appeared on a rising ground to the east of us. We could scarcely expect Arabs from that quarter; however, all our party made ready for an attack. Cawal Yusuf and myself, being the best mounted, rode towards them to reconnoitre. Then one or two horsemen advanced warily from the opposite party. We neared each other. Yusuf spied the well-known black turban, dashed forward with a shout of joy, and in a moment we were surrounded, and in the embrace of friends. Hussein Bey and Sheikh Nasr, with the Cawals and Yezidi elders, had ridden nearly forty miles through the night to meet and escort me, if needful, to Mosul! Their delight at seeing us knew no bounds; nor was I less touched by a display of gratitude and good feeling, equally unexpected and sincere.

They rode with us as far as Tel Eskoff, where the danger from the Arabs ceased, and then turned their hardy mares, still fresh after their long journey, towards Sheikhan. I was now once more with old friends. We had spent the first day of our journey, on leaving Mosul two years ago, in the house of Toma, the Christian Kiayah of Tel Eskoff; we now eat bread with him the last on our return. In the afternoon, as we rode towards Tel Kef, I left the high road with Hormuzd to drink water at some Arab tents. As we approached we were greeted with exclamations of joy, and were soon in the midst of a crowd of men and women, kissing our knees, and exhibiting other tokens of welcome. They were Jebours, who had been employed in the excavations. Hearing that we were again going to dig after old stones, they at once set about striking their tents to join us at Mosul or Nimroud.

As we neared Tel Kef we found groups of my old superintendents and workmen by the road side. There were fat Toma,

Mansour, Behnan, and Hannah, joyful at meeting me once more, and at the prospect of fresh service.   In the village we found Mr. Rassam (the vice-consul) and Khodja Toma, his dragoman, who had made ready the feast for us at the house of the Chaldæan bishop.   Next morning, as we rode the three last hours of our journey, we met fresh groups of familiar faces :—Merjan, with my old groom holding the stirrup ready for me to mount, the noble animal looking as beautiful, as fresh, and as sleek as when I last saw him, although two long years had passed ; former servants, Awad and the Sheikhs of the Jebours, even the very greyhounds who had been brought up under my roof.   Then as we ascend an eminence midway, walls, towers, minarets, and domes rise boldly from the margin of the broad river, cheating us into the belief, too soon to be dispelled, that Mosul is still a not unworthy representative of the great Nineveh.   As we draw near, the long line of lofty mounds, the only remains of mighty bulwarks and spacious gates, detach themselves from the low undulating hills : now the vast mound of Kouyunjik overtops the surrounding heaps ; then above it peers the white cone of the tomb of the prophet Jonah , many other well-remembered spots follow in rapid succession ; but we cannot linger.   Hastening over the creaking bridge of boats, we force our way through the crowded bazars, and alight at the house I had left two years ago.   Old servants take their places as a matter of course, and, uninvited, pursue their regular occupations as if they had never been interrupted.   Indeed it seemed as if we had but returned from a summer's ride ; two years had passed away like a dream.

I may in this place add a few words on part of the route pursued by Xenophon and the Ten Thousand during their memorable retreat, the identification of which had been one of my principal objects during our journey.   I have, in the course of my narrative, already pointed out one or two spots signalled by remarkable events on their march.

I must first state my conviction that the parasang, like its representative the modern farsang or farsakh of Persia, was not a measure of distance very accurately determined, but rather indicated a certain amount of time employed in traversing a given space.   Travellers are well aware that the Persian farsakh varies considerably according to the nature of the country, and the usual modes of conveyance adopted by its inhabitants.   In the plains of Khorassan and central Persia, where mules and horses are chiefly used by caravans, it is equal to about four miles, whilst in the

mountainous regions of Western Persia, where the roads are difficult and precipitous, and in Mesopotamia and Arabia, where camels
are the common beasts of burden, it scarcely amounts to three.
The farsakh and the hour are almost invariably used as expressing
the same distance. That Xenophon reckoned by the common
mode of computation of the country is evident by his employing,
almost always, the Persian " parasang " instead of the Greek
stadium ; and that the parasang was the same as the modern hour,
we find by the distance between Larissa (Nimroud) and Mespila
(Kouyunjik) being given as six parasangs, corresponding exactly with the number of hours assigned by the present inhabitants of the country, and by the authorities of the Turkish post,
to the same road. The six hours in this instance are equal to
about eighteen English miles.

The ford, by which the Greeks crossed the Great Zab (Zabates)
may, I think, be accurately determined. It is still the principal
ford in this part of the river, and must, from the nature of the
bed of the stream, have been so from the earliest periods. It is
about twenty-five miles from the confluence of the Zab and Tigris.*
A march of twenty-five stadia, or nearly three miles, in the direction
of Larissa, would have brought them to the Ghazir, or Bumadus ;
and this stream was, I have little doubt, the deep valley formed
by the torrent where Mithridates, venturing to attack the retreating army, was signally defeated.† This action took place
eight stadia beyond the valley; the Persian commander having
neglected to intercept the Greeks when endeavoring to cross
the difficult ravine, in which they would most probably have
been entangled. A short march of three parasangs, or hours ‡,
brought them to Larissa, the modern Nimroud. The Greeks
could not have crossed the Zab above the spot I have indicated,
as the bed of the river is deep, and confined within high rocky
banks. They might have done so *below* the junction of the
Ghazir, and a ravine worn by winter rains may correspond with
the valley mentioned by Xenophon, but I think the Ghazir far
more likely to have been the torrent bed viewed with so much

---

* Mr. Ainsworth would take the Greeks up to the modern ferry, where there
could never have been a ford, and which would have been some miles out of
their route. (Travels in the Track of the Ten Thousand.)
† Anab. book iii. ch. 4.
‡ Xenophon merely says that they marched the rest of the day. After the
action, they could scarcely have advanced more than three parasangs, or nine
miles.

alarm by the Greek commander, and the passage of which Mithridates might have disputed with some prospect of success.*

That Larissa and Mespila are represented by the ruins of Nimroud and Kouyunjik no one can reasonably doubt. Xenophon's description corresponds most accurately with the ruins and with the distance between them.

From Mespila the Greeks marched four parasangs and probably halted near the modern village of Batnai, between Tel Kef and Tel Eskof, an ancient site exactly four hours, by the usual caravan road, from Kouyunjik. Many ancient mounds around Batnai mark the remains of those villages, from which, after having repulsed the Persian forces under Tissaphernes and Orontas, the Greeks obtained an abundant supply of provisions. Instead of fording the Khabour near its junction with the Tigris, and thus avoiding the hills, they crossed them by a precipitous pass to the site of the modern Zakko. They reached this range in four days, traversing it on the fifth, probably by the modern caravan road. The distance from Batnai to Zakko, according to the Turkish post, is twenty hours. This would give between four and five hours, or parasangs, a day for the march of the Greeks, the distance they usually performed. They were probably much retarded during the last day, by having to fight their way over three distinct mountain ridges. It is remarkable that Xenophon does not mention the Khabour, although he must have crossed that river either by a ford or by a bridge † before reaching the plain. Yet the stream is broad and rapid, and the fords at all times deep. Nor does he allude to the Hazel, a confluent of the Khabour, to which he came during his first day's march, after leaving Zakko. These omissions prove that he does not give an accurate itinerary of his route.

Four days' march, the first of only sixty stadia, or about seven miles‡, brought the Greeks to the high mountains of Kurdistan, which, meeting the Tigris, shut out all further advance except by difficult and precipitous passes, already occupied by the Persians.

---

* In Chapter X. will be found some further remarks on this subject; many reasons, based upon personal experience, may be adduced for the probability of Xenophon's preferring the upper ford.

† He probably took the more difficult road over the pass, and not that round the spur, in order to cross the Khabour by a bridge or ferry. It must be remembered that it was winter, and that the rivers were consequently swollen.

‡ This halt, after so short a day's march, may have been occasioned by the Hazel. The distance corresponds with sufficient accuracy.

Xenophon, having dislodged the enemy from the first ridge, returned to the main body of the army, which had remained in the plain. This must have been near Fynyk, where the very foot of the Kurdish mountains is first washed by the river. The spot agrees accurately with Xenophon's description, as it does with the distance. "The Greeks," says he, "came to a place where the river Tigris is, both from its depth and breadth, absolutely impassable; no road appeared, the craggy mountains of the Carduchians hanging over the river." The offer of the Rhodian to cross the army on inflated skins, bound together to form a bridge, having been rejected, on account of the strong force assembled on the opposite side to dispute the passage, the Greeks marched back to the villages. The Persian prisoners informed Xenophon that four roads branched off from this spot: one to the south, by which the Greeks had retreated from Babylonia; the second eastwards, to Susa and Ecbatana, by the plain of Zakko, the modern Amadiyah, Suleimaniyah, and the foot of the great range of Zagros; a third to the west, crossing the Tigris, near Jezireh, and thence through Orfa, Aintab, Tarsus, and the Cilician gates to Lydia and Ionia; and a fourth across the mountains of the Carduchians, or Kurdistan. The tribes infesting this fourth road were represented to Xenophon as notorious for their courage and warlike habits. They only held intercourse with the inhabitants of the low country, when they were at peace with the governor residing in the plain, and such has been precisely the case with their descendants to this day. This route was, however, preferred, as it led into Armenia, a country from which they might choose their own road to the sea, and which abounded in villages and the necessaries of life.

The Greeks appear to have followed the route taken by Sultan Murad in his expedition against Baghdad, and, recently, by part of the Turkish forces sent against Beder Khan Bey; in fact, the great natural highway from the remotest period between eastern Armenia and Assyria. Beyond the Carduchian mountains there were, according to the prisoners, two roads into Armenia, one crossing the head waters of the principal branch of the Tigris, the other going round them; that is, leaving them to the left. These are the roads to this day followed by caravans, one crossing the plains of Kherzan to Diarbekir, and thence, by well-known mountain-passes to Kharput, the other passing through Bitlis. Xenophon chose the latter. The villages in the valleys and recesses of the mountains are still found around Funduk; and, on their first day's march over the Carduchian hills, the Greeks probably reached the neighbourhood of this village. There now remained about ten parasangs to the plain through which flows the eastern branch of the Tigris; but the country was difficult, and at this time of the year (nearly midwinter)*, the lower road along the river was impassable. The Greeks had, therefore, to force their way over a series of difficult passes, all stoutly defended by warlike tribes. They were consequently four days in reaching the Centritis, or eastern Tigris, the united waters of the rivers of Bitlis, Sert, and Bohtan. The stream was rapid, the water reaching to the breast, and the ford, owing to the unevenness of the bottom and the loose, slippery stones, exceedingly difficult; such, it will be remembered, we found to be the case near Tilleh. The opposite banks were, moreover, defended by the combined forces of the Armenians, Mygdonians, and Chaldæans. It was impossible to cross the river at this spot in the face of the enemy. At length a ford was discovered higher up, and Xenophon, by skilful strategy, effected the passage. This must have been at a short distance from Tilleh, as the river, narrowed between rocky banks, is no longer fordable higher up. The Greeks came upon the Centritis soon after leaving the Carduchian mountains.

The direct and most practicable road would now have been along the river banks to Bitlis†, but owing to the frequent incursions of the Carduchi, the villages in that direction had been abandoned, and the Greeks were compelled to turn to the westward, to find provisions and habitations. Still *there was no road* into Armenia, particularly at this time of year, for an army encumbered with baggage, except that through the Bitlis valley. The

* It is a matter of surprise that Cyrus should have chosen the very middle of summer for his expedition into Babylonia, and still more wonderful that the Greeks, unused to the intense heats of Mesopotamia, and encumbered with their heavy arms and armour, should have been able to brave the climate. No Turkish or Persian commander would in these days venture to undertake a campaign against the Arabs in this season of the year; for, besides the heat, the want of water would be almost an insurmountable obstacle. During their retreat, the Greeks had to encounter all the rigor of an Armenian winter; so that, during the few months they were under arms, they went through the most trying extremes of climate. The expedition of Alexander was also undertaken in the middle of summer. It must, however, be borne in mind, that Mesopotamia was probably then thickly peopled and well cultivated, and that canals and wells of water must have abounded.

† That by Sert passes over very precipitous mountains, and is only now taken by caravans, because it is more secure than the other, and leads through a town in which there is some trade.

remains of an ancient causeway are even now to be traced, and this probably has always been the great thoroughfare between western Armenia and the Assyrian plains. Xenophon consequently made nearly the same detour as I had made on my journey from Constantinople.

Six marches, of five parasangs each, brought them to the small river Teleboas. I am convinced that this river cannot be identified with the Kara Su, which would be at least between forty and fifty parasangs, or from eight to ten days' march, from Tilleh, supposing Xenophon to have made the smallest possible deviation to the west. I believe the Teleboas to have been the river of Bitlis.* After crossing the low country of Kherzan, well described by Xenophon as "a plain varied by hills of an easy ascent," the Greeks must necessarily have turned slightly to the eastward to reach the Bitlis valley, as inaccessible mountains stopped all further progress. My caravan was thirty-three hours in journeying from Bitlis to Tilleh, corresponding exactly with the six days' march of the Greeks. They probably came to the river somewhat below the site of the modern town, where it well deserves the epithet of " beautiful." It may have then had, as at this day, many villages near its banks. It will be observed that Xenophon says that *they came to*, not that they *crossed*, the Teleboas.

From this river they reached the Euphrates in six marches, making, as usual, five parasangs each day; in all, thirty parasangs, or hours. Now from the Kara Su to the Euphrates, even supposing the Greeks to have gone far to the eastward out of the direct route on the plain of Malaskert, there would scarcely be twenty parasangs, whereas the high road from Bitlis to Northern Armenia would lead in exactly thirty hours, or six marches, to the Euphrates, which it crosses near Karaghal. I believe, therefore, that, after issuing from the valley of Bitlis, Xenophon turned to the westward, leaving the lake of Wan a little to the right, though completely concealed from him by a range of low hills.† Skirting the western foot of the Nimroud Dagh range, he passed through a plain thickly inhabited, abounding in well-provisioned villages, and crossed here and there by ranges of hills. This country still tallies precisely with Xenophon's description.

---

* It must be borne in mind that the river of Bitlis joins the Sert Su before it falls into the main branch of the Tigris at Tilleh, and might therefore, under a different name, have appeared another river to Xenophon.

† Had he seen this large inland sea, he would probably have mentioned it.

We have not, I conceive, sufficient data in Xenophon's narrative to identify with any degree of certainty his route after crossing the Euphrates. We know that about twenty parasangs from that river the Greeks encamped near a hot spring, and this spring might be recognised in one of the many which abound in the country. It is most probable that the Greeks took the road still used by caravans through the plains of Hinnis and Hassan-Kalah, as offering the fewest difficulties. But what rivers are we to identify with the Phasis and Harpasus, the distance between the Euphrates and Phasis being seventy parasangs, and between the Phasis and Harpasus ninety-five, and the Harpasus being the larger of the two rivers? I cannot admit that the Greeks turned to the west, and passed near the site of the modern Erzeroom. There are no rivers in that direction to answer the description of Xenophon. Moreover, the Greeks came to the high mountain, and beheld the sea for the first time, at the distance of thirty-two parasangs from Trebizond. Had they taken either of the three modern roads from Erzeroom to the coast, and there are no others, they must have seen the Euxine in the immediate vicinity of Trebizond, certainly not more than six or eight parasangs from that city. I am on the whole inclined to believe, that either the Greeks took a very tortuous course after leaving the Euphrates, making daily but little actual progress towards the great end of their arduous journey, the sea coast, or that there is a considerable error in the amount of parasangs given by Xenophon; that the Harpasus must be the Tcherouk, and the Phasis either the Araxes or the Kur *; and that Mount Theches, the holy mountain from which the Greeks beheld the sea, was between Batoun and Trebizond, the army having followed the valley of the Tcherouk, but leaving it before reaching the site of the modern port on the Black Sea.

* In no way, however, would a *direct* line of march between these two rivers, nor between any other two rivers which can possibly answer to his description, tally with the distances given by Xenophon.

Mosul, from the North.

Subterranean Excavations at Kouyunjik

CHAP. IV

ON the morning after our arrival in Mosul, I rode at sunrise
to Kouyunjik. The reader may remember that, on my return
to Europe in 1847, Mr. Ross had continued the researches in that

PLAN
of
EXCAVATED CHAMBERS
of
KOUYUNJIK.
No 1

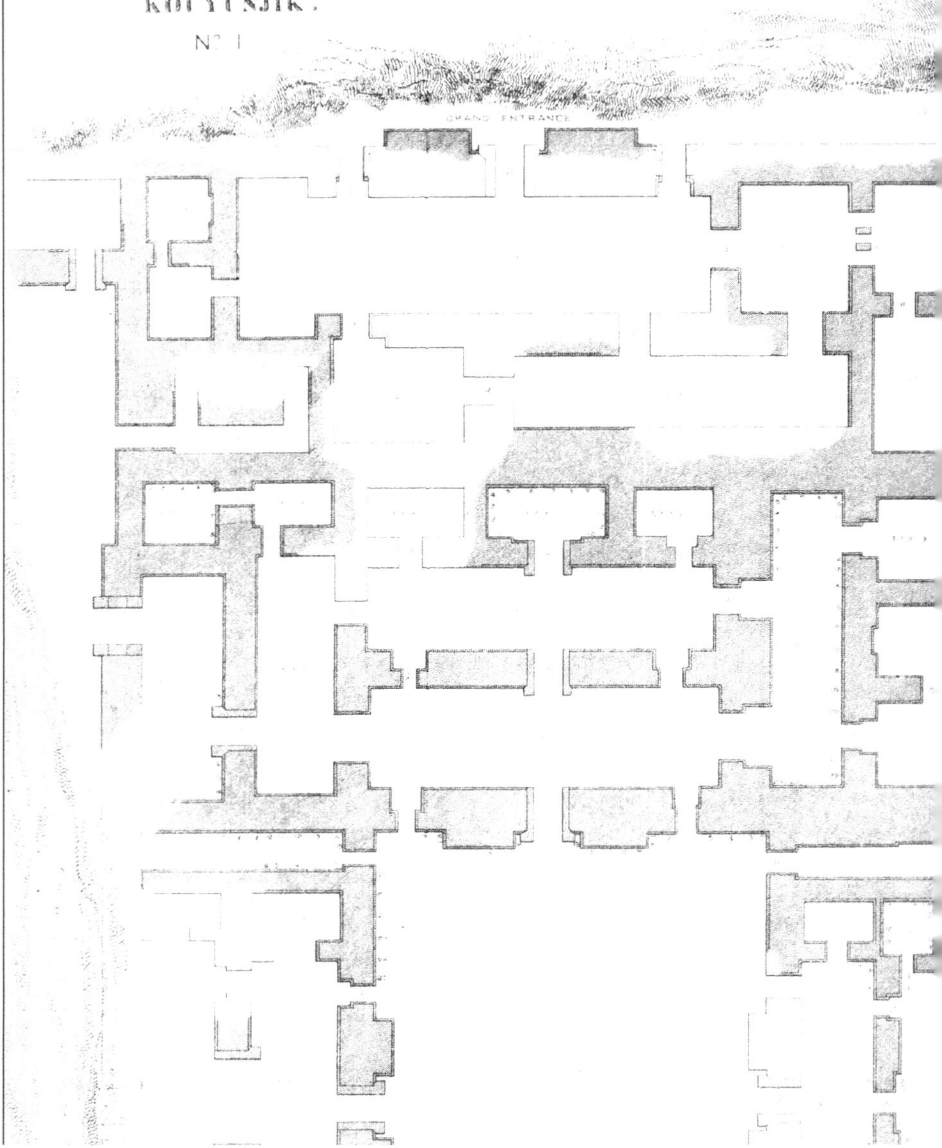

GRAND ENTRANCE

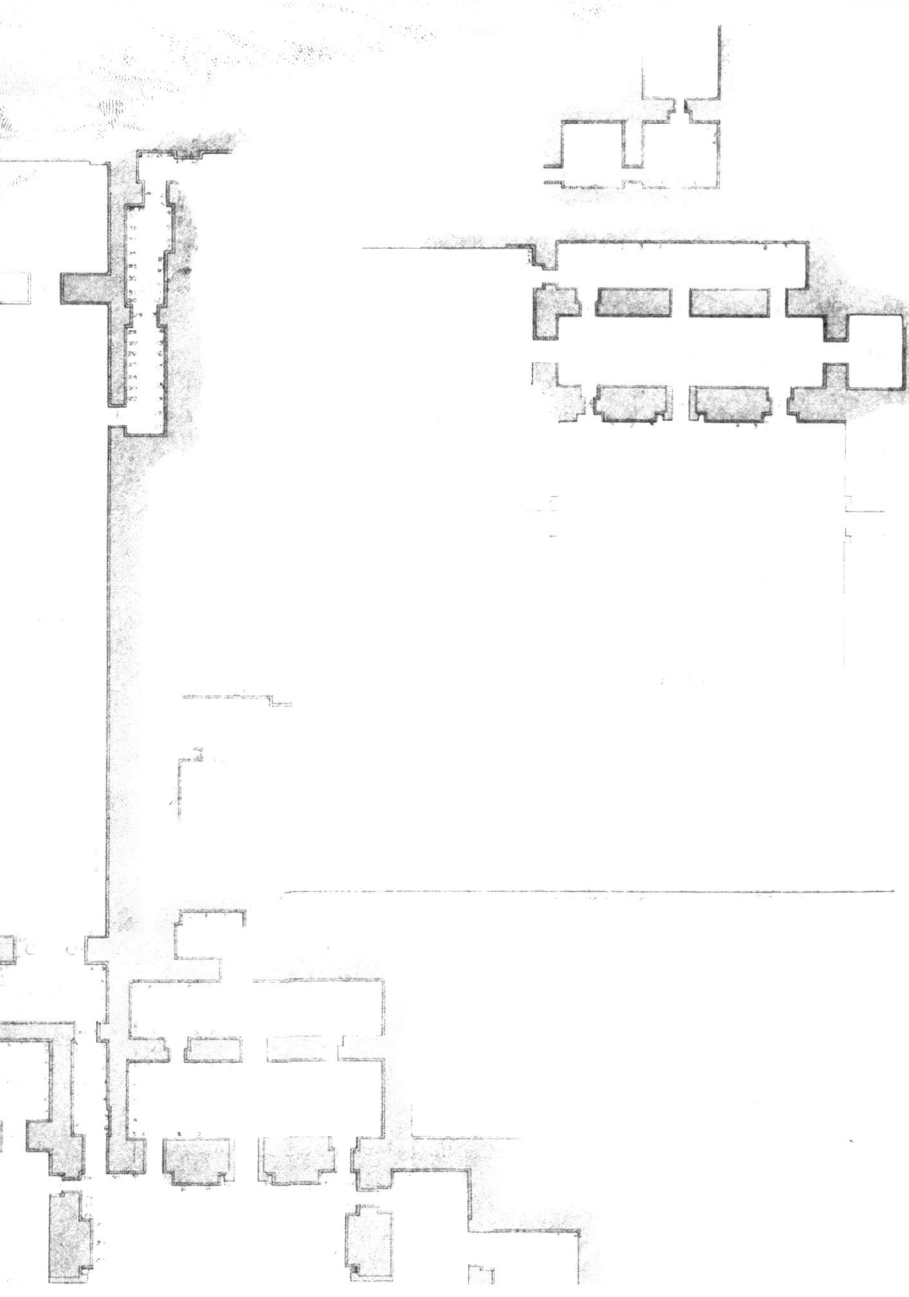

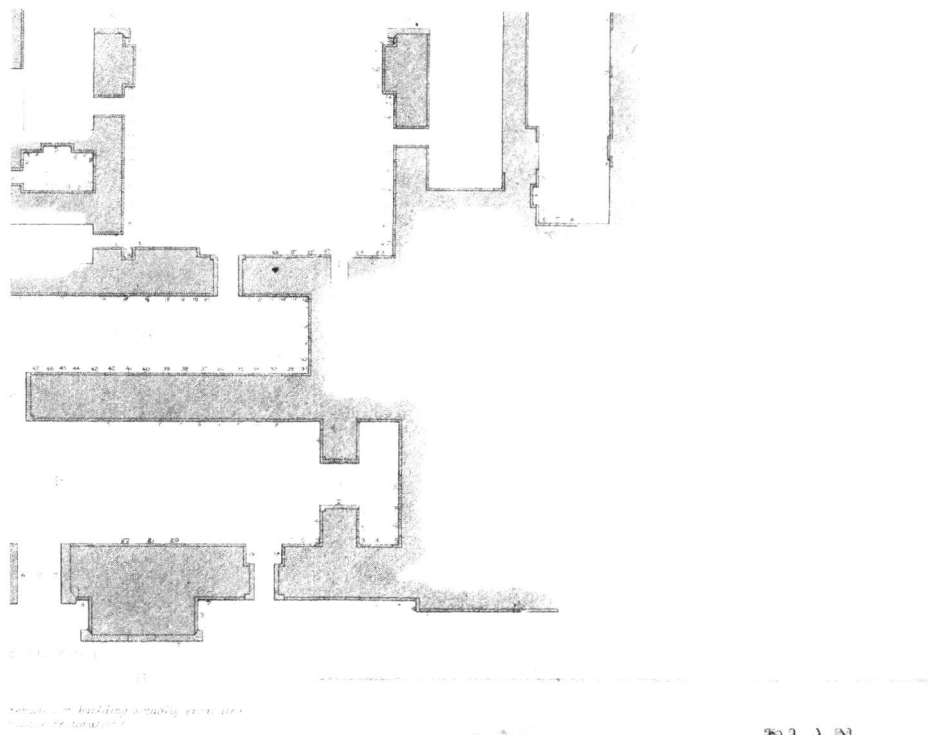

PLAN
of
THE MOUND OF KOUYUNJIK
by
Lieut. F. G. Edoswitt R.A.

mound, and had uncovered several interesting bas-reliefs, which I have already described from his own account of his discoveries.* That gentleman had, to my great regret, left Mosul. Since his departure the excavations had been placed under the charge of Mr. Rassam, the English vice-consul, who was directed by the Trustees of the British Museum to employ a small number of men, rather to retain possession of the spot, and to prevent interference on the part of others, than to carry on extensive operations. Toma Shishman, or "the Fat," was still the overseer of the workmen, and accompanied me on my first visit to the ruins.

But little change had taken place in the great mound since I had last seen it. It was yellow and bare, as it always is at this time of the year. Heaps of earth marked the site of former excavations, the chambers first discovered having been again completely buried with rubbish. Of the sculptured walls laid bare two years before no traces now remained. The trenches dug under Mr. Ross's directions, in the southern corner, opposite the town of Mosul, were still open. It was evident at a glance that the chambers he had entered did not, as he had been led to suppose, belong to a second palace. They formed part of the same great edifice once standing on this angle of the mound, and already partly explored. The style of the bas-reliefs, and of the inscriptions, marked them at once as of the same epoch as those previously discovered. They belonged to the same king, and also recorded his wars and his triumphs. The same great fire, too, which had raged in the rest of the building, turning the sculptured panelling to lime, defacing the ancient records, and reducing the edifice to a heap of ashes and rubbish, had done its work here. But four or five feet remained of the bas-reliefs once covering the walls of sun-dried bricks to the height of eight or nine, and even these fragments were generally too much defaced to admit of minute description.

The walls of two chambers had been laid bare. In one†, the lower part of a long series of sculptures was still partly preserved, but the upper had been completely destroyed, the very alabaster itself having disappeared. The bas-reliefs recorded the subjection by the Assyrian king of a nation inhabiting the banks of a river. The captive women are distinguished by long embroidered robes fringed with tassels, and the castles have a peculiar wedge-shaped

---

* See Nineveh and its Remains, vol. ii. p. 139.
† No. LI. Plan I

ornament on the walls.   The towns probably stood in the midst
of marshes, as they appear to be surrounded by canes or reeds, as

Castle near a River or Marsh (Kouyunjik).

well as by groves of palm trees.   The Assyrians having captured
the strong places by escalade, carried the inhabitants into captivity,
and drove away cattle, camels, and carts drawn by oxen.   Some
of the men bear large baskets of osier work, and the women vases
or cauldrons.   The king, standing in his chariot, attended by
his warriors, and preceded by an eunuch registering the number of
prisoners and the amount of the spoil, receives the conquered chiefs.
Not a vestige of inscription remains to record the name of the
vanquished people; but we may conjecture, from the river and
the palm trees, that they inhabited some district in southern Mesopotamia.   They were, probably, one of the numerous Arab tribes
who lived in the marshes formed by the Euphrates and Tigris, and
took advantage, as their descendants do to this day, of their almost
inaccessible position in the midst of vast swamps to be in continual
rebellion against the supreme government.   Many of these tribes, it
will hereafter be seen, are mentioned amongst the southern conquests of the king who built the palace.   In the southern wall of
this chamber was a doorway formed by plain, upright slabs of
a close-grained magnesian limestone, almost as hard as flint;
between them were two small, crouching lions, in the usual

alabaster. This entrance led into a further room, of which only a small part had been explored.* The walls were panelled with unsculptured slabs of the same compact limestone.

The sculptured remains hitherto discovered in the mound of Kouyunjik had been reached by digging down to them from the surface, and then removing the rubbish. After the departure of Mr. Ross, the accumulation of earth above the ruins had become so considerable, frequently exceeding thirty feet, that the workmen, to avoid the labor of clearing it away, began to tunnel along the walls, sinking shafts at intervals to admit light and air. The hardness of the soil, mixed with pottery, bricks, and remains of buildings raised at various times over the buried ruins of the Assyrian palace, rendered this process easy and safe with ordinary care and precaution. The subterraneous passages were narrow, and were propped up when necessary either by leaving columns of earth, as in mines, or by wooden beams. These long galleries, dimly lighted, lined with the remains of ancient art, broken urns projecting from the crumbling sides, and the wild Arab and hardy Nestorian wandering through their intricacies, or working in their dark recesses, were singularly picturesque.

Toma Shishman had removed the workmen from the southern corner of the mound, where the sculptures were much injured, and had opened tunnels in a part of the building previously explored, commencing where I had left off on my departure from Mosul.† I descended into the vaulted passages by an inclined way, through which the workmen issued from beneath to throw away the rubbish dug out from the ruins. At the bottom I found myself before a wall forming the southern side of the great Hall, discovered, though only partly explored, during my former researches. ‡ The sculptures, faintly seen through the gloom, were still well enough preserved to give a complete history of the subject represented, although, with the rest of the bas-reliefs of Kouyunjik, the fire had nearly turned them to lime, and had cracked them into a thousand pieces. The faces of the slabs had been entirely covered with figures, varying from three inches

---

* No. LIII. Plan I.

† At No. VI. same plan. The chambers marked with letters in the Plan of Kouyunjik in the 2d vol. of " Nineveh and its Remains," are distinguished, for convenience of general reference, by numbers in Plan I. of this work, which includes all those excavated during the first expedition, as well as those discovered during the second : the letters are, however, also inserted.

‡ No. VI. Plan I.

to one foot in height, carefully finished, and designed with great spirit.

In this series of bas-reliefs the history of an Assyrian conquest was more fully portrayed than in any other yet discovered, from the going out of the monarch to battle, to his triumphal return after a complete victory. The first part of the subject has already been described in my former work.* The king, accompanied by his chariots and horsemen, and leaving his capital in the Assyrian plains, passed through a mountainous and wooded district.† He does not appear to have been delayed by the siege of many towns or castles, but to have carried the war at once into the high country. His troops, cavalry and infantry, are represented in close combat with their enemies, pursuing them over hills and through valleys, beside streams, and in the midst of vineyards. The Assyrian horsemen are armed with the spear and the bow, using both weapons whilst at full speed : their opponents seem to be all archers. The vanquished turn to ask for quarter ; or, wounded, fall under the feet of the advancing horses, raising their hands imploringly to ward off the impending deathblow. The triumph follows. The king standing in his chariot, beneath the royal parasol, followed by long lines of dismounted warriors leading richly caparisoned horses, and by foot soldiers variously armed and accoutred, is receiving the captives and spoil taken from the conquered people. First approach the victorious warriors, throwing the heads of the slain into heaps before the registering officers. They are followed by others leading, and urging onwards with staves, the prisoners — men chained together, or bound singly in fetters, and women, some on foot, carrying their children on their shoulders, and leading them by the hand, others riding on mules. The procession is finished by asses, mules, and flocks of sheep. As on the bas-reliefs uncovered by Mr. Ross, there is unfortunately no inscription by which the name of the conquered people can be determined. We are left to conjecture the site of the country they inhabited from its natural features, rudely portrayed in the bas-reliefs, or from notices that may hereafter — on a better acquaintance with the cuneiform character — be found in the great inscriptions on the bulls containing the history of the wars of the

---

* Nineveh and its Remains, vol. ii. p. 134.

† The long lines of variously armed troops, described in my former work (vol. ii. p. 134.) as covering several slabs from top to bottom, form the army of the king marching to this campaign. Monuments of Nineveh, Plate 81.

Assyrian king. The mountains, valleys, and streams, the vines and dwarf oaks, probably indicate a region north of Assyria, in Armenia, Media, or Kurdistan, countries we know to have been invaded by the royal builder of the palace. The dress of the men consists of a short tunic; that of the women, of a shirt falling to the ankles, and cut low in front of the neck.*

In the side of the hall sculptured with these bas-reliefs was a wide portal, formed by a pair of gigantic human-headed bulls.† They had suffered, like all those previously discovered, from the fire, and the upper part, the wings and human head, had been completely destroyed. The lower half had, however, escaped, and the inscriptions were consequently nearly entire. Joined to the forepart of the bulls were four small figures, two on each side, and one above the other. They had long hair, falling in large and massive curls on their shoulders, wore short tunics descending to the knee, and held a pole topped by a kind of cone in one hand, raising the other as in act of adoration.‡ At right angles with the slabs bearing these sculptures were colossal figures carrying the oft-repeated cone and basket.

In this entrance a well, cut through the large pavement slab between the bulls, was afterwards discovered. It contained broken pottery, not one vase having been taken out whole, apparently human remains, and *some fragments of calcined sculptured alabaster*, evidently detached from the bas-reliefs on the walls. It is doubtful whether this well was sunk after the Assyrian ruins had been buried, or whether it had been from the earliest times a place of deposit for the dead. The remains of bas-reliefs found in it, at a considerable depth, show that it must have been filled up after the destruction of the Assyrian palace; and, as no such wells exist in similar entrances, I am inclined to believe that, like many others discovered during the excavations, it had been made by those who built on the mound above the ancient ruins. When sinking the shaft they probably met with the pavement slab, and cut through it. It appears to have been afterwards choked by the falling in of the rubbish through which it had been

---

* Two plates from these spirited sculptures are given in the 2d series of the Monuments of Nineveh, Plates 37. 38. They represent the battle, and part of the triumph.

† Entrance k. No. VI. Plan I.

‡ One such figure has been placed in the British Museum, and see 2d series of the Monuments of Nineveh, Plate 6.

carried, and hence the fragments of sculptured alabaster mixed
with the broken pottery.   Being unable to support its crumbling
sides, I was obliged to abandon the attempt after digging to the
depth of about fifteen feet.

A small doorway to the right of the portal formed by the
winged bulls, led into a further chamber*, in which an entrance
had been found into a third room †, whose walls had been completely uncovered. Its dimensions were 26 feet by 23, and it
had but this one outlet, flanked on either side by two colossal
figures, whose lower extremities alone remained, the upper part
of the slabs having been destroyed: one appeared to have been
eagle-headed, with the body of a man, and the other a monster,
with human head and the feet of a lion. The bas-reliefs round
the chamber represented the siege of a castle standing on an artificial mound, surrounded at its base by houses. The besieged
defended themselves on the walls and turrets with bows, spears,
and stones. The Assyrian army was composed of spearmen,
slingers, and bowmen, some of whom had already gained the
housetops. Male and female captives had been taken and heads
cut off; the victorious warriors according to custom, and probably to claim a reward ‡, bringing them to the registrars. The
led horses and body-guard of the king was still preserved, but
that part of the bas-relief containing the monarch himself, probably standing in his chariot, had been destroyed. In the back
ground were wooded mountains; vines and other trees formed a
distinct band in the middle of the slabs; and a river ran at the
foot of the mound. The dress of the male prisoners consisted
either of a long robe falling to the ankles, or of a tunic reaching to
the knees, over which was thrown an outer garment, apparently
made of the skins of animals, and they wore greaves laced up in
front. The women were clothed in a robe descending to the feet,
with an outer fringed garment thrown over the shoulders; a
kind of hood or veil covered the back of the head, and fell over
the neck. Above the castle was the fragment of an inscription
in two lines, containing the name of the city, of which unfortunately the first character is wanting. It reads: " *The city of*
. . . alammo I attacked and captured; I carried away its spoil." No

---

* No. XIII. Plan I.   † No. XIV. same plan.
    ‡ It is still the custom in Persia, and was so until lately in Turkey, for soldiers
to bring the heads of the slain to their officers after a battle, and to claim a small
pecuniary reward.

name, however, corresponding with it has yet been found in the royal annals, and we can only infer, from the nature of the country represented, that the place was in a mountainous district to the north of Assyria.* It is remarkable that in this chamber, as in others afterwards explored, some of the slabs (those adjoining the entrance) had been purposely defaced, every vestige of sculpture having been carefully removed by a sharp instrument.

Returning to the great hall, I found that a third outlet had been discovered, opening, however, to the west. This entrance had been guarded by six colossal figures, three on each side. The upper part of all of them had been destroyed. They appear to have been eagle-headed and lion-headed monsters.†

This doorway led into a narrow passage, one side of which had alone been excavated; on it was represented the siege of a walled city, divided into two parts by a river. One half of the place had been captured by the Assyrians, who had gained possession of the towers and battlements, but that on the opposite bank of the stream was still defended by slingers and bowmen. Against its walls had been thrown banks or mounds, built of stones, bricks, and branches of trees.‡ The battering-rams, covered with skins or hides looped together, had been rolled up these inclined ways, and had already made a breach in the fortifications. Archers and spearmen were hurrying to the assault, whilst others were driving off the captives, and carrying away the idols of the enemy. The dress of the male prisoners consisted of a plain under-shirt, an upper garment falling below the knees, divided in the front and buttoned at the neck, and laced greaves. Their hair and beards were shorter and less elaborately curled than those of the Assyrians. The women were distinguished by high rounded turbans, ornamented with plaits or folds. A veil fell from the back of this headdress over the shoulders. § No inscription remained to record the name of the vanquished nation. Their castles stood in a wooded and mountainous country, and their

---

* As much of the bas-reliefs as could be moved is now in the British Museum; see also 2d series of the Monuments of Nineveh, Plate 39.

† Entrance i. No. VI. Plan I.

‡ For an account of these mounds represented in the Assyrian sculptures, and the manner in which they illustrate various passages in Scripture, see my Nineveh and its Remains, vol ii. p. 367. and note.

§ Such is the costume of the women in ships in a bas-relief discovered during my former researches (see Nineveh and its Remains, vol. ii. p. 129. and Monuments of Nineveh, Plate 71.), and which, I have conjectured, may represent the capture of Tyre or Sidon.

peculiar costume, and the river passing through the centre of their chief city, may help hereafter to identify them.

The opposite side of this narrow chamber, or passage, was shortly afterwards uncovered. The bas-reliefs on its walls represented the king in his chariot, preceded and followed by his warriors. The only remarkable feature in the sculptures was the highly decorated trappings of the horses, whose bits were in the form of a horse at full speed.

Such were the discoveries that had been made during my absence. There could be no doubt whatever that all the chambers hitherto excavated belonged to one great edifice, built by one and the same king. I have already shown how the bas-reliefs of Kouyunjik differed from those of the older palaces of Nimroud, but closely resembled those of Khorsabad in the general treatment, in the costumes of the Assyrian warriors, as well as of the nations with whom they warred, and in the character of the ornaments, inscriptions, and details. Those newly uncovered were, in all these respects, like the bas-reliefs found before my departure, and upon which I had ventured to form an opinion as to the respective antiquity and origin of the various ruins hitherto explored in Assyria. The bas-reliefs of Nimroud, the reader may remember, were divided into two bands or friezes by inscriptions; the subject being frequently confined to one tablet, or slab, and arranged with some attempt at composition, so as to form a separate picture. At Kouyunjik the four walls of a chamber were generally occupied by one series of sculptures, representing a consecutive history, uninterrupted by inscriptions, or by the divisions in the alabaster panelling. Figures, smaller in size than those of Nimroud, covered from top to bottom the face of slabs, eight or nine feet high, and sometimes of equal breadth.

The sculptor could thus introduce more action, and far more detail, into his picture. He aimed even at conveying, by rude representations of trees, valleys, mountains, and rivers, a general idea of the natural features of the country in which the events recorded took place. A chamber thus generally contained the whole story of a particular war, from the going out of the king to his triumphal return. These pictures, including a kind of plan of the campaign, add considerably to the interest of the monuments, and allow us to restore much of the history of the period. They will probably also enable us to identify the sculptured records with the descriptive accounts contained in the great inscriptions carved upon the bulls, at the various entrances to the palace, and embracing a general chronicle of the reign of the king. At Kouyunjik there were probably few bas-reliefs, particularly those containing representations of castles and cities, that were not accompanied by a short epigraph or label, giving the name of the conquered king and country, and even the names of the principal prisoners, especially if royal personages. Unfortunately these inscriptions having been usually placed on the upper part of the slabs, which has very rarely escaped destruction, but few of them remain. These remarks should be borne in mind to enable the reader to understand the descriptions of the excavated chambers at Kouyunjik, which will be given in the following pages in the order that they were discovered.

I lost no time in making arrangements for continuing the excavations with as much activity as the funds granted to the Trustees of the British Museum would permit. Toma Shishman was placed over Kouyunjik; Mansour, Behnan (the marble cutter), and Hannah (the carpenter), again entered my service. Ali Rahal, a sheikh of the Jebours, who, hearing of my return, had hastened to Mosul, was sent to the desert to collect such of my old workmen from his tribe as were inclined to re-enter my service. He was appointed " sheikh of the mound," and duly invested with the customary robe of honor on the occasion.

The accumulation of soil above the ruins was so great, that I determined to continue the tunnelling, removing only as much earth as was necessary to show the sculptured walls. But to facilitate the labor of the workmen, and to avoid the necessity of their leaving the tunnels to empty their baskets, I made a number of rude triangles and wooden pulleys, by which the excavated rubbish could be raised by ropes through the shafts, sunk at intervals for this purpose, as well as to admit light and air. One or two passages then sufficed for the workmen to descend into the subterranean galleries.

Many of the Nestorians formerly in my service as diggers, having also heard of my intended return, had left their mountains, and had joined me a day or two after my arrival. There were Jebours enough in the immediate neighbourhood of the town to make up four or five gangs of excavators, and I placed parties at once in the galleries already opened, in different parts of Kouyunjik not previously explored, and at a high mound in the northwest walls, forming one side of the great inclosure opposite

Mosul — a ruin which I had only partially examined during my previous visit.*

During the spring of this year Colonel Williams, the British commissioner for the settlement of the disputed boundaries between Turkey and Persia, had visited Mosul on his way to Baghdad, and had kindly permitted Lieutenant Glascott, R.N., the engineer of the commission, to make a careful survey of Kouyunjik. His plan, into which the excavations subsequently made have been introduced, will show the position of the palace and the general form of the mound.† The shape of this great ruin is very irregular; nearly square at the S.W. corner, it narrows almost to a point at the N.E. The palace occupies the southern angle. At the opposite, or northern, extremity are the remains of the village of Kouyunjik, from which the mound takes its name.‡ From this spot a steep road leads to the plain, forming the only access to the summit of the mound for loaded animals or carts. Nearly midway between the ruined village and the excavations is a small whitewashed Mussulman tomb, surmounted by a dome, belonging to some sheikh, or holy man, whose memory and name have long passed away. A little beyond it, to the south-west, the level of the mound rises above that of any other part; in consequence probably of the ruins of ancient buildings, belonging to a period preceding the Arab conquest, though still erected over the older Assyrian edifices. Beyond it, to the north, the level is considerably below that part of the mound which covers the remains of the excavated palace. To the south of the tomb the platform suddenly sinks, leaving a semicircular ridge, resembling an amphitheatre. There are ravines on all sides of Kouyunjik, except that facing the Tigris. If not entirely worn by the winter rains, they have, undoubtedly, been deepened and increased by them. They are strewed with fragments of pottery, bricks, and sometimes stone and burnt alabaster, whilst the falling earth frequently discloses in their sides vast masses of solid brick masonry, which fall in when undermined by the rains. Through these ravines are carried the steep and narrow pathways leading to the top of the mound. As they reach far into the ruins, frequently laying bare

---

* See Nineveh and its Remains, vol. i. p. 144., for a description of the discoveries previously made in this mound.

† See General Plan of the mound of Kouyunjik, in corner of Plan I.

‡ "The little sheep." Kouyunjik is, however, generally known to the Arabs by the name of Armousheeyah.

the very foundations of the artificial platform of earth on which the edifices were erected, they afford the best places to commence experimental tunnels.

The Khauser winds round the eastern base of Kouyunjik, and leaving it near the angle occupied by the ruins of the palace, runs in a direct line to the Tigris. Although a small and sluggish stream, it has worn for itself a deep bed, and is only fordable near the mound immediately below the southern corner, where the direct road from Mosul crosses it, and at the northern extremity where a flour mill is turned by its waters. After rain it becomes an impetuous torrent, overflowing its banks, and carrying all before it. It then rises very suddenly, and as suddenly subsides. The Tigris now flows about half a mile from the mound, but once undoubtedly washed its base. Between them is a rich alluvium deposited by the river during its gradual retreat; it is always under cultivation, and is divided into corn fields, and melon and cucumber beds.* In this plain stands the small modern village of Kouyunjik, removed for convenience from its ancient site on the summit of the mound. Round the foot of the platform are thickly scattered fragments of pottery, brick, and stone, fallen from the ruins above.

In Mosul I had to call upon the governor, and renew my acquaintance with the principal inhabitants, whose good will was in some way necessary to the pleasant, if not successful, prosecution of my labors. Kiamil Pasha had been lately named to the pashalic. He was the sixth or seventh pasha who had been appointed since I had left, for it is one of the banes of Turkish administration that, as soon as an officer becomes acquainted with the country he is sent to govern, and obtains any influence over its inhabitants, he is recalled to make room for a new ruler. Kiamil had been ambassador at Berlin, and had visited several European courts. His manners were eminently courteous and polished; his intelligence,

---

* The river Tigris flows in this part of its course, and until it reaches Saimarrah, on the confines of Babylonia, through a valley varying from one to two miles in width, bounded on both sides by low limestone and conglomerate hills. Its bed has been undergoing a continual and regular change. When it reaches the hills on one side, it is thrown back by this barrier, and creeps gradually to the opposite side, leaving a rich alluvial soil quickly covered with jungle. This process it has been repeating, backwards and forwards, for countless ages, and will continue to repeat as long as it drains the great highlands of Armenia. At Nimroud it is now gradually returning to the base of the mound, which it deserted some three thousand years ago; but centuries must elapse before it can work its way that far.

and, what is of far more importance in a Turkish governor, his integrity, were acknowledged. His principal defects were great inactivity and indolence, and an unfortunate irritability of temper, leading him to do foolish and mischievous things, of which he generally soon found cause to repent. He offered a very favorable contrast to the Pasha who received me on my visit to Mosul in 1847, and who, by the way, notwithstanding a decree of the supreme council condemning him to death for his numerous misdeeds, but not carried into execution in consequence of the misdirected humanity of the Sultan, had been recently appointed to a comfortable pashalic in Asia Minor, far from consuls and other troublesome checks upon his tyranny and extortion. Our right to excavate was now too well established to admit of question, and my visit to the Pasha was rather one of friendship than of duty. I had known him at the capital, where he held a high post in the council of state, and at Belgrade, when governor there during troublous times.

Soon after my arrival, my old friends Sheikh Abd-ur-rahman, of the Abou Salman, and Abd-rubbou, chief of the Jebours, rode into the town to see me. The former complained bitterly of poverty: his claims upon Mohammed Pasha, although recognised by the government, had not been paid, and by the new system of local administration introduced into the pashalic since my departure, his old pasture grounds near Nimroud had been taken from his tribe, and made "miri," or public property. The Jebours, under Abd-rubbou, were encamping in the desert to the south of Mosul. He offered to accompany me to Kalah Sherghat, or to any other ruin I might wish to examine, and a silk robe cemented our former friendship.

I had scarcely settled myself in the town, when Cawal Yusuf came in from Baadri, with a party of Yezidi Cawals, to invite me, on the part of Hussein Bey and Sheikh Nasr, to the annual festival at Sheikh Adi. The invitation was too earnest to be refused, nor was I sorry to have this occasion of meeting the principal chiefs of the sect assembled together, of explaining to them what had occurred at Constantinople, and of offering them a few words of advice as to their future conduct. The Jebour workmen, too, had not yet moved their tents to Nimroud or Mosul, and the excavations had consequently not been actively resumed.

I was accompanied in this visit by my own party, with the addition of Mr. Rassam, the vice-consul, and his dragoman. We rode the first day to Baadri, and were met on the road by Hussein Bey and a large company of Yezidi horsemen. Sheikh Nasr had already gone to the tomb, to make ready for the ceremonies. The young chief entertained us for the night, and on the following morning, an hour after sunrise, we left the village for Sheikh Adi. At some distance from the sacred valley we were met by Sheikh Nasr, Pir Sino, the Cawals, the priests, and the chiefs. They conducted us to the same building in the sacred grove that I had occupied on my former visit. The Cawals assembled around us and welcomed our coming on their tambourines and flutes; and soon about us was formed one of those singularly beautiful and picturesque groups which I have attempted to describe in my previous account of the Yezidi festival.*

The Yezidis had assembled in less numbers this year than when I had last met them in the valley. Only a few of the best armed of the people of the Sinjar had ventured to face the dangers of the road now occupied by the Bedouins. Abde Agha and his adherents were fully occupied in defending their villages against the Arab marauders, who, although repulsed after we quitted Semil, were still hanging about the district, bent upon revenge. The Kochers, and the tribes of Dereboun, were kept away by the same fears. The inhabitants of Kherzan and Redwan were harrassed by the conscription. Even the people of Baasheikhah and Baazani had been so much vexed by a recent visit from the Pasha that they had no heart for festivities. His Excellency not fostering feelings of the most friendly nature towards Namik Pasha, the new commander-in-chief of Arabia, who was passing through Mosul on his way to the head-quarters of the army at Baghdad, and unwilling to entertain him, was suddenly taken ill and retired for the benefit of his health to Baasheikhah. On the morning after his arrival he complained that the asses by their braying during the night had allowed him no rest; and the asses were accordingly peremptorily banished from the village. The dawn of the next day was announced, to the great discomfort of his Excellency, who had no interest in the matter, by the cocks; and the irregular troops who formed his body-guard were immediately incited to a genera. slaughter of the race. The third night his sleep was disturbed by the crying of the children, who, with their mothers, were at once locked up, for the rest of his sojourn, in the cellars. On the fourth he was awoke at daybreak by the chirping of sparrows,

* Nineveh and its Remains, vol. i. ch. ix.

Valley and Tomb of Sheikh Adi.

and every gun in the village was ordered to be brought out
to wage a war of extermination against them. But on the fifth
morning his rest was sorely broken by the flies, and the enraged
Pasha insisted upon their instant destruction. The Kiayah, who,
as chief of the village, had the task of carrying out the Governor's
orders, now threw himself at his Excellency's feet, exclaiming,
"Your Highness has seen that all the animals here, praise be to
God, obey our Lord the Sultan; the infidel flies alone are rebellious to his authority. I am a man of low degree and small power,
and can do nothing against them; it now behoves a great Vizir
like your Highness to enforce the commands of our Lord and
Master." The Pasha, who relished a joke, forgave the flies, but
left the village.

I have already so fully described the general nature of the annual festival at Sheikh Adi, and the appearance of the valley on
that occasion, that I shall confine myself to an account of such
ceremonies as I was now permitted to witness for the first time.

About an hour after sunset, Cawal Yusuf summoned Hormuzd
and myself, who were alone allowed to be present, to the inner
yard, or sanctuary, of the Temple. We were placed in a room
from the windows of which we could see all that took place in
the court. The Cawals, Sheikhs, Fakirs, and principal chiefs
were already assembled. In the centre of the court was an iron
lamp, with four burners — a simple dish with four lips for the wicks,
supported on a sharp iron rod driven into the ground. Near it
stood a Fakir, holding in one hand a lighted torch, and in the other
a large vessel of oil, from which he, from time to time, re-
plenished the lamp, loudly invoking Sheikh Adi. The Cawals
stood against the wall on one side of the court, and commenced a
slow chant, some playing on the flute, others on the tambourine,
and accompanying the measure with their voices. The Sheikhs
and chiefs now formed a procession, walking two by two. At
their head was Sheikh Jindi. He wore a tall shaggy black cap,
the hair of which hung far over the upper part of his face. A
long robe, striped with horizontal stripes of black and dark red,
fell to his feet. A countenance more severe, and yet more imposing, than that of Sheikh Jindi could not well be pictured by
the most fanciful imagination. A beard, black as jet, waved low
on his breast; his dark piercing eyes glittered through ragged eye-
brows, like burning coals through the bars of a grate. The color

G

of his face was of the deepest brown, his teeth white as snow, and his features, though stern beyond measure, singularly noble and well formed. It was a by-word with us that Sheikh Jindi had never been seen to smile. To look at him was to feel that a laugh could not be born in him. As he moved, with a slow and solemn step, the flickering lamp deepening the shadows of his solemn and rugged countenance, it would have been impossible to conceive a being more eminently fitted to take the lead in ceremonies consecrated to the evil one. He is the *Peesh-namaz,* " the leader of prayer," to the Yezidi sect. Behind him were two venerable sheikhs. They were followed by Hussein Bey and Sheikh Nasr, and the other chiefs and Sheikhs came after. Their long

Sheikh Nasr High Priest of the Yezidis

robes were all of the purest white. As they walked slowly round, sometimes stopping, then resuming their measured step, they chanted prayers in glory and honor of the Deity. The Cawals accompanied the chant with their flutes, beating at intervals the tambourines. Round the burning lamp, and within the circle formed by the procession, danced the Fakirs in their black dresses, with solemn pace timed to the music, raising and swinging to and fro their arms after the fashion of Eastern dancers, and placing themselves in attitudes not less decorous than elegant. To hymns in praise of the Deity succeeded others in honor of Melek Isa and Sheikh Adi. The chants passed into quicker strains, the tambourines were beaten more frequently, the Fakirs became more active in their motions, and the women made the loud *tahlel*, the ceremonies ending with that extraordinary scene of noise and excitement that I have attempted to describe in relating my first visit. When the prayers were ended, those who marched in procession kissed, as they passed by, the right side of the doorway leading into the temple, where a serpent is figured on the wall; but not, as I was assured, the image itself, which has no typical or other meaning, according to Sheikh Nasr and Cawal Yusuf. Hussein Bey then placing himself on the step at this entrance, received the homage of the Sheikhs and elders, each touching the hand of the young chief with his own, and raising it to his lips. All present, afterwards, gave one another the kiss of peace.

The ceremonies having thus been brought to a close, Hussein Bey and Sheikh Nasr came to me, and led me into the inner court. Carpets had been spread at the doorway of the temple for myself and the two chiefs; the Sheikhs, Cawals, and principal people of the sect, seated themselves, or rather crouched, against the walls. By the light of a lamp, dimly breaking the gloom within the temple, I could see Sheikh Jindi unrobing. During the prayers, priests were stationed at the doorway, and none were allowed to enter except a few women and girls: the wives and daughters of sheikhs and cawals had free access to the building, and appeared to join in the ceremonies. The Vice-Consul and Khodja Toma were now admitted, and took their places with us at the upper end of the court. Cawal Yusuf was then called upon to give a full account of the result of his mission to Constantinople, which he did with the same detail, and almost in the same words, that he had used so frequently during our journey. After he had concluded, I endeavored to point out to the chiefs that by the new concessions made to them, liberty of conscience and the enjoyment of property were, if not completely secured, at least fully recognised as their right, and that the great burdens to which the Yezidis had long been exposed were abolished. Their children could no longer be taken as slaves, and the Sultan had even ordered the liberation of those who were already in bondage.* Henceforward none would suffer torture or death for their religion's sake. Whatever their objections to the conscription and military service, it was but reasonable that, as subjects of the Sultan, and as exempt from the capitation tax paid by Christians, they should be placed under the same laws as Mussulmans, and should serve the state. Such practices and food as were repugnant to them, the Grand Vizir had promised should not be forced upon those who were enrolled in the regular army. For the first time the Yezidis had been in direct communication with the Sultan's ministers, and had been formally recognised as one of the sects of the empire. They were to justify the good intentions of the Porte towards them by proving themselves loyal and faithful subjects. But, above all, they were to eschew internal quarrels, and to maintain peace and unity among the tribes, by which means alone they could defy their enemies. Their industry had already raised them above their Mohammedan and Christian neighbours, and now that additional protection was extended to them they might fairly hope to be wealthy and prosperous. It was finally agreed that letters of thanks, sealed by all the chiefs of the Yezidis, should be sent to the Grand Vizir, Reshid Pasha, for the reception given to the Yezidi deputation, and to Sir Stratford Canning for his generous intercession in their behalf.

The private and domestic affairs of the sect were then discussed, and various reforms proposed. The mode of contracting marriages required some change. The large sums of money demanded by parents for their daughters had been the cause that many girls remained unmarried, a state of things rarely found in Eastern countries, and the source of loud complaints amongst the younger members of the community. Rassam suggested that the price paid to the father should be reduced, or he should encourage elopements, and give the fugitives the benefit of his

* During my subsequent residence in Mosul, I was able, with the assistance of Mr. Rassam, the vice-consul, who always exerted himself zealously and most disinterestedly in the cause of humanity, to take from the very harem of the Cadi, a Yezidi girl, who had been torn from her parents some time before, and had been compelled to embrace the Mohammedan religion. Such an unusual proceeding had a great effect in the town.

protection. The proposed alternative caused much merriment; but one of the old Sheikhs of Baazani at once consented to take 300 piasters (about 2*l.* 10*s.*) for his daughter, instead of 3000, which he had previously asked. This led to several betrothals on the spot, amidst much mirth and great applause on the part of such young Cawals as were anxious to get married. It was nearly midnight before the assembly broke up. We then went into the outer court, where dances were kept up until late in the morning, by the light of torches; all the young men and women joining in the Debka.

Soon after sunrise on the following morning the Sheikhs and Cawals offered up a short prayer in the court of the temple, but without any of the ceremonies of the previous evening. Some prayed in the sanctuary, frequently kissing the threshold and holy places within the building. When they had ended they took the green cloth covering from the tomb of Sheikh Adi, and, followed by the Cawals playing on their tambourines and flutes, walked with it round the outer court. The people flocked about them, and reverently carried the corner of the drapery to their lips, making afterwards a small offering of money. After the cover had been again thrown over the tomb, the chiefs and priests seated themselves round the inner court. The Fakirs and Sheikhs especially devoted to the service of the sanctuary, who are called Kotcheks, now issued from the kitchens of the temple bearing large platters of smoking *harisa**, which they placed on the ground. The company collected in hungry groups round the messes, and whilst they were eating, the *Kotcheks* standing by called upon them continually in a loud voice to partake of the hospitality of Sheikh Adi. After the empty plates had been removed, a collection was made towards the support of the temple and tomb of the saint. It is also customary for all families who come to the annual festival to send some dish as an

---

* A mixture of bruised wheat, chopped meat, milk and curds, boiled into a thick pulpy mass, over which melted butter is poured. It is a favorite dish in Syria and Mesopotamia, and is cooked by families on great festivals, or on certain days of the year, in consequence of vows made during sickness or in travel. On these occasions it is sent round to friends, and distributed amongst the poor. The wealthy sprinkle it with cinnamon and sugar, and it is then agreeable to the taste, and palatable enough. It is sold early in the morning in the bazars of many Eastern towns.

offering to Sheikh Nasr. He merely tastes these contributions to show his acceptance of them, and they are then shared by the servants of the sanctuary.

These ceremonies occupied us until nearly mid-day; we then sat by the fountain in the valley, and the men and women danced before us, the boys climbing into the trees and hanging on the boughs to see the dancers. Sugar, dates, and raisins were afterwards scrambled amongst the children. The men soon took part in the amusement. A party of Kurds, bringing grapes from the mountains to sell at the festival, were maliciously pointed out as good objects for a joke. The hint was no sooner given than they, their donkeys, and their grapes, were all rolled into one heap under a mountain of human beings. The Kurds, who were armed, resisted manfully; and, ignorant of our intentions, might have revenged themselves on their assailants, but were soon restored to good humour when they found that they were to receive ample compensation for their losses and personal injuries. A fat *bakkal*, a peddling dealer in nuts, raisins, and dates from Mosul, was then thrown with all his stores into a pond, and was well-nigh drowned by the crowd of boys who dived into the reservoir on the chance of sharing in the contents of his panniers. The young chief mingled heartily in the sport, stripping off his gay robes and inciting the people to mischief. There was general laughing in the valley, and the Yezidis will long remember these days of simple merriment and happiness.

In the afternoon the wives and daughters of the chiefs and Cawals called upon me. The families of the Cawals, evidently descended from the same stock, are remarkable for the beauty both of the men and women, all of whom are strikingly like one another. Their complexion is, perhaps, too dark, but their features are regular and admirably formed. The dresses of the girls were elegant, and as rich as the material they could obtain would allow. Some wove flowers into their hair, others encircled their black turbans with a single wreath of myrtle, a simple and elegant ornament. They all wore many strings of coins, amber, coral, agate, and glass beads round their necks, and some had the black skull cap completely covered with gold and silver money. A kind of apron of grey or yellowish check, like a Scotch plaid, tied over one shoulder, and falling in front over the silk dress, is a peculiar feature in the costume of the Yezidi girls, and of some Christians from the same district. Unmarried women have the

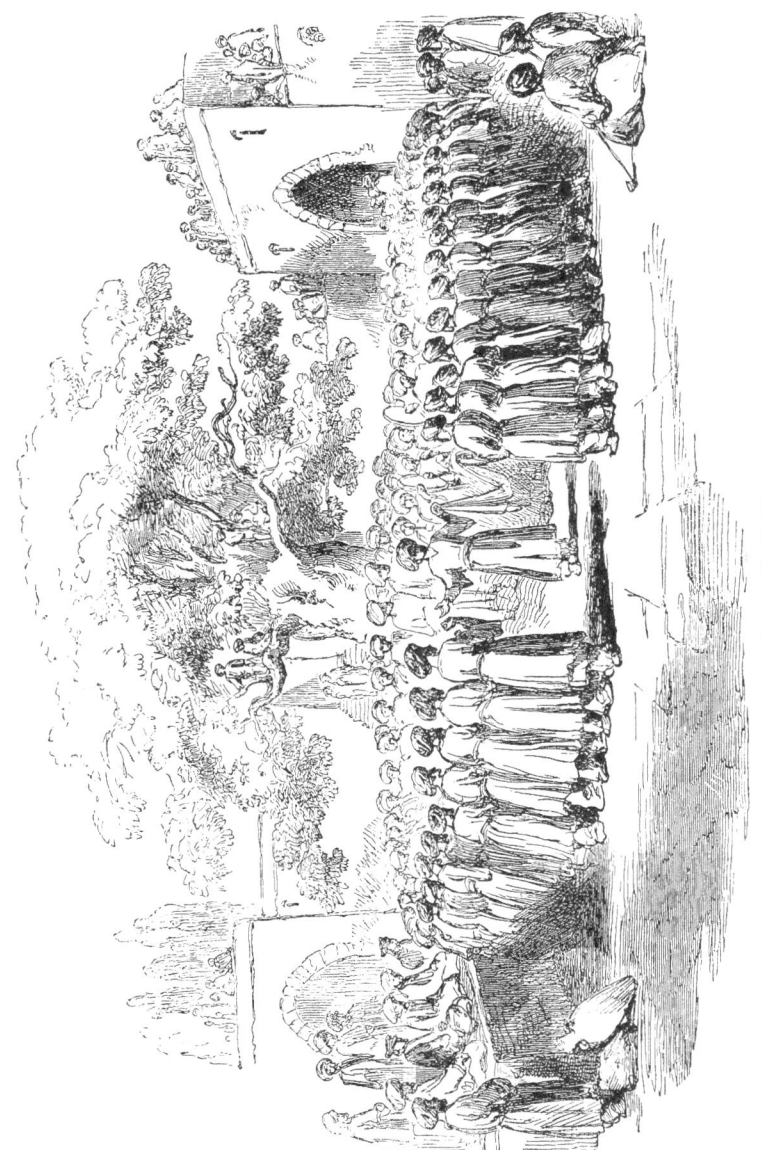

Yezidi Dance at Sheikh Adi

neck bare, the married conceal it with a white kerchief, which passes under the chin, and is tied on the top of the head. The brightest colors are worn by the girls, but the matrons are usually clothed in plain white. The females of the Cawal families always wear black turbans and skull caps. Cawal Yusuf, to show how the Frank ladies he had seen at Constantinople were honored by their husbands, made his young wife walk arm in arm with him before us, to the great amusement of the bystanders.

At night the same religious ceremonies were repeated in the temple, and I was allowed to sleep in the room overlooking the inner court from whence I had witnessed them on the previous evening. After all had retired to rest, the Yezidi Mullah recited, in a low chanting tone, a religious history, or discourse, consisting of the adventures and teachings of a certain Mirza Mohammed. He stood before the burning lamp, and around him were stretched at full length on the stone pavement, and covered by their white cloaks, the sleeping Sheikhs and Cawals. The scene was singularly picturesque and impressive.

Next morning I visited, with Mr. Rassam and Mr. Cooper, the rock-sculptures of Bavian, which are not more than six miles from the valley of Sheikh Adi in the same range of hills; but I will defer a description of these remarkable monuments until I come to relate my second journey to the spot.

The Kaidi, a Yezidi tribe, perform, at the annual festival, the following curious ceremony, said to be of great antiquity, which we witnessed on the day of our departure from Sheikh Adi. They ascend, in company with all those who have fire-arms, the rocks overhanging the temple, and, placing small oak twigs into the muzzles of their guns, discharge them into the air. After having kept up a running fire for nearly half an hour, they descend into the outer court and again let off their pieces. When entering the inner court they go through a martial dance before Hussein Bey, who stands on the steps of the sanctuary amidst the assembled priests and elders. The dance being ended, a bull, presented by the Yezidi chief, is led out from the temple. The Kaidi rush upon the animal with shouts, and, seizing it, lead it off in triumph to Sheikh Mirza, one of the heads of the sect, from whom they also receive a present, generally consisting of sheep. During these ceremonies the assembled crowd of men, women, and children form groups on the steep sides of the ravine, some standing on the well-wooded terraces, others on projecting rocks and ledges, whilst the boys clamber into the high trees, from whence they can obtain a view of the proceedings. The women make the *tahlel* without ceasing, and the valley resounds with the deafening noise. The long white garments fluttering amongst the trees, and the gay costumes of some of the groups, produce a very beautiful and novel effect.

The Kaidi were formerly a powerful tribe, sending as many as six hundred matchlock-men to the great feast. They have been greatly reduced in numbers and wealth by wars and oppression.

Cawal Yusuf had promised, on the occasion of the festival, to show me the sacred book of the Yezidis. He accordingly brought a volume to me one morning, accompanied by the secretary of Sheikh Nasr, the only Yezidi, as far as I am aware, who could read it. It consisted of a few tattered leaves, of no ancient date, containing a poetical rhapsody on the merits and attributes of Sheikh Adi, who is identified with the Deity himself, as the origin and creator of all things, though evidently distinguished from the Eternal Essence by being represented as seeking the truth, and as reaching through it the highest place, which he declares to be attainable by all those who like him shall find the truth. I will, however, give a translation of this singular poem, for which I am indebted to Mr. Hormuzd Rassam.*

THE RECITATION (OR POEM) OF SHEIKH ADI — PEACE BE UPON HIM!

1. My understanding surrounds the truth of things,
2. And my truth is mixed up in me.
3. And the truth of my descent is set forth by itself †;
4. And when it was known it was altogether in me.‡
5. All who are in the universe are under me,

---

* The year after my visit to Sheikh Adi this poem was shown, through Mr. C. Rassam, to the Rev. Mr. Badger, who has also given a translation of it in the first volume of his "Nestorians and their Rituals." The translation in the text was, however, made before Mr. Badger's work was published. That gentleman is mistaken in stating that "Sheikh Adi is one of the names of the Deity in the theology of the Yezidis," and "that he is held by them to be the good deity," for in the fifty-eighth verse the Sheikh is expressly made to say, "The *All-merciful* has distinguished me with names;" and the Yezidis always admit him to be but a great prophet, or *Vicegerent* of the Almighty.

† Or, "I am come of myself."

‡ According to Mr. Badger, "I have not known evil to be with me," but the verse seems to have reference to the Sheikh's self-existence.

6. And all the habitable parts and the deserts *,
7. And every thing created is under me. †
8. And I am the ruling power preceding all that exists.
9. And I am he who spake a true saying.
10. And I am the just judge, and the ruler of the earth (Bat'ha).
11. And I am he whom men worship in my glory,
12. Coming to me and kissing my feet.
13. And I am he who spread over the heavens their height.
14. And I am he who cried in the beginning (or in the wilderness, Al bidaee).
15. And I am the Sheikh, the one and only one.
16. And I am he who of myself revealeth all things.
17. And I am he to whom came the book of glad tidings,
18. From my Lord who burneth (or cleaveth) the mountains.
19. And I am he to whom all created men come,
20. In obedience to kiss my feet.
21. I bring forth fruit from the first juice of early youth,
22. By my presence, and turn towards me my disciples. ‡
23. And before his light the darkness of the morning cleared away.
24. I guide him who asketh for guidance.
25. And I am he that caused Adam to dwell in Paradise,
26. And Nimrod to inhabit a hot burning (or hell) fire.
27. And I am he who guided Ahmed the Just,
28. And led him into my path and way.
29. And I am he unto whom all creatures
30. Come unto for my good purposes and gifts.§
31. And I am he who visited all the heights (or, who hath all majesty),
32. And goodness and charity proceed from my mercy.
33. And I am he who made all hearts to fear
34. My purpose, and they magnified the power and majesty of my awfulness.‖
35. And I am he to whom the destroying lion came,

---

* Or, " And who are in distress and in a thicket."
† Or, " And in every good action I take delight."
‡ The Rev. Mr. Badger translates the 21st and 22d verses differently : —
      " I am the mouth, the moisture of whose spittle
      Is as honey, wherewith I constitute my confidants ;"
referring to the mode of initiation amongst Mussulman dervishes, who drink a bowl of milk into which a Sheikh has spat.
§ Or, " Mine are all created, or existing things ;
      They are my gifts, and for my purposes."
‖ " And I am he that entereth the heart in my zeal,
    And I shine through the power of my awfulness and majesty."
Mr. Badger.

36. Raging, and I shouted against him and he became stone.
37. And I am he to whom the serpent came,
38. And by my will I made him dust.
39. And I am he who struck the rock and made it tremble,
40. And made to burst from its side the sweetest of waters.
41. And I am he who sent down the certain truth.
42. From me (is) the book that comforteth the oppressed.
43. And I am he who judged justly;
44. And when I judged it was my right.
45. And I am he who made the springs to give water,
46. Sweeter and pleasanter than all waters.
47. And I am he that caused it to appear in my mercy,
48. And by my power I called it the pure (or the white).
49. And I am he to whom the Lord of Heaven hath said,
50. Thou art the Just Judge, and the ruler of the earth (Bat hai).
51. And I am he who disclosed some of my wonders.
52. And some of my virtues are manifested in that which exists
53. And I am he who caused the mountains to bow,
54. To move under me, and at my will.
55. And I am he before whose awful majesty the wild beasts cried :
56. They turned to me worshipping, and kissed my feet.
57. And I am Adi Es-shami (or, of Damascus), the son of Moosafir.*
58. Verily the All-Merciful has assigned unto me names,
59. The heavenly throne, and the seat, and the seven (heavens) and the earth.†
60. In the secret of my knowledge there is no God but me.
61. These things are subservient to my power.
62. And for which state do you deny my guidance.‡
63. Oh men ! deny me not, but submit ;
64. In the day of Judgment you will be happy in meeting me.
65. Who dies in my love I will cast him
66. In the midst of Paradise by my will and pleasure ;
67. But he who dies unmindful of me,
68. Will be thrown into torture in misery and affliction.§
69. I say that I am the only one and the exalted ;
70. I create and make rich those whom I will.

* There is some doubt about this passage; Mr. Badger has translated it,
      " I am Adi of the mark, a wanderer."

Guided by the spirit of the passage, I prefer, however, Mr. Rassam's version
which agrees with the common tradition amongst the Yezidis, with whom Sheikh
Moosafir is a venerated personage. His mother was a woman of Busrah. He
was never married.

  † " And my seat and throne are the wide-spread earth." — *Mr. Badger.*
  ‡ Or, " O mine enemies, why do you deny me ? "
  § Or, " Shall be punished with my contempt and rod." — *Mr. Badger.*

71. Praise be to myself, and all things are by my will.
72. And the universe is lighted by some of my gifts.
73. I am the King who magnifies himself;
74. And all the riches of creation are at my bidding.
75. I have made known unto you, O people, some of my ways,
76. Who desireth me must forsake the world.
77. And I can also speak the true saying.
78. And the garden on high is for those who do my pleasure.
79. I sought the truth, and became a confirming truth;
80. And by the like truth shall they possess the highest place like me.

This was the only written work that I was able to obtain from the Yezidis; their cawals repeated several prayers and hymns to me, which were purely laudatory of the Deity, and unobjectionable in substance. Numerous occupations during the remainder of my residence in Assyria prevented me prosecuting my inquiries much further on this subject. Cawal Yusuf informed me that before the great massacre of the sect by the Bey of Rahwanduz they possessed many books which were lost during the general panic, or destroyed by the Kurds. He admitted that this was only a fragmentary composition, and by no means "the Book" which contained the theology and religious laws of the Yezidi. He even hinted that the great work did still exist, and I am by no means certain that there is not a copy at Baasheikhah or Baazani. The account given by the Cawal seems to be confirmed by the allusion made in the above poem to the "Book of Glad Tidings," and "the Book that comforteth the oppressed," which could scarcely have been inserted for any particular purpose, such as to deceive their Mohammedan neighbours.

I have given in an appendix three chants of the Yezidis, which were noted down by M. Lowy as Cawal Yusuf played on his flute when with me at Constantinople.* Two of them are not without originality and melody.

I will here add a few notes concerning the Yezidis and their faith to those contained in my former work; they were chiefly obtained from Cawal Yusuf.

They believe that Christ will come to govern the world, but that after him Sheikh Medi will appear, to whom will be given special jurisdiction over those speaking the Kurdish language,

---

* The flute of the Yezidis consists of a reed blown at one end. The tone is exceedingly sweet and mellow, and some of their melodies very plaintive.

including the Yezidis (this is evidently a modern interpolation derived from Mussulman sources, perhaps invented to conciliate the Mahommedans).

All who go to heaven must first pass an expiatory period in hell, but no one will be punished eternally. Mohammedans they exclude from all future life, but not Christians. (This may have been said to avoid giving offence.)

The Yezidis will not receive converts to their faith; circumcision is optional. When a child is born near enough to the tomb of Sheikh Adi, to be taken there without great inconvenience or danger, it should be baptized as early as possible after birth. The Cawals in their periodical visitations carry a bottle or skin filled with the holy water, to baptize those children who cannot be brought to the shrine.

There are forty days fast in the spring of the year, but they are observed by few; one person in a family may fast for the rest.* They should abstain during that period as completely as the Chaldæans from animal food. Sheikh Nasr fasts rigidly for one month in the year, eating only once in twenty-four hours and immediately after sunset.

Only one wife is strictly lawful, although the chief takes more; but concubines are not forbidden. The wife may be turned away for great misconduct, and the husband, with the consent of the Sheikhs, may marry again; but the discarded wife never can. Even such divorces ought only to be given in cases of adultery; for formerly, when the Yezidis administered their own temporal laws, the wife was punished with death, and the husband of course was then released.

The religious, as well as the political, head of all Yezidis, wherever they may reside, is Hussein Bey, who is called the Kalifa, and he holds this position by inheritance. As he is young and inexperienced, he deputes his religious duties to Sheikh Nasr. He should be the *Peesh-Namaz*, or leader of the prayers, during sacred ceremonies; but as a peculiar dress is worn on this occasion, and the Bey is obliged to be in continual intercourse with the Turkish authorities, these robes might fall into their hands, and they are, therefore, entrusted to Sheikh Jindi, who officiates for the

---

* This reminds me of the Bedouins, who, when they come into a town in a party, send one of their number to the mosque to pray for his companions as well as himself.

young chief.* Sheikh Nasr is only the chief of the Sheikhs of the
district of Sheikhan. The Cawals are all of one family, and are
under the orders of Hussein Bey, who sends them periodically to
collect the voluntary contributions of the various tribes. The
amount received by them is divided into two equal parts, one of
which goes to the support of the tomb of Sheikh Adi, and half of
the other to Hussein Bey, the remainder being equally shared by
the Cawals. Neither the priests nor Hussein Bey ever shave their
beards. They ought not to marry out of their own order, and
though the men do not observe this rule very strictly, the women
are never given in marriage to one out of the rank of the priesthood. Hussein Bey ought to take his wife from the family of
Chul Beg.

After death, the body of a Yezidi, like that of a Mohammedan,
is washed in running water, and then buried with the face turned
towards the north star. A Cawal should be present at the ceremony, but if one cannot be found, the next who visits the neighbourhood should pray over the grave. I have frequently seen
funeral parties of Yezidis in their villages. The widow dressed in
white, throwing dust over her head, which is also well smeared
with clay, and accompanied by her female friends, will meet the
mourners dancing, with the sword or shield of her husband in one
hand, and long locks cut from her own hair in the other.

I have stated that it is unlawful amongst the Yezidis to know
how to read or write. This, I am assured, is not the case, and
their ignorance arises from want of means and proper teachers.
Formerly a Chaldæan deacon used to instruct the children.

Cawal Yusuf mentioned accidentally, that, amongst the Yezidis,
the ancient name for God was Azed, and from it he derived the
name of his sect. He confirmed to me the fact of the small
Ziareh at Sheikh Adi being dedicated to the sun, who, he says,
is called by the Yezidis " Wakeel el Ardth " (the Lieutenant or
Governor of the world). They have no particular reverence for
fire ; the people pass their hands through the flame of the lamps
at Sheikh Adi, merely because they belong to the tomb. Their
Kublah, he declared, was the polar star and not the east.

On my way to Mosul from Sheikh Adi, I visited the ruins of
Jerraiyah, where excavations had been again carried on by one of

---

* Ali Bey, Hussein Bey's father was initiated in the performance of all the
ceremonies of the faith.

my agents. No ancient buildings were discovered. The principal mound is lofty and conical in shape, and the base is surrounded by smaller mounds, and irregularities in the soil which
denote the remains of houses. I had not leisure during my residence in Assyria to examine the spot as fully as it may deserve.

Yezidi Cawals

Mound of Nimroud.

CHAP. V.

WE were again in Mosul by the 12th of October. The Jebours, my old workmen, had now brought their families to the town. I directed them to cross the river, and to pitch their tents over the excavations at Kouyunjik, as they had formerly done around the trenches at Nimroud. The Bedouins, unchecked in their forays by the Turkish authorities, had become so bold, that they ventured to the very walls of Mosul, and on the opposite bank of the Tigris had plundered the cattle belonging to the inhabitants of the village of the tomb of Jonah. On one occasion I saw an Arab horseman of the desert dart into the high road, seize a mule, and drive it off from amidst a crowd of spectators. This state of things made it necessary to have a strong party on the ruins for self-defence. The Jebours were, however, on good terms with the Bedouins, and had lately encamped amongst them. Indeed, it was suspected, that whilst Abd-rubbou and his tribe were more than usually submissive in their dealings with the local government, they were the receivers of goods carried off by their friends, their intercourse with the town enabling them to dispose of such property to the best advantage in the market-place.

About one hundred workmen, divided into twelve or fourteen parties, were employed at Kouyunjik. The Arabs, as before, removed the earth and rubbish, whilst the more difficult labor with the pick was left entirely to the Nestorian mountaineers. My old friend, Yakoub, the Rais of Asheetha, made his appearance one morning, declaring that things were going on ill in the mountains; and that, although the head of a village, he hoped to spend the winter more profitably and more pleasantly in my service. He was accordingly named superintendent of the Tiyari workmen, for whom I built mud huts near the foot of the mound.

The work having been thus began at Kouyunjik, I rode with Hormuzd to Nimroud for the first time on the 18th of October. It seemed but yesterday that we had followed the same track. We stopped at each village, and found in each old acquaintances ready to welcome us. From the crest of the hill half way, the first view of Nimroud opened upon us; the old mound, on which I had gazed so often from this spot, and with which so many happy recollections were bound up, rising boldly above the Jaif, the river winding through the plain, the distant wreaths of smoke marking the villages of Naifa and Nimroud. At Selamiyah we sought the house of the Kiayah, where I had passed the first winter whilst excavating at Nimroud; but it was now a house of mourning. The good old man had died two days before, and the wails of the women, telling of a death within, met our ears as we approached the hovel. Turning from the scene of woe, we galloped over the plain, and reached Nimroud as the sun went down. Saleh Shahir, with the elders of the village, was there to receive us. I dismounted at my old house, which was still standing, though somewhat in ruins, for it had been the habitation of the Kiayah during my absence. Toma Shishman had, however, been sent down the day before, and had made such preparations for our reception as the state of the place would permit. To avoid the vermin swarming in the rooms, my tent was pitched in the courtyard, and I dwelt entirely in it.

The village had still, comparatively speaking, a flourishing appearance, and had not diminished in size since my last visit. The *tanzimat*, or reformed system of local administration, had been introduced into the pashalic of Mosul, and although many of its regulations were evaded, and arbitrary acts were still occasionally committed, yet on the whole a marked improvement had taken place in the dealings of the authorities with the subjects of the Sultan. The great cause of complaint was the want of security. The troops under the command of the Pasha were not sufficient in number to keep the Bedouins in check, and there was scarcely a village in the low country which had not suffered more or less from their depredations. Nimroud was particularly exposed to their incursions, and the inhabitants lived in continual agitation and alarm.

The evening was spent with the principal people of the village, talking with them of their prospects, taxes, harvests, and the military conscription, now the great theme of discontent in Southern Turkey, where it had been newly introduced.

By sunrise I was amongst the ruins. The mound had undergone no change. There it rose from the plain, the same sun-burnt yellow heap that it had stood for twenty centuries. The earth and rubbish, which had been heaped over the excavated chambers and sculptured slabs, had settled, and had left uncovered in sinking the upper part of several bas-reliefs. A few colossal heads of winged figures rose calmly above the level of the soil, and with two pairs of winged bulls, which had not been reburied on account of their mutilated condition, was all that remained above ground of the north-west palace, that great storehouse of Assyrian history and art. Since my departure the surface of the mound had again been furrowed by the plough, and ample crops had this year rewarded the labors of the husbandman. The ruins of the south-west palace were still uncovered. The Arabs had respected the few bas-reliefs which stood against the crumbling walls, and Saleh Shahir pointed to them as a proof of the watchfulness of his people during my long absence.

Collecting together my old excavators from the Shemutti and Jehesh (the Arab tribes who inhabit Nimroud and Naifa), and from the tents of a few Jebours who still lingered round the village to glean a scanty subsistence after the harvest, I placed workmen in different parts of the mound. The north-west palace had not been fully explored. Most of the chambers which did not contain sculptured slabs, but were simply built of sundried bricks, had been left unopened. I consequently directed a party of workmen to resume the excavations where they had been formerly abandoned.* New trenches were also opened in the ruins of the centre palace, where, as yet, no sculptures had been discovered in

* To the south of Chamber X. Plan III. "Nineveh and its Remains," vol. i. p. 62.

their original position against the walls. The high conical mound forming the north-west corner of Nimroud, the pyramid as it has usually been called, had always been an object of peculiar interest, which want of means had hitherto prevented me fully examining. With the exception of a shaft, about forty feet deep, sunk nearly in the centre, and passing through a solid mass of sundried bricks, no other opening had been made into this singular ruin. I now ordered a tunnel to be carried into its base on the western face, and on a level with the conglomerate rock upon which it rested.

Whilst riding among the ruins giving directions to the workmen, we had not escaped the watchful eyes of the Abou-Salman Arabs, whose tents were scattered over the Jaif. Not having heard of my visit, and perceiving horsemen wandering over the mound, they took us for Bedouin marauders, and mounting their ever-ready mares, sallied forth to reconnoitre. Seeing Arabs galloping over the plain I rode down to meet them, and soon found myself in the embrace of Schloss, the nephew of Sheikh Abd-ur-Rahman. We turned together to the tents of the chief, still pitched on the old encamping ground. The men, instead of fighting with Bedouins, now gathered round us in the *muzeef**, and a sheep was slain to celebrate my return. The Sheikh himself was absent, having been thrown into prison by the Pasha for refusing to pay some newly-imposed taxes. I was able to announce his release, at my intercession, to his wife, who received me as his guest. The Sheikh of the Haddedeen Arabs, hearing that I was at the Abou-Salman camp, rode over with his people to see me. His tents stood on the banks of the Tigris, and he had united with Abd-ur-Rahman for mutual defence against the Bedouins.

As we returned to Nimroud in the evening, we stopped at a small encampment in the Jaif, and buried beneath a heap of old felts and sacks found poor Khalaf-el-Hussein, who had, in former times, been the active and hospitable Sheikh of my Jebour workmen at the mound. The world had since gone ill with him. Struck down by fever, he had been unable to support himself and his family by labor, or other means open to an Arab. He was in great poverty, and still helpless from disease. He rose up as we rode to his tent, and not having heard of our arrival was struck with astonishment and delight as he saw Hormuzd and myself at its entrance. We gave him such help as was in our power, and he declared that the prospect of again being in my service would soon prove the best remedy for his disease.

* The *muzeef* is that part of an Arab tent in which guests are received.

As I ascended the mound next morning I perceived a group of travellers on its summit, their horses picketted in the stubble. Ere I could learn what strangers had thus wandered to this remote region, my hand was seized by the faithful Bairakdar. Beneath, in an excavated chamber, wrapped in his travelling cloak, was Rawlinson deep in sleep, wearied by a long and harassing night's ride. For the first time we met in the Assyrian ruins, and besides the greetings of old friendship there was much to be seen together, and much to be talked over. The fatigues of the journey had, however, brought on fever, and we were soon compelled, after visiting the principal excavations, to take refuge from the heat of the sun in the mud huts of the village. The attack increasing in the evening, it was deemed prudent to ride into Mosul at once, and we mounted our horses in the middle of the night.

During two days Col. Rawlinson was too ill to visit the excavations at Kouyunjik. On the third we rode together to the mound. After a hasty survey of the ruins we parted, and he continued his journey to Constantinople and to England, to reap the laurels of a well-earned fame.*

I had now nearly all my old adherents and workmen about me. The Bairakdar, who had hastened to join me as soon as he had heard of my return, was named principal cawass, and had the general management of my household. One Latiff Agha, like the Bairakdar, a native of Scio, carried off as a slave after the massacre, and brought up in the Mussulman creed, was appointed an overseer over the workmen. He had been strongly recommended to me by the British consul at Kaiseriyah, and fully justified in my service by his honesty and fidelity the good report I had received of him.

My readers would be wearied were I to relate, day by day, the progress of the excavations, and to record, as they were gradually made, the discoveries in the various ruins. It will give a more complete idea of the results of the researches to describe the

---

* Shortly after Col. Rawlinson's departure, Capt. Newbold, of the East India Company's service, spent a few days with me at Mosul. Although, alas! I can no longer recall to his recollection the happy hours we passed together, let me pay a sincere tribute to the memory of one who, in spite of hopeless disease, and sufferings of no common kind, maintained an almost unrivalled sweetness of disposition, and never relaxed from the pursuit of knowledge and the love of science. Those who enjoyed his intimacy, and profited by his learning, will know that this testimony to his worth is not the exaggerated praise of partial friendship.

sculptured walls of a whole chamber when entirely explored, instead of noting, one by one, as dug out, bas-reliefs which form but part of the same subject. I will, therefore, merely mention that, during the months of October and November, my time was spent between Kouyunjik and Nimroud, and that the excavations were carried on at both places without interruption. Mr. Cooper was occupied in drawing the bas-reliefs discovered at Kouyunjik, living in Mosul, and riding over daily to the ruins. To Mr. Hormuzd Rassam, who usually accompanied me in my journeys, were confided, as before, the general superintendence of the operations, the payment of the workmen, the settlement of disputes, and various other offices, which only one, as well acquainted as himself with the Arabs and men of various sects employed in the works, and exercising so much personal influence amongst them, could undertake. To his unwearied exertions, and his faithful and punctual discharge of all the duties imposed upon him, to his inexhaustible good humour, combined with necessary firmness, to his complete knowledge of the Arab character, and the attachment with which even the wildest of those with whom we were brought in contact regarded him, the Trustees of the British Museum owe not only much of the success of these researches, but the economy with which I was enabled to carry them through. Without him it would have been impossible to accomplish half what has been done with the means placed at my disposal.

The Arab workmen, as I have already observed, lived in tents amongst the ruins. The overseers of the works of Kouyunjik resided either in the village near the foot of the mound, or in Mosul, and crossed the river every morning before the labors of the day began. The workmen were divided into several classes, and their wages varied according to their respective occupations, as well as according to the time of year. They were generally paid weekly by Hormuzd. The diggers, who were exposed to very severe labor, and even to considerable risk, received from two piastres and a half to three piastres (from 5d. to 6d.) a-day ; those who filled the baskets from two piastres to two and a half; and the general workmen from one and a half to two piastres. The earth, when removed, was sifted by boys, who earned about one piastre for their day's labor. These wages may appear low, but they are amply sufficient for the support of a family in a country where the camel-load of wheat (nearly 480 lbs.) is sold for about four shillings, and where no other protection from the inclemencies of the weather is needed than a linen shirt and the black folds of an Arab tent.*

The Kouyunjik workmen were usually paid in the subterraneous galleries, some convenient space where several passages met being chosen for the purpose; those of Nimroud generally in the village. A scene of wild confusion ensued on these occasions, from which an inexperienced observer might argue a sad want of order and method. This was, however, but the way of doing business usual in the country. When there was a difference of opinion, he who cried the loudest gained the day, and after a desperate struggle of voices matters relapsed into their usual state, every one being perfectly satisfied. Screaming and gesticulation with Easterns by no means signify ill will, or even serious disagreement. Without them, except of course amongst the Turks, who are staid and dignified to a proverb, the most ordinary transactions cannot be carried on, and they are frequently rather symptoms of friendship than of hostility. Sometimes the Arabs employed at Kouyunjik would cross the river to Mosul to receive their pay. They would then walk through the town in martial array, brandishing their weapons and chanting their war cries in chorus, to the alarm of the authorities and the inhabitants, who generally concluded that the place had been invaded by the Bedouins. It was Mr. Hormuzd Rassam's task to keep in check these wild spirits.

By the end of November several entire chambers had been excavated at Kouyunjik, and many bas-reliefs of great interest had been discovered. The four sides of the hall, part of which has already been described†, had now been explored.‡ In the centre of each side was a grand entrance, guarded by colossal human-headed bulls. § This magnificent hall was no less than 124

---

* At Mosul, a bullock, very small certainly when compared with our high-fed cattle, is sold for forty or fifty piastres, 8s. or 10s. ; a fat sheep for about 4s.; a lamb for 2s. or 2s. 6d. Other articles of food are proportionally cheap. The camel-load of barley was selling at my departure for ten or twelve piastres (2s. or 2s. 6d.). A common horse is worth from 3l. to 5l.; a donkey about 10s.; a camel about the same as a horse.

† See p. 70.

‡ It will be borne in mind that it was necessary to carry tunnels round the chambers, and along the walls, leaving the centre buried in earth and rubbish, a very laborious and tedious operation with no more means at command than those afforded by the country.

§ All these entrances were formed in the same way as that in the south-eastern side, described p. 72., namely, by a pair of human-headed bulls, flanked on each side by a winged giant, and two smaller figures one above the other.

feet in length by 90 feet in breadth, the longest sides being those to the north and south. It appears to have formed a centre, around which the principal chambers in this part of the palace were grouped. Its walls had been completely covered with the most elaborate and highly finished sculptures. Unfortunately all the bas-reliefs, as well as the gigantic monsters at the entrances, had suffered more or less from the fire which had destroyed the edifice; but enough of them still remained to show the subject, and even to enable me in many places to restore it entirely.

The narrow passage leading from the great hall at the south-west corner had been completely explored. Its sculptures have already been described.* It opened into a chamber 24 feet by 19, from which branched two other passages.† The one to the west was entered by a wide doorway, in which stood two plain spherical stones about three feet high, having the appearance of the bases of columns, although no traces of any such architectural ornament could be found. This was the entrance into a broad and spacious gallery, about 218 feet long and 25 wide.‡ A tunnel at its western end, cut through the solid wall, as there was no doorway on this side of the gallery, led into the chambers excavated by Mr. Ross§, thus connecting them with the rest of the building. Opposite this tunnel the gallery turned to the right, but was not explored until long after. From this part of the excavations an inclined way, dug from the surface of the mound, was used by the Arabs in descending to the subterraneous works.

I have already described the bas-reliefs representing the conquest of a mountainous country on the southern side of the great hall. ‖ The same subject was continued on the western wall, without much variety in the details. But on the northern, the sculptures differed from any others yet discovered, and from their interest and novelty merit a particular notice. They were in some cases nearly entire, though much cracked and calcined by fire, and represented the process of transporting the great human-headed bulls to the palaces of which they formed so remarkable a feature. But before giving a particular description of them, I must return to the long gallery to the west of the great hall¶, as the sculptures still preserved in it form part of and complete this important series.

---

* P. 74.   † Nos. XLVIII. and XLII. Plan I.
‡ No. XLIX. same Plan.   § Nos. LI. and LII. same Plan.
‖ P. 71. I assume the building to be due north and south, although it is not so. It faces nearly north-east and south-west.
¶ No. XLIX. Plan I.

The slabs on one side of this gallery had been entirely destroyed, except at the eastern end; and from the few which still remained, every trace of sculpture had been carefully removed by some sharp instrument. Along the opposite wall (that to the right on leaving the great hall) only eight bas-reliefs still stood in their original position, and even of these only the lower part was preserved. Detached fragments of others were found in the rubbish, and from them I ascertained that the whole gallery had been occupied by one continuous series, representing the different processes adopted by the Assyrians in moving and placing various objects used in their buildings, and especially the human-headed bulls, from the first transport of the huge stone in the rough from the quarry, to the raising of these gigantic sculptures in the gateways of the palace-temples. On these fragments were seen the king in his chariot, superintending the operations, and workmen carrying cables, or dragging carts loaded with coils of ropes, and various implements for moving the colossi. Enough, however, did not remain to restore any one series of bas-reliefs, but fortunately, on the slabs still standing, was represented the first process, that of bringing the stone from the quarry, whilst those on the northern walls of the great hall furnished many of the subjects which were here wanting. Amongst the scattered fragments was the figure of a lion-headed man raising a sword*, which does not appear to have belonged to this gallery, unless it had been used to break the monotony of one long line of elaborate bas-reliefs representing nearly the same subject. Similar figures only occur at entrances in the ruins of Kouyunjik.

I will commence, then, by a description of the sculptures still standing in their original position in the gallery. A huge block of stone (probably of the alabaster used in the Assyrian edifices), somewhat elongated in form so as to resemble an obelisk in the rough†, is lying on a low flat-bottomed boat floating on a river. It has probably been towed down the Tigris from some quarry, and is to be landed near the site of the intended palace, to be carved

---

* This sculpture is now in the British Museum. The opposite lithograph, from a sketch by the able pencil of the Rev. S. C. Malan, will show in what state these fragments were discovered.

† It is just possible that this object may really represent an obelisk, similar to that brought, according to Diodorus Siculus (lib. ii. c. 1.), by Semiramis, from Armenia to Babylon; but I think it far more probable, for several reasons, that it is a block in the rough from the quarry, to be sculptured into the form of a winged bull.

Sketched on the spot by S. C. Malan   John Murray, Albemarle Street, 1852.   N. Chevalier, lith

Excavations at Kouyunjik.

by the sculptor into the form of a colossal bull. It exceeds the boat considerably in length, projecting beyond both the head and stern, and is held by upright beams fastened to the sides of the vessel, and kept firm in their places by wooden wedges. Two cables are passed through holes cut in the stone itself, and a third is tied to a strong pin projecting from the head of the boat. Each cable is held by a large body of men, who pull by means of small ropes fastened to it and passed round their shoulders. Some of these trackers walk in the water, others on dry land. The number altogether represented must have been nearly 300, about 100 to each cable, and they appear to be divided into distinct bands, each distinguished by a peculiar costume. Some wear a kind of em-

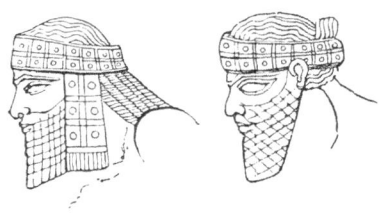

Head-dress of Captives employed by Assyrians in moving Bull (Kouyunjik).

broidered turban, through which their long hair is gathered behind; the heads of others are encircled by a fringed shawl, whose ends hang over the ears and neck, leaving the hair to fall in long curls upon the shoulders. Many are represented naked, but the greater number are dressed in short chequered tunics, with a long fringe attached to the girdle. They are urged on by taskmasters armed with swords and staves. The boat is also pushed by men wading through the stream. An overseer, who regulates the whole proceedings, is seated astride on the fore-part of the stone. His hands are stretched out in the act of giving commands. The upper part of all the bas-reliefs having unfortunately been destroyed, it cannot be ascertained what figures were represented above the trackers; probably Assyrian warriors drawn up in martial array, or may be the king himself in his chariot, accompanied by his body-guard, and presiding over the operations.[*]

The huge stone having been landed, and carved by the Assyrian sculptor into the form of a colossal human-headed bull, is to be moved from the bank of the river to the site it is meant to occupy permanently in the palace-temple. This process is represented on the walls of the great hall. From these bas-reliefs, as well as from discoveries to be hereafter mentioned, it is therefore evident that the Assyrians sculptured their gigantic figures before, and not

---

[*] For the details of these interesting bas-reliefs, I must refer my readers to Plates 10 and 11. in the 2nd series of the Monuments of Nineveh.

after, the slabs had been raised in the edifice, although all the details and the finishing touches were not put in, as it will be seen, until they had been finally placed.* I am still, however, of opinion, that the smaller bas-reliefs were entirely executed after the slabs had been attached to the walls.

In the first bas-relief I shall describe, the colossal bull rests horizontally on a sledge similar in form to the boat containing the rough block from the quarry, but either in the carving the stone has been greatly reduced in size, or the sledge is much larger than the boat, as it considerably exceeds the sculpture in length. The bull faces the spectator, and the human head rests on the fore part of the sledge, which is curved upwards and strengthened by a thick beam, apparently running completely through from side to side. The upper part, or deck, is otherwise nearly horizontal; the under, or keel, being slightly curved throughout. Props, probably of wood, are placed under different parts of the sculpture to secure an equal pressure. The sledge was dragged by cables, and impelled by levers. The cables are four in number; two fastened to strong projecting pins in front, and two to similar pins behind. They are pulled by small ropes passing over the shoulders of the men, as in the bas-reliefs already described. The numbers of the workmen may of course be only conventional, the sculptor introducing as many as he found room for on the slab. They are again distinguished by various costumes, being probably captives from different conquered nations, and are urged on by task-masters. The sculpture moves over rollers, which, as soon as left behind by the advancing sledge, are brought again to the front by parties of men, who are also under the control of overseers armed with staves. Although these rollers materially facilitated the motion, it would be almost impossible, when passing over rough ground, or if the rollers were jammed, to give the first impetus to so heavy a body by mere force applied to the cables. The Assyrians, therefore, lifted, and consequently eased, the hinder part of the sledge with huge levers of wood, and in order to obtain the necessary fulcrum they carried with them during the operations wedges of different sizes. Kneeling workmen are represented in the bas-reliefs inserting an additional wedge to raise the fulcrum. The lever itself was worked by ropes, and on a detached fragment,

---

* In my former work (vol. ii. p. 255.) I had stated that *all* the Assyrian sculptures were carved in their places against the walls of the buildings.

discovered in the long gallery, men were seen seated astride upon it to add by their weight to the force applied.

On the bull itself are four persons, probably the superintending officers. The first is kneeling, and appears to be clapping his hands, probably beating time, to regulate the motions of the workmen, who unless they applied their strength at one and the same moment would be unable to move so large a weight. Behind him stands a second officer with outstretched arm, evidently giving the word of command. The next holds to his mouth, either a speaking-trumpet, or an instrument of music. If the former, it proves that the Assyrians were acquainted with a means of conveying sound, presumed to be of modern invention. In form it undoubtedly resembles the modern speaking-trumpet, and in no bas-relief hitherto discovered does a similar object occur as an instrument of music. The fourth officer, also standing, carries a mace, and is probably stationed behind to give directions to those who work the levers. The sledge bearing the sculpture is followed by men with coils of ropes and various implements, and drawing carts laden with cables and beams. Even the landscape is not neglected; and the country in which these operations took place is indicated by trees, and by a river. In this stream are seen men swimming on skins; and boats and rafts, resembling those still in use in Assyria, are impelled by oars with wedge-shaped blades.

A subject similar to that just described is represented in another series of bas-reliefs, with even fuller details. The bull is placed in the same manner on the sledge, which is also moved by cables and levers. It is accompanied by workmen with saws, hatchets, pick-axes, shovels, ropes, and props, and by carts carrying cables and beams. Upon it are three officers directing the operations, one holding the trumpet in his hands, and in front walk four other overseers. Above the sledge and the workmen are rows of trees, and a river on which are circular boats resembling in shape the " kufas," now used on the lower part of the Tigris, and probably, like them, built of reeds and ozier twigs, covered with square pieces of hide.* They are heavily laden with beams and implements required for moving the bulls. They appear to have been near the sledge when dragged along the bank of the river, and were impelled by four oars similar to those above described. Near the

---

* Such appear to have been the boats described by Herodotus (lib. i. c. 194.). The modern " kufa " is covered with bitumen.

boats, astride on inflated skins in the water, are fishermen angling with hook and line.*

Workmen carrying Ropes, Saws, and other Implements for moving Bull (Kouyunjik)

On a fallen slab, forming part of the same general series, is the king standing in a richly decorated chariot, the pole of which, curved upwards at the end, and ornamented with the head of a horse, is raised by eunuchs. From the peculiar form of this chariot and the absence of a yoke, it would seem to have been intended purposely for such occasions as that represented in the bas-relief, and to have been a kind of moveable throne drawn by men and not by horses.† Behind the monarch, who holds a kind of flower, or ornament in the shape of the fruit of the pine, in one hand, stand

Stag (Kouyunjik)

two eunuchs, one raising a parasol to shade him from the sun, the other cooling him with a fan. He appears to have been superintending the transport of one of the colossal sculptures, and his chariot is preceded and followed by his bodyguard armed with maces. In the upper part of the slab is a jungle of high reeds, or canes, in which are seen a wild sow with its young,

* This bas-relief is now in the British Museum, and see Plate 12., 2nd series of Monuments of Nineveh.

† A throne on wheels, with a yoke, carried by two eunuchs, is represented in a bas-relief at Khorsabad. Botta, Plate 17.

and a stag and two hinds. These animals are designed with great spirit and truth.*

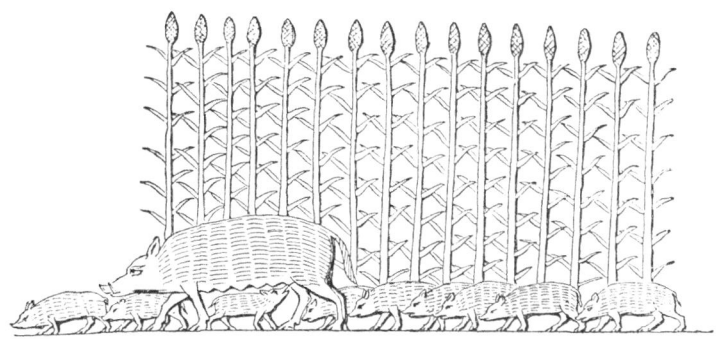

Wild Sow and Young, amongst Reeds (Kouyunjik).

The next series of bas-reliefs represents the building of the artificial platforms on which the palaces were erected, and the Assyrians moving to their summit the colossal bulls. † The king is again seen in his chariot drawn by eunuchs, whilst an attendant raises the royal parasol above his head. He overlooks the operations from that part of the mound to which the sledge is being dragged, and before him stands his body-guard, a long line of alternate spearmen and archers, resting their arms and shields upon the ground. Above him are low hills covered with various trees, amongst which may be distinguished by their fruit the vine, the fig, and the pomegranate. At the bottom of the slab is represented either a river divided into two branches and forming an island, as the Tigris does to this day opposite Kouyunjik, or the confluence of that stream and the Khauser, which then probably took place at the very foot of the mound. On the banks are seen men raising water by a simple machine, still generally used for irrigation in the East, as well as in Southern Europe, and called in Egypt a *shadoof.* It consists of a long pole, balanced on a shaft of masonry, and turning on a pivot; to one end is attached a stone, and to the other a bucket, which, after being lowered into the water and filled, is easily raised by the help of the opposite weight. Its contents are then emptied into a conduit communicating with the various

---

* See Plate 12. 2d series of Monuments of Nineveh.

† See Plates 14 and 15. 2d series of Monuments of Nineveh.

watercourses running through the fields. In the neighbourhood of Mosul this mode of irrigation is now rarely used, the larger skins raised by oxen affording a better supply, and giving, it is considered, less trouble to the cultivator.*

The process of building the artificial mound adjoined the subject just described.† Men, apparently engaged in making bricks, are crouching and kneeling round a square space, probably representing the pit whence the clay for this purpose was taken. Unfortunately this part of the subject, on the only two slabs on which it occurs, has been so much defaced, that its details cannot be ascertained with certainty. These brickmakers are between two mounds, on which are long lines of workmen going up and down. Those who toil upwards carry large stones, and hold on their backs by ropes baskets filled with bricks, earth, and rubbish. On reaching the top of the mound they relieve themselves of their burdens, and return again to the foot for fresh loads in the order they went up.

It would appear that the men thus employed were captives and malefactors, for many of them are in chains, some singly, others bound together by an iron rod attached to rings in their girdles. The fetters, like those of modern criminals, confine the legs, and are supported by a bar fastened to the waist, or consist of simple shackles round the ankles. They wear a short tunic, and a conical cap, somewhat resembling the Phrygian bonnet, with the curved crest turned backwards, a costume very similar to that of the tribute bearers on the Nimroud obelisk. Each band of workmen is followed and urged on by task-masters armed with staves.

The mound, or artificial platform, having been thus built, not always, as it has been seen, with regular layers of sundried bricks, but frequently in parts with mere heaped-up earth and rubbish‡, the next step was to drag to its summit the colossal figures prepared for the palace. As some of the largest of these sculptures were full twenty feet square, and must have weighed between forty and fifty tons, this was no easy task with such means as the Assyrians possessed. The only aid to mere manual strength

---

* I have described the mode of irrigation now generally employed by the Mesopotamian Arabs, in my "Nineveh and its Remains," vol. ii. p. 353.

† Part of this bas-relief is in the British Museum, and see 2d series of Monuments of Nineveh, Plates 14 & 15. The whole series occupied about twenty-five slabs in the N.E. walls of the great hall, from No. 43. to No. 68. Plan I. Unfortunately some of the slabs had been entirely destroyed.

‡ Subsequent excavations at Kouyunjik and Nimroud fully verified this fact.

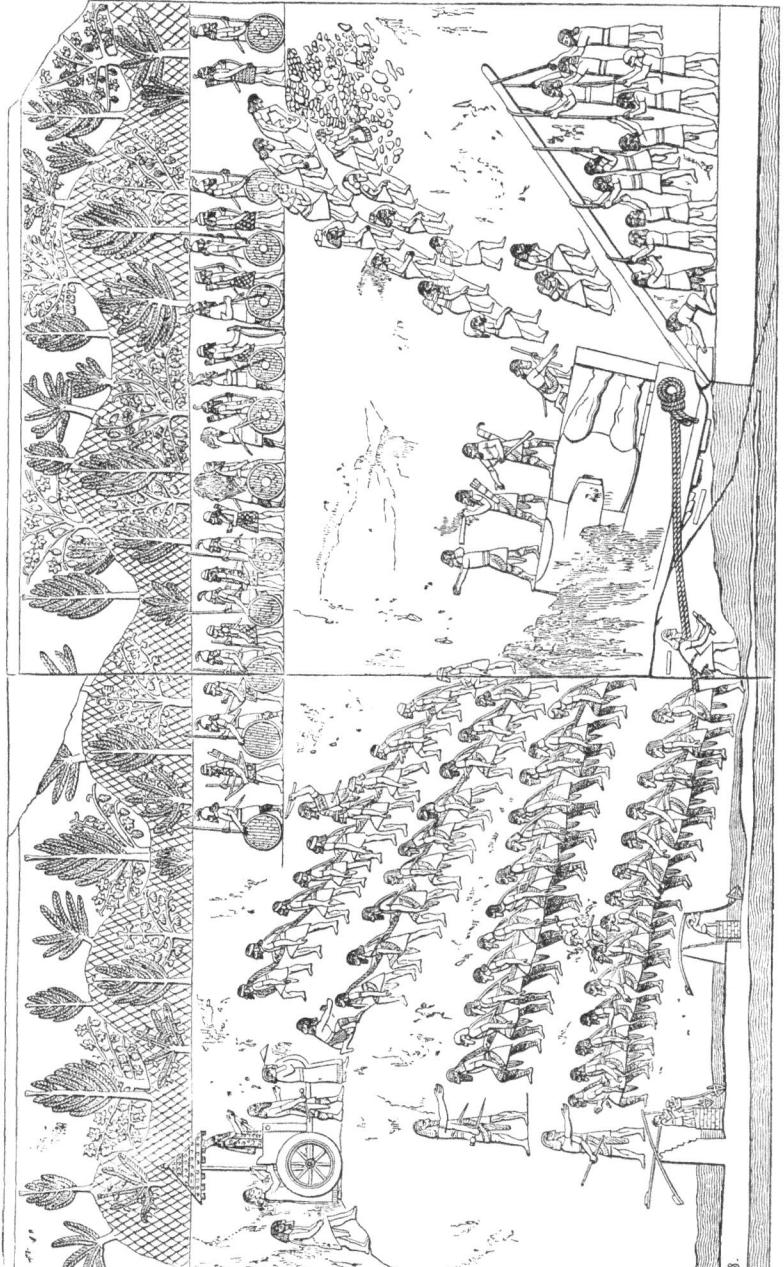

King superintending Removal of colossal Bull (Kouyunjik).

was derived from the rollers and levers. A sledge was used similar to that already described, and drawn in the same way. In the bas-relief representing the operation, four officers are seen on the bull, the first apparently clapping his hands to regulate the motions of those who draw, the second using the trumpet, the third directing the men who have the care of the rollers, and the fourth kneeling down on the edge of the back part of the sculpture to give orders to those who use the lever. Two of the groups of workmen are preceded by overseers, who turn back to encourage them in their exertions; and in front of the royal chariot, on the edge of the mound, kneels an officer, probably the chief superintendent, looking towards the king to receive orders direct from him.

Behind the monarch, on an adjoining slab, are carts bearing the cables, wedges, and implements required in moving the sculpture. A long beam or lever is slung by ropes from the shoulders of three men, and one of the great wedges is carried in the same way. In the upper compartment of this slab is a stream issuing from the foot of hills wooded with vines, fig-trees, and pomegranates. Beneath stands a town or village, the houses of which have domes and high conical roofs, probably built of mud, as in parts

Village with conical Roofs near Aleppo

of northern Syria. The domes have the appearance of dish-covers with a handle, the upper part being topped by a small circular projection, perhaps intended as an aperture to admit light and air.

This interesting series is completed by a bas-relief, showing, it would seem, the final placing of the colossal bull. The figure no longer lies horizontally on the sledge, but is raised by men with ropes and forked wooden props. It is kept in its erect position by beams, held together by cross bars and wedges *,

---

* It may be remarked, that precisely the same framework was used for moving the great sculptures in the British Museum.

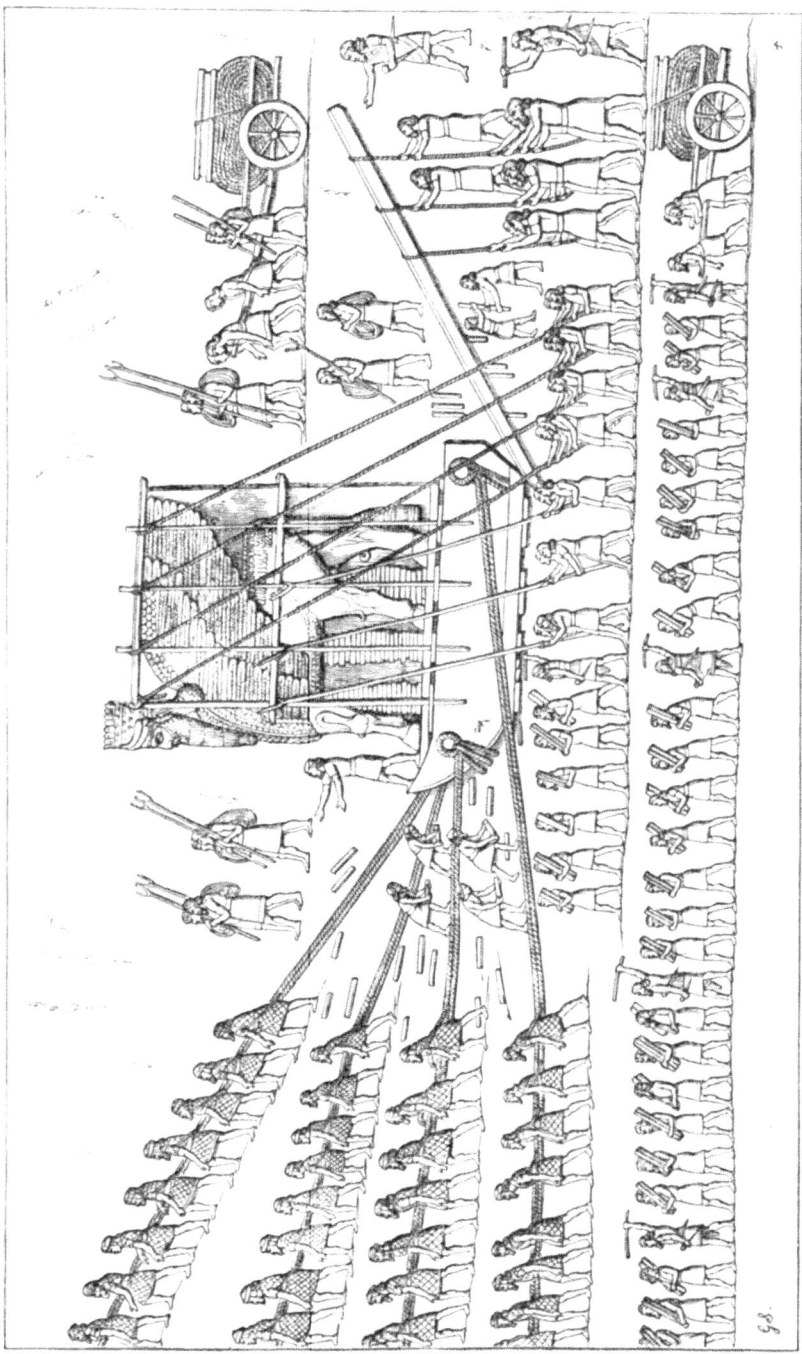

Assyrians placing a human-headed Bull (partly restored from a Bas-relief at Kouyunjik).

and is further supported by blocks of stone, or wood, piled up under the body. On the sledge, in front of the bull, stands an officer giving directions with outstretched hands to the workmen. Cables, ropes, rollers, and levers are also employed on this occasion to move the gigantic sculpture. The captives are distinguished by the peculiar turbans before described.* Unfortunately the upper part of all the slabs has been destroyed, and much of the subject is consequently wanting.

We have thus represented, with remarkable fidelity and spirit†, the several processes employed to place these colossi where they still stand, from the transport down the river of the rough block to the final removal of the sculptured figure to the palace. From these bas-reliefs we find that the Assyrians were well acquainted with the lever and the roller, and that they ingeniously made use of the former by carrying with them wedges, of different dimensions, and probably of wood, to vary the height of the fulcrum. When moving the winged bulls and lions now in the British Museum from the ruins to the banks of the Tigris, I used almost the same means.‡ The Assyrians, being unable to construct a wheeled cart of sufficient strength to carry so great a weight, employed a sledge, probably built of some hard wood obtained from the mountains. It seems to have been nearly solid, or to have been filled with beams, or decked, as the sculpture is raised above its sides. Unless the levers were brought from a considerable distance they must have been of poplar, no other beams of sufficient length existing in the country. Although weak, and liable to break with much strain, I found them strong enough for purposes of the same kind. The Assyrians, like the Egyptians, had made considerable progress in rope twisting, an art now only known in its rudest state in the same part of the East. The cables appear to be of great length and thickness, and ropes of various dimensions are represented in the sculptures.§

---

* See woodcut, p. 105.

† Although in these bas-reliefs, as in other Assyrian sculptures, no regard is paid to perspective, the proportions are very well kept. I must refer my readers to the 2d series of the Monuments of Nineveh for detailed drawings of these highly interesting sculptures.

‡ See woodcut in the Abridgment of my "Nineveh and its Remains" (p. 297.), which may be compared with the Assyrian bas-reliefs, to show the difference between the ancient and modern treatment of a subject almost identic.

§ There appears to be a curious allusion to ropes and cables of different sizes, and to their use for such purposes as that described in the text in Isaiah, v. 18. " Woe unto them that draw iniquity with *cords* of vanity and sin as it were

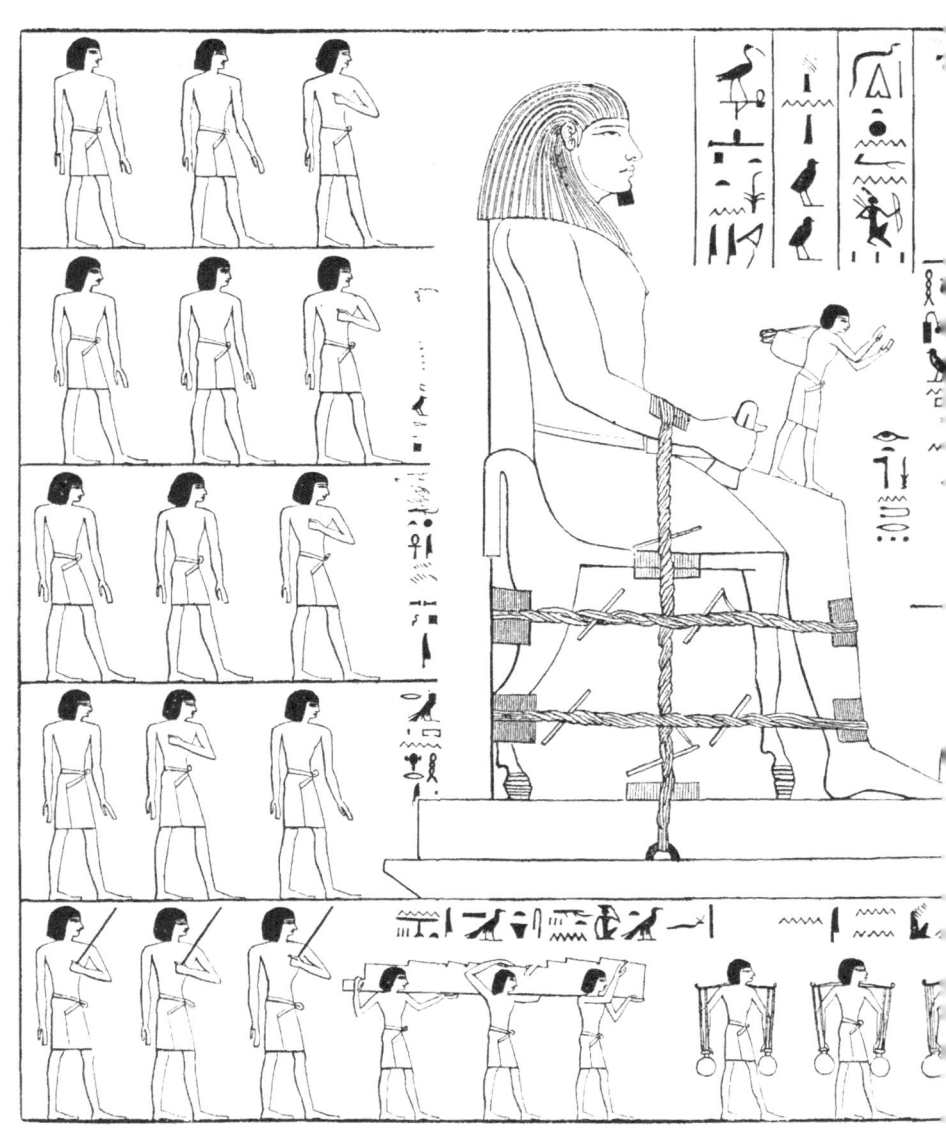

EGYPTIANS MOV.

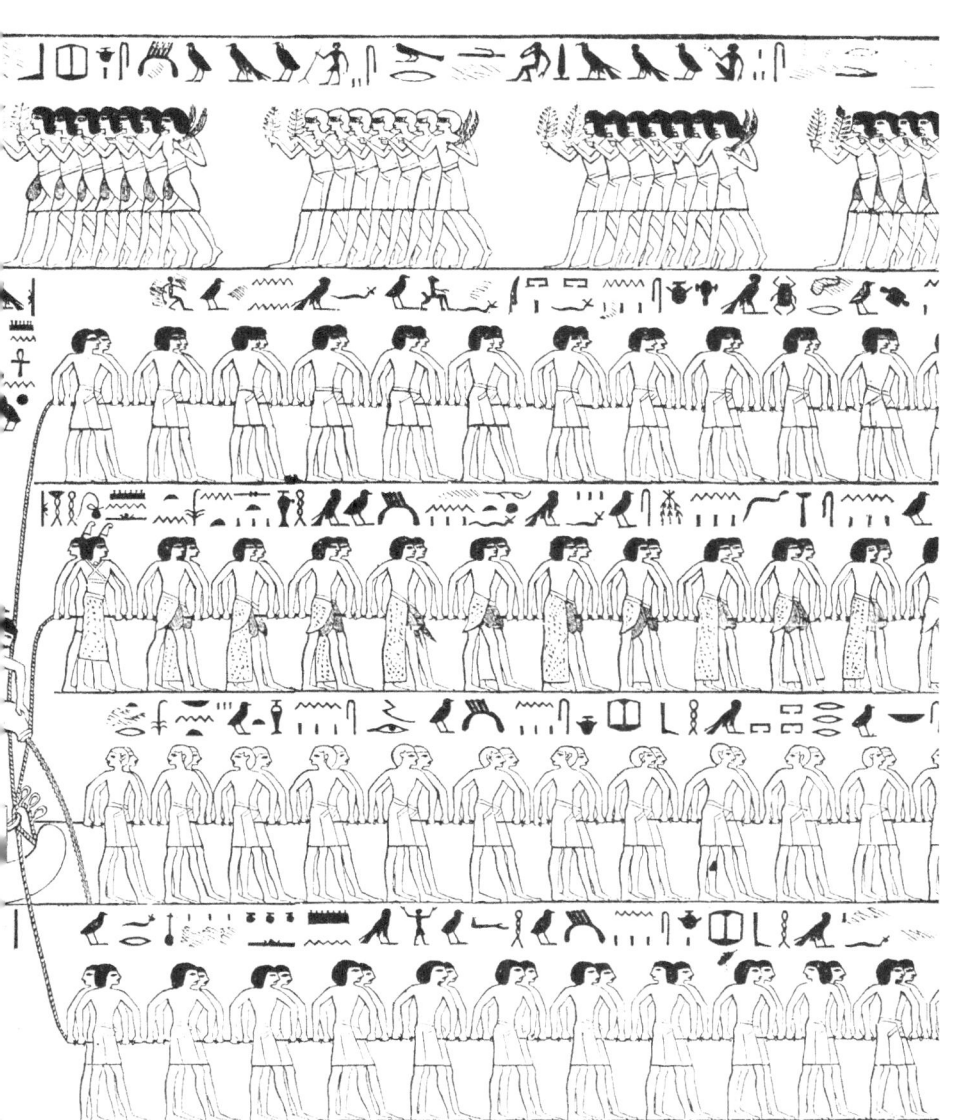

COLOSSUS FROM THE QUARRIES, IN A GROTTO AT EL BERSHEH. (From a drawing by Sir

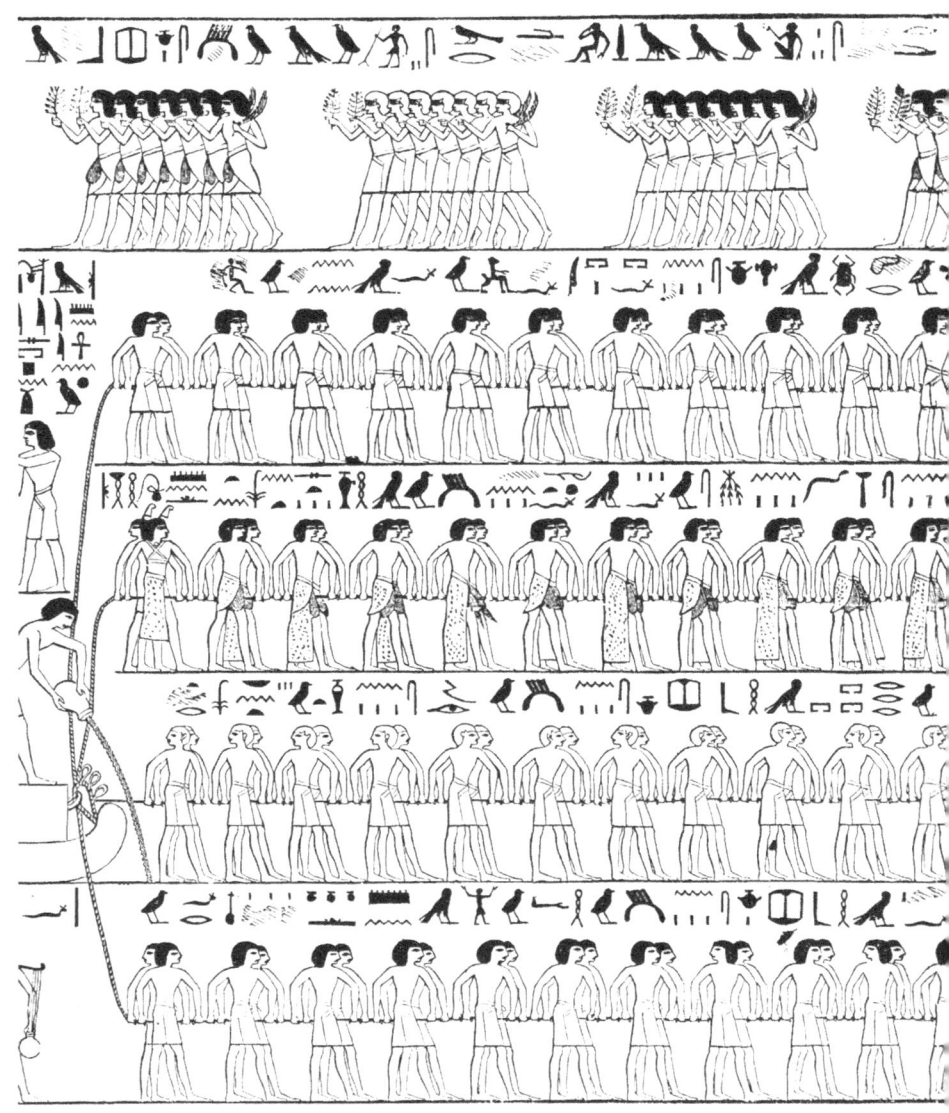

A COLOSSUS FROM THE QUARRIES, IN A GROTTO AT EL BERSHEH. (From a drawing by

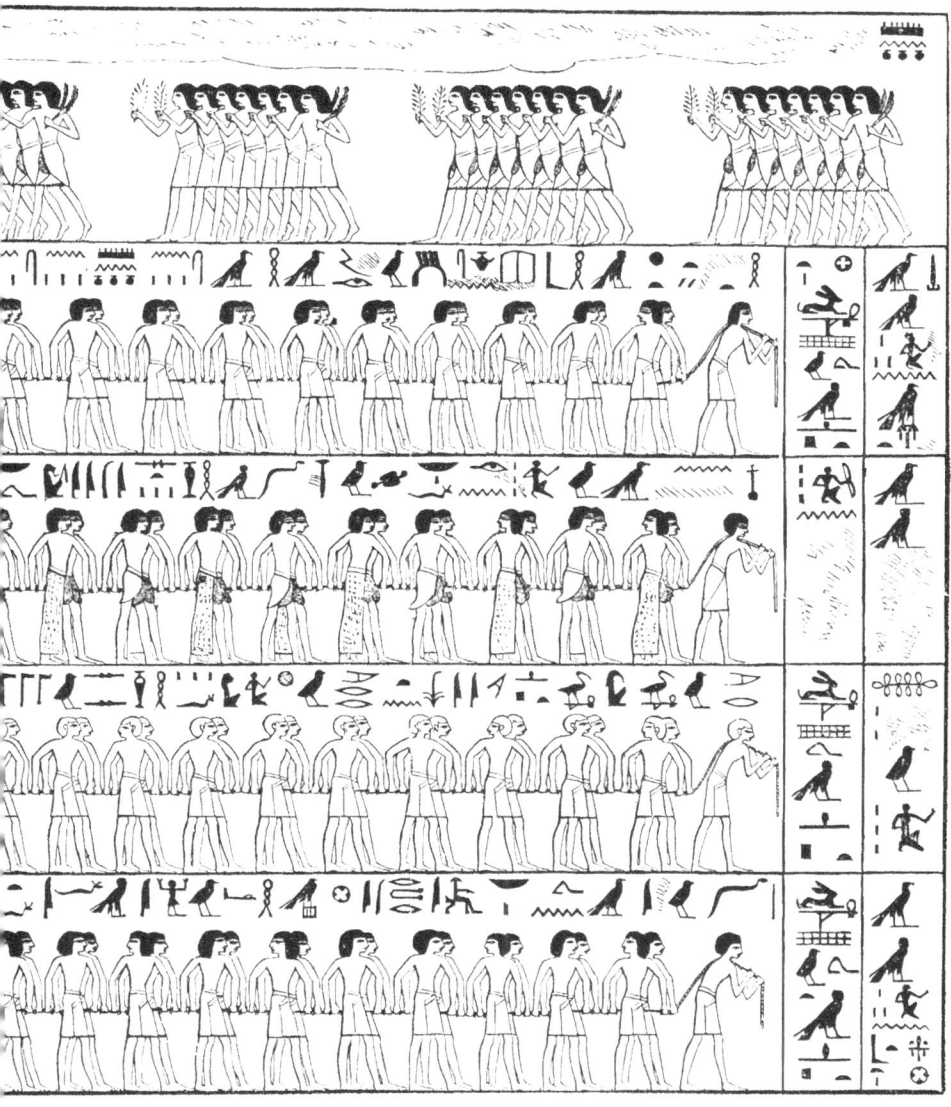

I have given, for the sake of comparison, a woodcut of the well-known painting in an Egyptian grotto at El Bersheh of the moving of a colossal figure.* It will show how the Egyptians and Assyrians represented nearly a similar subject, and in what way these nations differed in their mode of artistic treatment.   The Egyptian colossus is placed upon a sledge not unlike that of the Assyrian bas-reliefs in form, though smaller in comparison with the size of the figure, which appears in this case to have been about twenty-four feet high. † The ropes, four in number, as in the Kouyunjik sculptures, are all fastened to the fore part of the sledge, and are pulled by the workmen without the aid of smaller cords.   The absence of levers and rollers is remarkable, as the Egyptians must have been well acquainted with the use of both, and no doubt employed them for moving heavy weights.‡   On the statue, as in the Assyrian bas-reliefs, stands an officer who claps his hands in measured time to regulate the motions of the men, and from the front of the pedestal another pours some liquid, probably grease, on the ground to facilitate the progress of the sledge, which would scarcely be needed were rollers used.§   As in Assyria, the workmen included slaves and captives, who were accompanied by bands of armed men.

As this curious representation is believed to be of the time of Osirtasen II., a king of the seventeenth dynasty, who reigned, according to some, about sixteen centuries before Christ, it is far more ancient than any known Assyrian monument.   The masses of solid stone moved by the Egyptians also far exceeded in weight any sculpture that has yet been discovered in Assyria, or any monolith on record connected with that empire; with the exception, perhaps, of the celebrated obelisk which, according to Diodorus Siculus, was brought by Semiramis from Armenia to Babylon.‖

with a *cart rope.*" A most interesting collection of ancient Egyptian cordage of almost every kind has lately been purchased by the French Government from Clot Bey, and is now in the Louvre.

* This woodcut has been taken from a drawing by Sir Gardner Wilkinson, who has kindly allowed me to use it.   It is more correct in its details than that given in his work on the Ancient Egyptians, vol. iii. p. 328.

† Wilkinson, vol. iii. p. 327.

‡ Herodotus particularly mentions levers in his account of the transport of the monolith of Sais (lib. ii. 175.).

§ This looks as if the sledge were moving on an inclined way of boards constructed for the purpose.

‖ A colossus of granite of Rameses II., at the Memnonium, weighed when entire, according to Sir Gardner Wilkinson, 887 tons; and the stupendous mono-

It is a singular fact, that whilst the quarries of Egypt bear witness of themselves to the stupendous nature of the works of the ancient inhabitants of the country, and still show on their sides engraved records of those who made them, no traces whatever, notwithstanding the most careful research, have yet been found to indicate from whence the builders of the Assyrian palaces obtained their large slabs of alabaster. That they were in the immediate neighbourhood of Nineveh there is scarcely any reason to doubt, as strata of this material, easily accessible, abound, not only in the hills but in the plains. This very abundance may have rendered any particular quarry unnecessary, and blocks were probably taken as required from convenient spots, which have since been covered by the soil. The alabaster now used at Mosul is cut near the Sinjar gate, to the north-west of the town. The blocks are rarely larger than can be carried on the backs of horses. These quarries also supply Baghdad, where this material is much prized for the pavement of baths and serdaubs, or underground summer apartments.

There can be no doubt, as will hereafter be shown, that the king represented as superintending the building of the mounds and the placing of the colossal bulls is Sennacherib himself, and that the sculptures celebrate the building at Nineveh of the great palace and its adjacent temples described in the inscriptions as the work of this monarch. The bas-reliefs were accompanied in most instances by short epigraphs in the cuneiform character, containing a description of the subject with the name of the city to which the sculptures were brought. The great inscriptions on the bulls at the entrances of Kouyunjik record, it would seem, not only historical events, but, with great minuteness, the manner in which the edifice itself was erected, its general plan, and the various materials employed in decorating the halls, chambers, and roofs. When completely deciphered they will perhaps enable us to restore, with some confidence, both the general plan and elevation of the building.

Unfortunately only fragments of these epigraphs have been preserved. From them it would appear that the transport of more than one object was represented on the walls. Besides bulls and sphynxes in stone are mentioned figures in some kind of wood, perhaps of olive, like " the two cherubims of olive tree, each ten cubits

lith in the temple of Latona, at Buto, which, according to Herodotus, took 2000 men during three entire years to move to its place, upwards of 5000. (Wilkinson's Ancient Egyptians, vol. iii. p. 331.)

high," in the temple of Solomon.* Over the king superintending the removal of one of these colossi is the following short inscription thus translated by Dr. Hincks :—

"Sennacherib, king of Assyria, the great figures of bulls, which in the land of Belad, were made for his royal palace at Nineveh, he transported *thither*." (?)

The land of Belad, mentioned in these inscriptions, appears to have been a district in the immediate vicinity of Nineveh, and probably on the Tigris, as these great masses of stone would have been quarried near the river for the greater convenience of moving them to the palace. The district of Belad may indeed have been that in which the city itself stood.

Over the representation of the building of the mound there were two epigraphs, both precisely similar, but both unfortunately much mutilated. As far as they can be restored, they have thus been interpreted by Dr. Hincks :—

"Sennacherib, king of Assyria. Hewn stones, *which*, as the gods† willed, were found in the land of Belad, for the *walls* (?) (or foundations, the word reads '*shibri*') of my palace, *I caused the inhabitants of foreign countries* (?) and the people of the forests (Kershani)‡, the great bulls for the gates of my palace to *drag* (?) (or bring)."

If this inscription be rightly rendered, we have direct evidence that captives from foreign countries were employed in the great public works undertaken by the Assyrian kings, as we were led to infer, from the variety of costume represented in the bas-reliefs, and from the fetters on the legs of some of the workmen. The Jews themselves, after their captivity, may have been thus condemned to labor, as their forefathers had been in Egypt, in erecting the monuments of their conquerors; and we may, perhaps, recognise them amongst the builders portrayed in the sculptures. Two distinct objects appear to be mentioned in these epigraphs,—unhewn, or merely squared, stones for walls or foundations, and the colossal bulls for the entrances; unless some of the

---

* 1 Kings, vi. 23. I shall hereafter compare the edifices built by Solomon with the Assyrian palaces, and point out the remarkable illustrations of the Jewish temple afforded by the latter.

† A peculiar deity is mentioned who probably presided over the earth, but his name is as yet unknown; it is here denoted by a monogram.

‡ Compare the Hebrew חֹרֶשׁ, khersh, a thick wood, or, perhaps, חָרָשׁ, a stone-cutter, or a workman in stone or wood.

small stones carried on the backs of the workmen are intended by the former, we find only the colossi represented in the bas-reliefs.

From the long gallery, which appears to have been panelled with bas-reliefs, describing the removal of more than one object employed in the construction of the palace, we have unfortunately only three fragments of inscriptions without the sculptured representations of the events recorded. The most perfect is interesting on more than one account. According to Dr. Hincks it is to be translated : —

" Sennacherib, king of Assyria . . . . (some object, the nature not ascertained) of wood, which from the Tigris I caused to be brought up (*through* ?) the Kharri, or Khasri, on sledges (or boats), I caused to be carried (or to mount)."

The name of the river in this inscription very nearly resembles that of the small stream which sweeps round the foot of the great mound of Kouyunjik. In the woodcut of the king superintending the removal of the bull *, it will be perceived that two rivers, a smaller running into a larger, appear to be rudely represented. They correspond with the actual position of the Tigris and Khauser beneath Kouyunjik. It is possible, therefore, that the latter stream was deepened or enlarged, so as to enable the Assyrians to float heavy masses close to the mound; and from the bas-relief it would appear that the bull was moved from the very edge of the water up the artificial declivity. At that time, however, the Tigris was nearer to the palace than it now is to the ruins, its course having varied considerably at different periods ; but its ancient bed is still indicated by recent alluvial deposits.

In the fragment of another epigraph, we have mention of some objects also of wood " brought from Mount Lebanon, and taken up (to the top of the mound) from the Tigris." These may have been beams of cedar, which, it will be hereafter seen, were extensively used in the Assyrian palaces. It is highly interesting thus to find the inhabitants of Nineveh fetching their rare and precious woods from the same spot that king Solomon had brought the choicest woodwork of the temple of the Lord and of his own palaces.

On a third fragment similar objects are described as coming from or up the same Kharri or Khasri.

I have mentioned that the long gallery containing the bas-relief representing the moving of the great stone, led out of a chamber,

* *Ante*, page 111.

whose walls had been completely uncovered.* The sculptures upon them were partly preserved, and recorded the conquest of a city standing on a broad river, in the midst of mountains and forests. The Assyrians appear to have entered the enemy's country by a valley, to have forded the stream frequently, and to have continued during their march along its banks. Warriors on foot led their horses, and dragged the chariots over precipitous rocks. On each side of the river were wooded hills, with small streams flowing amongst vineyards. As they drew near to the city, the Assyrians cut down the woods to clear the approaches. Amongst the branches of a tree exceeding the others in size, and standing immediately beneath the walls, were birds and two nests containing their young. The sculptor probably introduced these accessories to denote the season of the year. The river appeared to flow through or behind the city. Long low walls with equidistant towers, the whole surmounted by cornices and angular battlements, stood on one side of the stream. Within the walls were large square buildings, curiously ornamented, and whose windows, immediately beneath the roof, were formed by small pillars with capitals in the form of the Ionic volute. The doors, except the entrance to the castle which was arched, were square, and, in some instances, surmounted by a plain cornice. That part of the city standing on the opposite side of the river, seemed to consist of a number of detached forts and houses, some of which had also open balustrades to admit the light. Flames issued from the dwellings, and on the towers were men apparently cutting down trees growing within the walls. Assyrian warriors, marching in a long line, carried away the spoil from the burning city. Some were laden with arms; others with furniture, chairs, stools, couches, and tables of various forms, ornamented with the heads and feet of animals. They were probably of metal, perhaps of gold or silver. The couches, or beds, borne by two men, had a curved head. Some of the chairs had high backs, and the tables resembled in shape the modern camp-stool.

The last bas-relief of the series represented the king seated within a fortified camp, on a throne of elaborate workmanship, and having beneath his feet a footstool of equally elegant form. He was receiving the captives, who wore long robes falling to their ankles. Unfortunately no inscription remained by which we might identify the conquered nation. It is probable, from

* No. XLVIII. Plan I. See Monuments of Nineveh, 2d series, Plate 40.

the nature of the country represented, that they inhabited some district in the western part of Asia Minor or in Armenia, in which direction, as we shall hereafter see, Sennacherib more than once carried his victorious arms. The circular fortified walls enclose tents, within which are seen men engaged in various domestic occupations.

It will be remembered that excavations had been resumed in a lofty mound in the north-west line of walls forming the enclosure round Kouyunjik. It was apparently the remains of a gate leading into this quarter of the city, and part of a building, with fragments of two colossal winged figures *, had already been discovered in it. By the end of November the whole had been explored, and the results were of considerable interest. As the mound rises nearly fifty feet above the plain, we were obliged to tunnel along the walls of the building within it, through a compact mass of rubbish, consisting almost entirely of loose bricks. Following the rows of low limestone slabs, from the south side of the mound, and passing through two halls or chambers, we came at length to the opposite entrance. This gateway, facing the open country, was formed by a pair of majestic human-headed bulls, fourteen feet in length, still entire, though cracked and injured by fire. They were similar in form to those of Khorsabad and Kouyunjik, wearing the lofty head-dress, richly ornamented with rosettes, and edged with a fringe of feathers peculiar to that period. Wide spreading wings rose above their backs, and their breasts and bodies were profusely adorned with curled hair. Behind them were colossal winged figures of the same height, bearing the pine cone and basket. Their faces were in full, and the relief was high and bold. More knowledge of art was shown in the outline of the limbs and in the delineation of the muscles, than in any sculpture I have seen of this period. The naked leg and foot were designed with a spirit and truthfulness worthy of a Greek artist.† It is, however, remarkable that the four figures were unfinished, none of the details having been put in, and parts being but roughly outlined. They stood as if the sculptors had been interrupted by some public calamity, and had left their work incomplete. Perhaps the murder of Senna-

---

* Nineveh and its Remains, vol. i. p. 146.

† The bulls and winged figures resembled those from Khorsabad, now in the great hall at the British Museum, but far exceeded them in beauty and grandeur, as well as in preservation. As nearly similar figures had thus already been sent to England, I did not think it advisable to remove them.

cherib by his sons, as he worshipped in the house of Nisroch his god, put a sudden stop to the great undertakings he had commenced in the beginning of his reign.

The sculptures to the left, on entering from the open country, were in a far more unfinished state than those on the opposite side. The hair and beard were but roughly marked out, square bosses being left for carving the elaborate curls. The horned cap of the human-headed bull was, as yet, unornamented, and the wings merely outlined. The limbs and features were hard and angular, still requiring to be rounded off, and to have expression given to them by the finishing touch of the artist. The other two figures were more perfect. The curls of the beard and hair (except on one side of the head of the giant) and the ornaments of the head-dress had been completed. The limbs of the winged deity and the body and legs of the bull had been sufficiently finished to give a bold and majestic character to the figures, which might have been rather lessened than improved by the addition of details. The wings of the giant were merely in outline. The sculptor had begun to mark out the feathers in those of the bull, but had been interrupted after finishing one row and commencing a second.* No inscription had yet been carved on either sculpture.

The entrance formed by these colossal bulls was fourteen feet and a quarter wide. It was paved with large slabs of limestone, still bearing the marks of chariot wheels. The sculptures were buried in a mass of brick and earth, mingled with charcoal and charred wood ; for "the gates of the land had been set wide open unto the enemy, and the fire had devoured the bars."† They were lighted from above by a deep shaft sunk from the top of the mound. It would be difficult to describe the effect produced, or the reflections suggested by these solemn and majestic figures, dimly visible amidst the gloom, when, after winding through the dark, underground passages, you suddenly came into their presence. Between them Sennacherib and his hosts had gone forth in all their might and glory to the conquest of distant lands, and had returned rich with spoil and captives, amongst whom may have been the handmaidens and wealth of Israel. Through them, too, the Assyrian monarch had entered his capital in shame, after his last and fatal defeat. Then the lofty walls, now but long lines

* See Plate 3. of the 2d series of the Monuments of Nineveh. The giant is correctly represented in its unfinished state in this plate, but the artist by mistake has filled up the details in the wings of the bulls.

† Nahum, iii. 13.

of low, wave-like mounds, had stretched far to the right and to the left — a basement of stone supporting a curtain of solid brick masonry, crowned with battlements and studded with frowning towers.

This entrance may have been arched like the castle gates of the bas-reliefs, and the mass of burnt bricks around the sculptures may be the remains of the vault.   A high tower evidently rose above this gate, which formed the great northern access to this quarter of Nineveh.

Behind the colossal figures, and between the outer and inner face of the gateway, were two chambers, nearly 70 feet in length by 23 in breadth.   Of that part of the entrance which was within the walls, only the fragments of winged figures, discovered during my previous researches, now remained.*   It is probable, however, that a second pair of human-headed bulls once stood there.   They may have been " the figures of animals," described to Mr. Rich as having been casually uncovered in this mound, and which were broken up nearly fifty years ago to furnish materials for the repair of a bridge. †

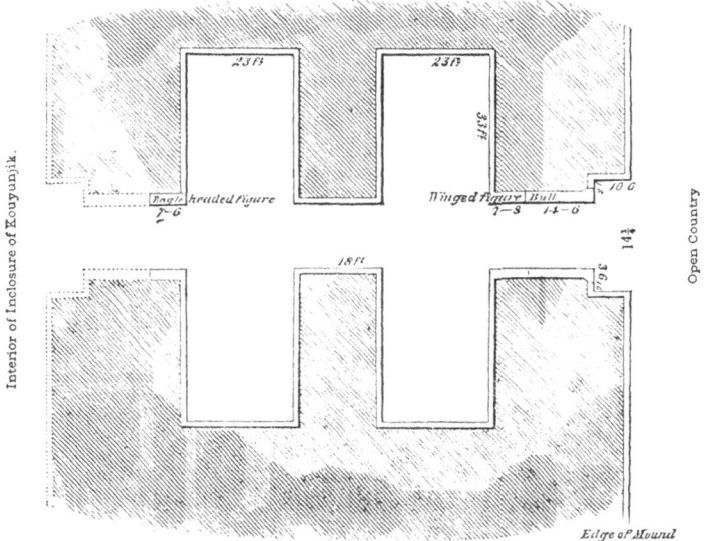

Plan of Northern Entrance to Inclosure of Kouyunjik.

The whole entrance thus consisted of two distinct chambers and three gateways, two formed by human-headed bulls, and a third between them simply panelled with low limestone slabs like the

---

* See Nineveh and its Remains, vol. i. p. 143.

† See Rich's Residence in Kurdistan and Nineveh, vol. ii. p. 39.

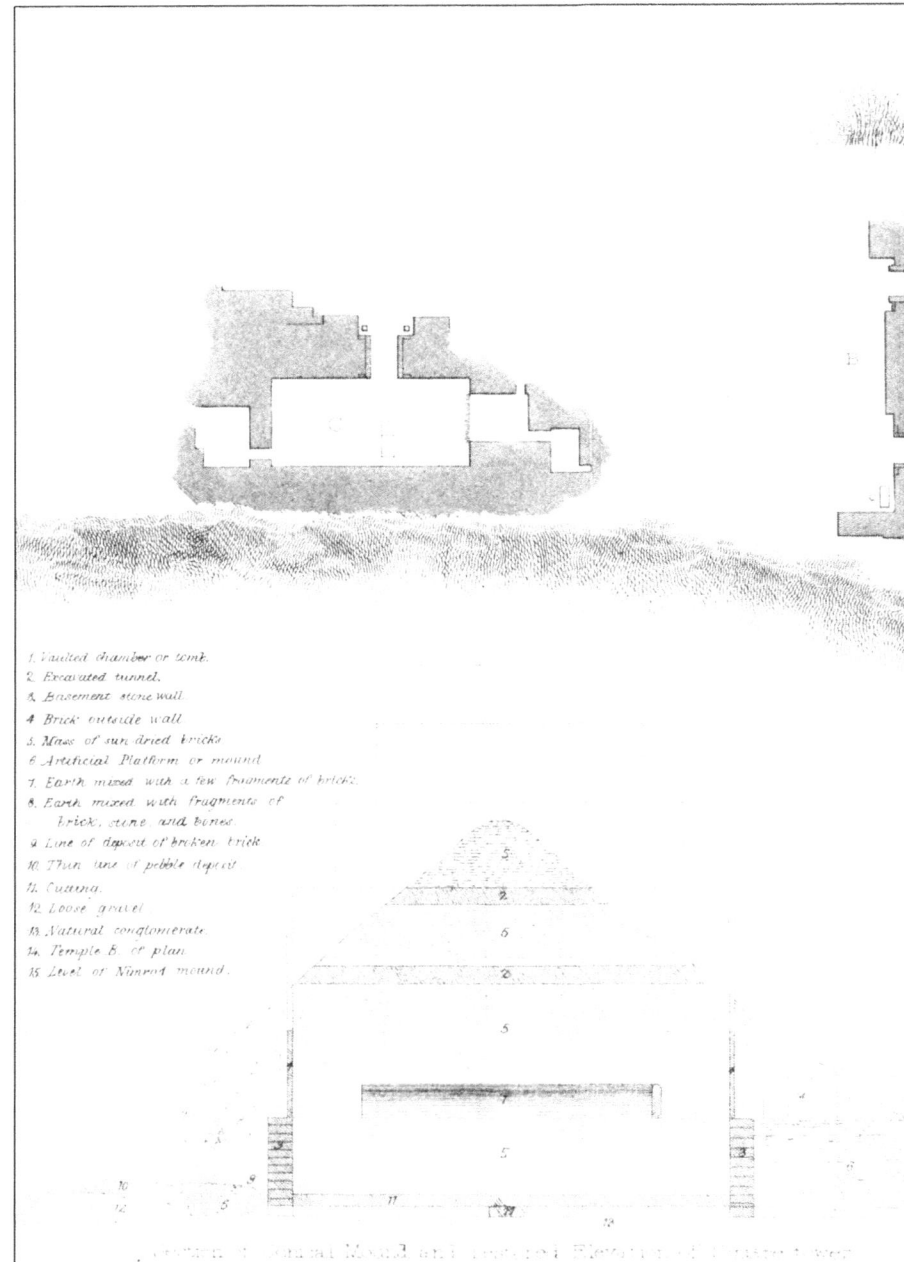

1. Vaulted chamber or tomb.
2. Excavated tunnel.
3. Basement stone wall.
4. Brick outside wall.
5. Mass of sun dried bricks.
6. Artificial Platform or mound.
7. Earth mixed with a few fragments of bricks.
8. Earth mixed with fragments of
   brick, stone, and bones.
9. Line of deposit of broken brick.
10. Thin line of pebble deposit.
11. Cutting.
12. Loose gravel.
13. Natural conglomerate.
14. Temple B or plan.
15. Level of Nimrod mound.

John Murray, Albemarle St. 1852

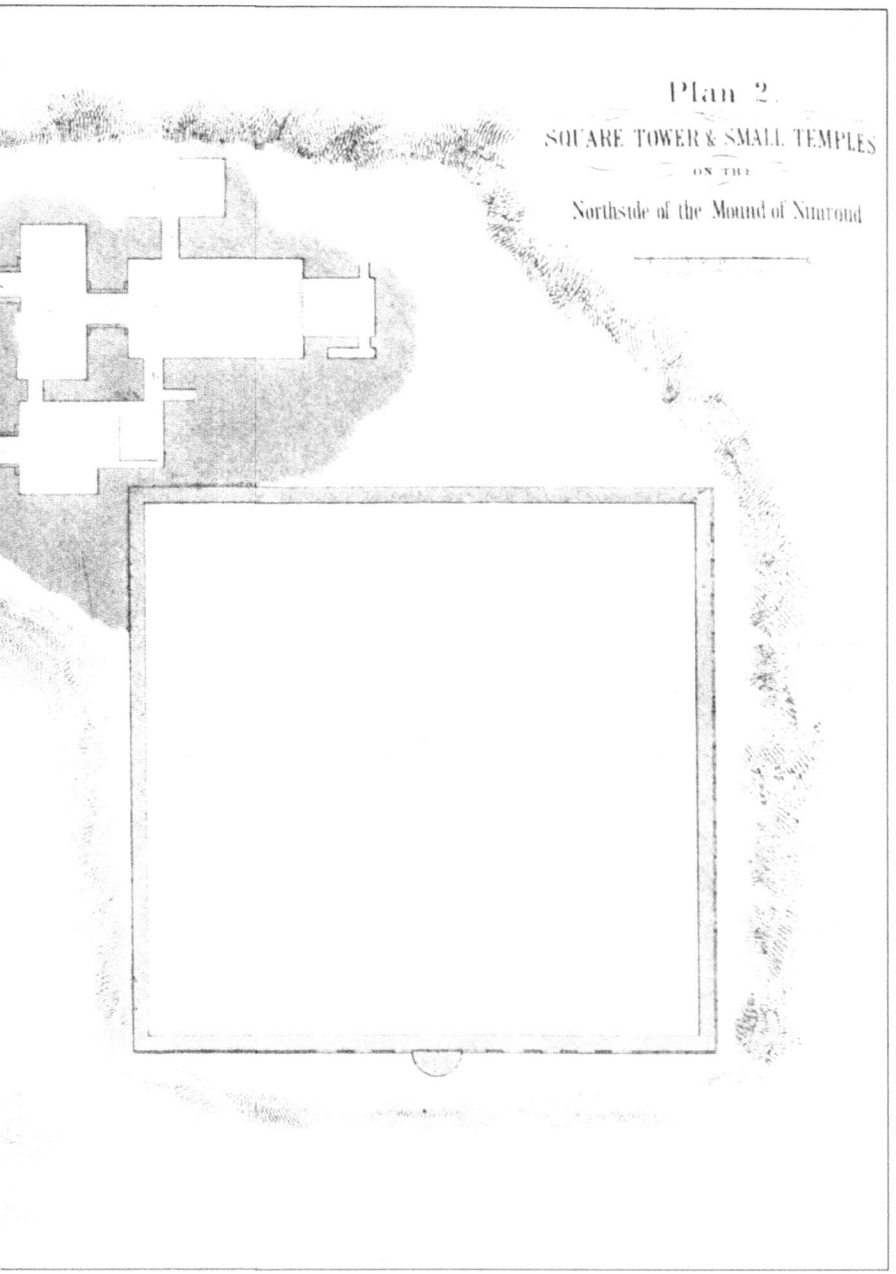

Plan 2.

SQUARE TOWER & SMALL TEMPLES

ON THE

Northside of the Mound of Nimroud

chambers. Its original height, including the tower, must have been full one hundred feet. Most of the baked bricks found amongst the rubbish bore the name of Sennacherib, the builder of the palace of Kouyunjik. A similar gateway, but without any remains of sculptured figures, and panelled with plain alabaster slabs, was subsequently discovered in the inner line of walls forming the eastern side of the quadrangle, where the road to Baashiekhah and Baazani leaves the ruins.

At Nimroud discoveries of very considerable importance were made in the high conical mound at the north-west corner. Desirous of fully exploring that remarkable ruin, I had employed nearly all the workmen in opening a tunnel into its western base. After penetrating for no less than eighty-four feet through a compact mass of rubbish, composed of loose gravel, earth, burnt bricks, and fragments of stone, the excavators came to a wall of solid stone masonry. The manner in which this structure had been buried is so curious, that I have given a section of the different strata through which the tunnel passed.* I have already observed that the edifice covered by this high mound was originally built upon the natural rock, a bank of hard conglomerate rising about fifteen feet above the plain, and washed in days of yore by the waters of the Tigris. Our tunnel was carried for thirty-four feet on a level with this rock, which appears to have been covered by a kind of flooring of sun-dried bricks, probably once forming a platform in front of the building. It was buried to the distance of thirty feet from the wall, by baked bricks broken and entire, and by fragments of stone, remains of the superstructure once resting upon the basement of still existing stone masonry. This mass of rubbish was about thirty feet high, and in it were found bones apparently human, and a yellow earthen jar rudely colored with simple black designs.† The rest of this part of the mound consisted of earth, through which ran two thin lines of extraneous deposit, one *of pebbles,* the other of fragments of brick and pottery. I am totally at a loss to account for their formation.

I ordered tunnels to be carried along the basement wall in both directions, hoping to reach some doorway or entrance, but it was found to consist of solid masonry, extending nearly the whole length of the mound. Its height was exactly twenty feet, which, sin-

---

* See section of conical mound, Plan II.

† These relics may have belonged to tombs made in the mound after the edifice had fallen into ruins.

Tunnel along Eastern Basement Wall (Nimroud).

Tunnel along Western Basement Wall (Nimroud).

gularly enough, coincides with that assigned by Xenophon to the
stone basement of the wall of the city (Larissa).* It was finished
at the top by a line of gradines, forming a kind of ornamental
battlement, similar to those represented on castles in the sculpture. ⌐⌐⌐ These gradines had fallen, and some of them
were discovered in the rubbish.† The stones in this structure were
carefully fitted together, though not united with mortar, unless the
earth which filled the crevices was the remains of mud used, as it
still is in the country, as a cement. They were bevelled with a
slanting bevel, and in the face of the wall were eight recesses or
false windows, four on each side of a square projecting block
between gradines.

The basement, of which this wall proved to be only one face,
was not excavated on the northern and eastern side until a later
period, but I will describe all the discoveries connected with this
singular building at once. The northern side was of the same
height as, and resembled in its masonry, the western. It had a
semicircular hollow projection in the centre, sixteen feet in diameter, on the east side of which were two recesses, and on the
west four, so that the two ends of the wall were not uniform.
That part of the basement against which the great artificial mound
or platform abutted, and which was consequently concealed by it,
that is, the eastern and southern sides, was of simple stone masonry
without recesses or ornament. The upper part of the edifice,
resting on the stone substructure, consisted of compact masonry of
burnt bricks, which were mostly inscribed with the name of the
founder of the centre palace (the obelisk king), the inscription
being in many instances turned outwards.

It was thus evident that the high conical mound forming the
north-west corner of the ruins of Nimroud, was the remains of a
square tower, and not of a pyramid, as had previously been conjectured. The lower part, built of solid stone masonry, had with-
stood the wreck of ages, but the upper walls of burnt brick, and
the inner mass of sun-dried brick which they encased, falling outwards, and having been subsequently covered with earth and vegetation, the ruin had taken the pyramidal form that loose materials
falling in this manner would naturally assume.

It is very probable that this ruin represents the tomb of Sarda-

---

* Anab. lib. iii. c. 4.

† Part of a wall, precisely similar in construction, still exists on one side of
the great mound of Kalah Sherghat. (Nineveh and its Remains, vol. ii. p. 61.)

napalus, which, according to the Greek geographers, stood at the entrance of the city of Nineveh. It will hereafter be seen that it is not impossible the builder of the north-west palace of Nimroud was a king of that name, although it is doubtful whether he can be identified with the historical Sardanapalus. Subsequent discoveries proved that he must himself have raised the stone substructure, although his son, whose name is found upon the bricks, completed the building. It was, of course, natural to conjecture that some traces of the chamber in which the royal remains were deposited, were to be found in the ruin, and I determined to examine it as fully as I was able. Having first ascertained the exact centre of the western⋅ stone basement, I there forced a passage through it. This was a work of some difficulty, as the wall was 8 ft. 9 in. thick, and strongly built of large rough stones. Having, however, accomplished this step, I carried a tunnel completely through the mound, at its very base, and on a level with the natural rock, until we reached the opposite basement wall, at a distance of 150 feet. Nothing having been discovered by this cutting, I directed a second to be made at right angles to it, crossing it exactly in the centre, and reaching from the northern to the southern basement; but without any discovery. At the point where they intersected, and therefore precisely in the centre of the building, I dug down through the solid conglomerate to the depth of five feet, but without finding any traces whatever of an ancient disturbance of the soil. I was unable to make further excavations in this part of the ruin, on account of the enormous mass of superincumbent earth, and the great risk to which the men were exposed from its falling in.

The next cutting was made in the centre of the mound, on a line with the top of the stone basement wall, which was also the level of the platform of the north-west palace. The workmen soon came to a narrow gallery, about 100 feet long, 12 feet high, and 6 feet broad, which was blocked up at the two ends without any entrance being left into it. It was vaulted with sun-dried bricks, a further proof of the use of the arch at a very early period, and the vault had in one or two places fallen in. No remains whatever were found in it, neither fragments of sculpture or inscription, nor any smaller relic. There were, however, undoubted traces of its having once been broken

---

* The walls, as well as the vault, were of sun-dried bricks. It is curious that between one row of bricks was a layer of reeds, as in the Babylonian ruins; the only instance of this mode of construction yet met with in Assyria.

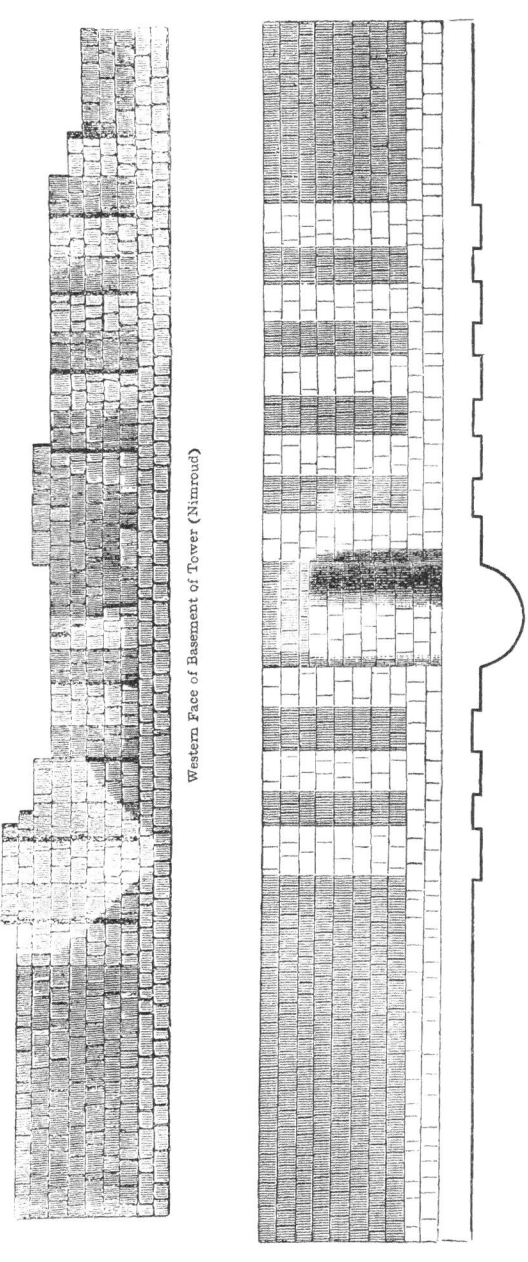

Western Face of Basement of Tower (Nimroud)

Northern Face of Basement of Tower (Nimroud).

into on the western side, by digging into the face of the mound
after the edifice was in ruins, and consequently, therefore, long
after the fall of the Assyrian empire. There was an evident
depression in the exterior of the mound, which could be perceived
by an observer from the plain, and the interior vault had been
forced through. The remains which it may have contained,
probably the embalmed body of the king, with vessels of precious
metals and other objects of value buried with it, had been carried
off by those who had opened the tomb at some remote period, in
search of treasure. They must have had some clue to the precise
position of the chamber, or how could they have dug into the
mound exactly at the right spot? Had this depositary of the dead
escaped earlier violation, who can tell with what valuable and important relics of Assyrian art or Assyrian history it might have
furnished us? I explored, with feelings of great disappointment,
the empty chamber, and then opened other tunnels, without further
results, in the upper parts of the mound.

It was evident that the long gallery or chamber I have described
was the place of deposit for the body of the king, if this were
really his tomb. The tunnels and cuttings in other parts of the
mound only exposed a compact and solid mass of sun-dried brick
masonry. I much doubt, for many reasons, whether any sepulchre
exists in the rock beneath the foundations of the tower, though, of
course, it is not impossible that such may be the case.*

From the present state of the ruin it is difficult to conjecture the
exact original form and height of this edifice. There can be no doubt
that it was a vast square tower, and it is not improbable that it may
have terminated in a series of three or more gradines, like the obelisk
of black marble from the centre palace now in the British Museum.
It is this shape that I have ventured to give it, in a general
restoration of the platform of Nimroud and its various edifices.†

* Col. Rawlinson, remarks in his memoir on the "Outlines of Assyrian
History" (published by the Royal Asiatic Society in 1852), that "the great
pyramid at Nimroud was erected by the son of the builder of the north-west
palace;" and as the Greeks name that monument the tomb of Sardanapalus, he
believes that "a shaft sunk into the centre of the mound, and carried down to the
foundations, would lay bare the original sepulchre. The difficulties (he adds) of such
an operation have hitherto prevented its execution, but the idea is not altogether
abandoned." He appears thus, curiously enough, to be ignorant of the excavations in that ruin described in the text, although he had just visited Nimroud.
The only likely place not yet examined would be *beneath* the very foundations.

† In the frontispiece to the 2d series of the Monuments of Nineveh. I am
indebted to Mr. Fergusson, who was good enough to make the original drawing,
for this restoration so ably executed by Mr. Baines.

Like the palaces, too, it was probably painted on the outside with various mythic figures and devices, and its summit may have been crowned by an altar, on which the Assyrian king offered up his great sacrifices, or on which was fed the ever-burning sacred fire. But I will defer any further remarks upon this subject until I treat of the architecture of the Assyrians.

As the ruin is 140 feet high, the building could scarcely have been much less than 200, whilst the immense mass of rubbish surrounding and covering the base shows that it might have been considerably more.

During the two months in which the greater part of the discoveries described in this chapter were made, I was occupied almost entirely with the excavations, my time being spent between Nimroud and Kouyunjik. The only incidents worth noting were a visit from Hussein Bey, Sheikh Nasr, and the principal chiefs of the Yezidis, and a journey taken with Hormuzd to Khorsabad and the neighbouring ruins.

The heads of the Yezidi sect came to Mosul to settle some differences with the Turkish authorities about the conscription. They lodged in my house. Sheikh Nasr had only once before ventured into the town, and then but for a few hours. To treat them with due honor I gave an entertainment, and initiated them into the luxuries of Turkish cookery. We feasted in the Iwan, an arched hall open to the courtyard, which was lighted up at night with *mashaals*, or bundles of flaming rags saturated with bitumen, and raised in iron baskets on high poles, casting a flood of rich red light upon surrounding objects. The Yezidis performed their dances to Mosul music before the chiefs. Suddenly the doors were thrown open, and a band of Arabs, stripped to the waist, brandishing their weapons and shouting their warcry, rushed into the yard. The Yezidis believed that they had been betrayed. The young chief drew his sword; and even Sheikh Nasr, springing to his feet, prepared to defend himself. Their fears, however, gave way to a hearty laugh, when they learnt that the intruders were a band of my workmen, who had been instigated by Mr. Hormuzd Rassam thus to alarm my guests.

Wishing to visit Baasheikhah, Khorsabad, and other ruins at the foot of the range of low hills of the Gebel Makloub, I left Nimroud on the 26th of November with Hormuzd and the Bairakdar. Four hours' ride brought us to some small artificial mounds near the village of Lak, about three miles to the east of the high road

K

to Mosul. Here we found a party of workmen excavating under
one of the Christian superintendents. Nothing had been discovered except fragments of pottery and a few bricks bearing the
name of the Kouyunjik king. As the ruins, from their size, did
not promise other results, I sent the men back to Mosul. We
reached Khorsabad after riding for nearly eight hours over a rich
plain, capable of very high cultivation, though wanting in water,
and still well stocked with villages, between which we startled large
flocks of gazelles and bustards. I had sent one of my overseers
there some days before to uncover the platform to the west of
the principal edifice, a part of the building I was desirous of examining. Whilst clearing away the rubbish, he had discovered
two bas-reliefs sculptured in black stone. They represented a
hunting scene. On one slab, broken into several pieces, was an
eunuch discharging an arrow at a flying bird, probably a pigeon
or partridge. He was dressed in a fringed robe, confined at
the waist by a girdle, and a short sword hung from his shoulder
by a broad and richly ornamented belt. The ends of his bow
were in the shape of the heads of birds. Behind the archer were
two figures, one carrying a gazelle over his shoulder and a hare
in his hand, the other wearing an embroidered tunic, and armed
with a bow and arrows. In the back ground were trees, and birds
flying amongst them.* On the second slab were huntsmen carrying birds, spears, and bows.

These bas-reliefs were executed with much truth and spirit.
They belonged to a small building, believed to be a temple,
entirely constructed of black marble, and attached to the palace.
It stood upon a platform 165 feet in length and 100 in width,
raised about 6 feet above the level of the flooring of the chambers, and ascended from the main building by a flight of broad
steps. This platform, or stylobate, is remarkable for a cornice
in grey limestone carried round the four sides, — one of the few
remains of exterior decoration in Assyrian architecture, with which
we are acquainted. It is carefully *built* of separate stones, placed
side by side, each forming part of the section of the cornice. Mr.
Fergusson observes†, with reference to it, " at first sight it
seems almost purely Egyptian ; but there are peculiarities in
which it differs from any found in that country, especially in the
curve being continued beyond the vertical tangent, and the conse-

---

* See Plate 32. of the 2d series of the Monuments of Nineveh. This bas-
relief, which has been perfectly repaired, is now in the British Museum
† Palaces of Nineveh and Persepolis restored, p. 223

quent projection of the torus giving a second shadow.　Whether the effect of this would be pleasant or not in a cornice placed so

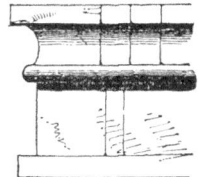

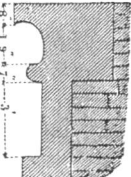

Elevation of Stylobate of Temp.e　　　　　Section of Stylobate of Temple.

high that we must look up to it is not quite clear; but below the level of the eye, or slightly above it, the result must have been more pleasing than any form found in Egypt, and where sculpture is not added might be used with effect anywhere."

Many fragments of bas-reliefs in the same black marble, chiefly parts of winged figures, had been uncovered; but this building has been more completely destroyed than any other part of the palace of Khorsabad, and there is scarcely enough rubbish even to cover the few remains of sculpture which are scattered over the platform.

The sculptures in the palace itself had rapidly fallen to decay, and of those which had been left exposed to the air after M. Botta's departure scarcely any traces remained.　Some, however, had been covered up and partly preserved by the falling in of the high walls of earth forming the sides of the trenches.　Here and there a pair of colossal bulls, still guarding the portals of the ruined halls, raised their majestic but weather-beaten human heads above the soil.　In one or two unexplored parts of the ruins my workmen had found inscribed altars or tripods, similar to that in the Assyrian collection of the Louvre, and bricks ornamented with figures and designs in color, showing that they had belonged to walls painted with subjects resembling those sculptured on the alabaster panels.

Since my former visit to Khorsabad, the French consul at Mosul had sold to Col. Rawlinson the pair of colossal human-headed bulls and winged figures, now in the great hall of the British Museum.*

---

* These sculptures were purchased by the Trustees of the British Museum from Col. Rawlinson.　Owing to that carelessness and neglect, of which there has been so much cause to complain in all that concerns the transport of the Assyrian antiquities to this country, they have suffered very considerable injury since their discovery.　They were sawn into many pieces for facility of transport by my marble-cutter Behnan, superintended by Mr. Rassam.

They had stood in a propylæum, about 900 feet to the south-east of the palace, within the quadrangle, but not upon the artificial mound. In form this small building appears to have been nearly the same as the gateway, in the walls of Kouyunjik*, and like it was built of brick and panelled with low limestone slabs. From the number of enamelled bricks discovered in the ruins it is probable that it was richly decorated in color.†

Trenches had also been opened in one of the higher mounds in the line of walls, and in the group of ruins at the S. W. corner of the quadrangle, but no discoveries of any interest had been made. The centre of the quadrangle was now occupied by a fever-breeding marsh formed by the waters of the Khauser.

We passed the night at Futhliyah, a village built at the foot of the Gebel Makloub, about a mile and a half from Khorsabad. A small grove of olive trees renders it a conspicuous object even from Mosul, whence it looks like a dark shadow on the tawny plain. Although once containing above two hundred houses it has now but sixty. It formerly belonged to the Mosul spahis, or military fief-holders, and is still claimed by them, although the government has abolished such tenures. We lodged in a well-built stone kasr, or large house, fast falling into ruins, belonging to the Alaï Bey, or chief of the spahis. Selim Bey, one of the former tenants of the land, still lingered about the place, gathering together such small revenues in money and in kind as he could raise amongst the more charitable of the inhabitants. He came to me in the morning, and gave me the history of the village and of its owners.

Near Futhliyah, and about two miles from the palace of Khorsabad, is a lofty conical Tel visible from Mosul, and from most parts of the surrounding country. It is one of those isolated mounds so numerous in the plains of Assyria, which do not appear to form part of any group of ruins, and the nature of which I have been unable to determine. Its vicinity to Khorsabad led me to believe that it might have been connected with those remains, and might have been raised over a tomb. By my directions deep trenches were opened into its sides, but only fragments of pottery were discovered. The place is, however, worthy of a more complete examination than the time and means at my disposal would permit.

From Futhliyah we rode across the plain to the large village of Baazani, chiefly inhabited by Yezidis. There we found Hussein

Bey, Sheikh Nasr, and a large party of Cawals assembled at the
house of one Abd-ur-rahman Chelibi, a Mussulman gentleman of
Mosul, who had farmed the revenues of the place.

Near Baazani is a group of artificial mounds of no great size.
The three principal have been used as burying-places by the
Yezidis, and are covered with their graves and white conical
tombs. Although no difficulties would have been thrown in my
way had I wished to excavate in these ruins, they did not appear
to me of sufficient importance to warrant an injury to the feelings
of these poor people by the desecration of the resting-places of
their dead. Having examined them, therefore, and taken leave
of the chiefs, I rode to the neighbouring village of Baasheikhah,
only separated from Baazani by a deep watercourse, dry except
during the rains. Both stand at the very foot of the Gebel
Makloub. Immediately behind them are craggy ravines worn by
winter torrents. In these valleys are quarries of the kind of alabaster used in the Assyrian palaces, but I could find no remains to show that the Assyrians had obtained their great slabs
from them, although they appear to be of ancient date. They are
now worked by the Yezidis, who set apart the proceeds for Sheikh
Nasr, as the highpriest of the tomb of Sheikh Adi. The stone
quarried from them is used for the houses both of Baazani and
Baasheikhah, which consequently have a more cleanly and substantial appearance than is usually the case in this part of
Turkey. Indeed, both villages are flourishing, chiefly owing to
the industry of their Yezidi inhabitants, and their cultivation of
several large groves of olive trees, which produce the only olive
oil in the country. Mixed with the Yezidis are some families of
Jacobite Christians, who live in peace and good understanding
with their neighbours.

I have already mentioned, in my former work*, the Assyrian
ruin near Baasheikhah. It is a vast mound, little inferior in size
to Nimroud, irregular in shape, uneven in level, and furrowed
by deep ravines worn by the winter rains. Standing, as it does,
near abundant quarries of the favorite sculpture-material of the
Assyrians, and resembling the platforms of Kouyunjik or Khorsabad, there was every probability that it contained the remains of
an edifice like those ruins. There are a few low mounds scattered
around it, but no distinct line of walls forming an inclosure. During
the former excavations only earthen jars, and bricks, inscribed

---

* Nineveh and its Remains, vol. i. p. 52.

with the name of the founder of the centre palace at Nimroud, had been discovered.* A party of Arabs and Tiyari were now opening trenches and tunnels in various parts of the mound, under the superintendence of Yakoub Rais of Asheetha. The workmen had uncovered, on the west side of the ruin near the surface, some large blocks of yellowish limestone apparently forming a flight of steps; the only other antiquities of any interest found during the excavations were a few bricks bearing the name of the early Nimroud king, and numerous fragments of earthenware, apparently belonging to the covers of some earthen vessels, having the guilloche and honeysuckle alternating with the cone and tulip, as on the oldest monuments of Nimroud, painted upon them in black upon a pale-yellow ground.†

It is remarkable that no remains of more interest have been discovered in this mound, which must contain a monument of considerable size and antiquity. Although the trenches opened in it were numerous and deep, yet the ruin has not yet probably been sufficiently examined. It can scarcely be doubted that on the artificial platform, as on others of the same nature, stood a royal palace, or some monument of equal importance.

* The fragment of sculpture brought me by a Christian overseer, employed during the former expedition, was, I have reason to believe, obtained at Khorsabad.

† Now in the British Museum. They appear to belong to several distinct objects, probably the covers to some funeral or other vases. See Plate 55. of 2d series of the Monuments of Nineveh.

Cart with Ropes and Workmen carrying Saws, Picks and Shovels, for moving Colossal Bull (Kouyunjik).

Bulls with historical Inscriptions of Sennacherib (Kouyunjik)

CHAP. VI.

DURING the month of December, several discoveries of the greatest
interest and importance were made, both at Kouyunjik and Nimroud. I will first describe the results of the excavations in the
ruins opposite Mosul.

I must remind the reader that, shortly before my departure for
Europe in 1848, the forepart of a human-headed bull of colossal
dimensions had been uncovered on the east side of the Kouyunjik
Palace.* This sculpture then appeared to form one side of an

* Nineveh and its Remains, vol. ii. p. 137.

entrance or doorway, and it is so placed in the plan of the ruins accompanying my former work.* The excavations had, however, been abandoned before any attempt could be made to ascertain the fact. On my return, I had directed the workmen to dig out the opposite sculpture. A tunnel, nearly 100 feet in length, was accordingly opened at right angles to the bull first discovered, but without coming upon any other remains than a pavement of square limestone slabs which stretched without interruption as far as the excavation was carried. I consequently discontinued the cutting, as it was evident that no entrance could be of so great a width, and as there were not even traces of building in that direction.

The workmen having been then ordered to uncover the bull which was still partly buried in the rubbish, it was found that adjoining it were other sculptures, and that it formed part of an exterior façade. The upper half of the next slab had been destroyed, but the lower still remained, and enabled me to restore the figure of the Assyrian Hercules strangling the lion, similar to that discovered between the bulls in the propylæa of Khorsabad, and now in the Louvre. The hinder part of the animal was still preserved. Its claws grasped the huge limbs of the giant, who lashed it with the serpent-headed scourge. The legs, feet, and drapery of the god were in the boldest relief, and designed with great truth and vigor. Beyond this figure, in the same line, was a second bull. The façade then opened into a wide portal, guarded by a pair of winged bulls, twenty feet long, and probably, when entire, more than twenty feet high. Forming the angle between them and the outer bulls were gigantic winged figures in low relief †, and flanking them were two smaller figures, one above the other.‡ Beyond this entrance was a group similar to and corresponding with that on the opposite side, also leading to a smaller entrance into the palace, and to a wall of sculptured slabs; but here all traces of building and sculpture ceased, and we found ourselves near the edge of the water-worn ravine.

Thus a façade of the south-east side of the palace, forming apparently the grand entrance to the edifice, had been discovered. Ten colossal bulls, with six human figures of gigantic

---

* Nineveh and its Remains, vol. ii. — plan of Kouyunjik.

† Nos. 4. and 9. Grand entrance, S.E. side, Plan I. These figures were those of winged priests, or deities, carrying the fir-cone and basket.

‡ Nos. 5. and 8. Same entrance. The small figures resembled No. 2. in Plate 6. of 2d series of Monuments of Nineveh.

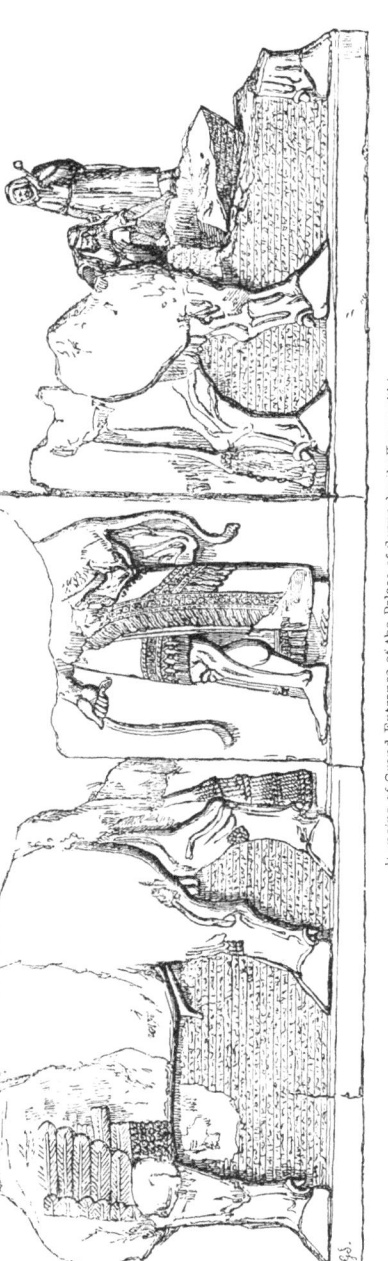

Remains of Grand Entrance of the Palace of Sennacherib (Kouyunjik)

Existing Remains at Khorsabad, showing original State of Grand Entrance at Kouyunjik.

proportions, were here grouped together, and the length of the whole, without including the sculptured walls continued beyond the smaller entrances, was 180 feet.* Although the bas-reliefs to the right of the northern gateway had apparently been purposely destroyed with a sharp instrument, enough remained to allow me to trace their subject. They had represented the conquest of a district, probably part of Babylonia, watered by a broad river and wooded with palms, spearmen on foot in combat with Assyrian horsemen, castles besieged, long lines of prisoners, and beasts of burden carrying away the spoil. Amongst various animals brought as tribute to the conquerors, could be distinguished a lion led by a chain. There were no remains whatever of the superstructure which once rose above the colossi, guarding this magnificent entrance; but I shall hereafter more particularly describe the principal decorations and details of Assyrian architecture, and shall endeavor to restore, as far as the remains still existing will permit, the exterior and interior of the palaces of Nineveh.

The bulls, as I have already observed, were all more or less injured. The same convulsion of nature—for I can scarcely attribute to any human violence the overthrow of these great masses — had shattered some of them into pieces, and scattered the fragments amongst the ruins. Fortunately, however, the lower parts of all, and, consequently, the inscriptions, had been more or less preserved. To this fact we owe the recovery of some of the most precious records with which the monuments of the ancient world have rewarded the labors of the antiquary.

On the great bulls forming the centre portal of the grand entrance, was one continuous inscription, injured in parts, but still so far preserved as to be legible almost throughout. It contained 152 lines. On the four bulls of the façade were two inscriptions, one inscription being carried over each pair, and the two being of precisely the same import. These two distinct records contain the annals of six years of the reign of Sennacherib, besides numerous particulars connected with the religion of the Assyrians, their gods, their temples, and the erection of their palaces, all of the highest interest and importance.

In my first work I had pointed out the evidence, irrespective of the inscriptions, which led me to identify the builder of the great

---

* The frontispiece to this volume will convey to the reader some idea of this magnificent façade when entire. This restoration, for which I am mainly indebted to Mr. Fergusson, has been made with a careful regard to the exact proportions.

palace of Kouyunjik with Sennacherib.* Dr. Hincks, in a memoir on the inscriptions of Khorsabad, read in June, 1849, but published in the " Transactions of the Royal Irish Academy "†, in 1850, was the first to detect the name of this king in the group of arrowheaded characters at the commencement of nearly all the inscriptions, and occurring on all the inscribed bricks from the ruins of this edifice. Subsequent discoveries confirmed this identification, but it was not until August, 1851, that the mention of any actual event recorded in the Bible, and in ancient profane history, was detected on the monuments, thus removing all further doubt as to the king who had raised them.

Shortly after my return to England my copies of these inscriptions having been seen by Colonel Rawlinson, he announced, in the Athenæum of the 23rd August, 1851, that he had found in them notices of the reign of Sennacherib, " which placed beyond the reach of dispute his historic identity," and he gave a recapitulation of the principal events recorded on the monuments, the greater part of which are known to us through history either sacred or profane. These inscriptions have since been examined by Dr. Hincks, and translated by him independently of Colonel Rawlinson. He has kindly assisted me in giving the following abridgment of their contents.‡

The inscriptions begin with the name and titles of Sennacherib.

---

* I had also shown the probability that the palace of Khorsabad owed its erection to a monarch of this dynasty, in a series of letters published in the Malta Times, as far back as 1843.

† Vol. xxii. p. 34. I take this opportunity of attributing to their proper source the discoveries of the names of Nebuchadnezzar and Babylon, inadvertently assigned to others in my "Nineveh and its Remains." We owe these, with many others of scarcely less importance, to the ingenuity and learning of Dr. Hincks. (Literary Gazette, June 27. 1846.)

‡ I must here remind the reader that any new discoveries in the cuneiform inscriptions referred to in the text are to be attributed to Dr. Hincks. The translation made by Col. Rawlinson, and published by the Royal Asiatic Society, was compiled from three distinct records of the same monarch, — the inscriptions on the bulls, on a large barrel-shaped terra-cotta cylinder, known as Bellino's cylinder, now in the British Museum, and on an hexagonal cylinder in the same material, in the possession of the late Col. Taylor. The first annals extend over six years of Sennacherib's reign, the second over only two, and the last, the fullest and most detailed, but unfortunately said to be lost, over eight. It will be perceived that Dr. Hincks's version differs somewhat from that published by Col. Rawlinson; and it must be observed that he was unable to refer to the more complete records, of which a cast in paper is in the Colonel's possession. He has availed himself of Bellino's cylinder to complete the annals of the first two years of the reign of the Assyrian king.

It is to be remarked that he does not style himself " King [or rather High Priest] of Babylon," as his father had done in the latter part of his reign, from which it may be inferred that at the time of engraving the record he was not the immediate sovereign of that city, although its chief may have paid tribute to him, and, no doubt, acknowledged his supremacy. He calls himself " the subduer of kings from the upper sea of the setting sun (the Mediterranean) to the lower sea of the rising sun (the Persian Gulf)." In the first year of his reign he defeated Merodach Baladan, a name with which we are familiar, for it is this king who is mentioned in the Old Testament as sending letters and a present to Hezekiah *, when the Jewish monarch in his pride showed the ambassadors " the house of his precious things, the silver and the gold, and the spices, and the precious ointment, and all the house of his armour, and all that was found in his treasures: there was nothing in his house, nor in all his dominions that Hezekiah showed them not;" an act of vain boasting which led to the reproof of the prophet Isaiah, and to his foretelling that all this wealth, together with the descendants of its owner, should be carried away as spoil to the very city from which these ambassadors came. Merodach Baladan is called king of Kar-Duniyas, a city and country frequently mentioned in the Assyrian inscriptions, and comprising the southernmost part of Mesopotamia, near the confluence of the Tigris and Euphrates, together with the districts watered by those two rivers, to the borders of Susiana. This king, with the help of his Susianian allies, had recently recovered Babylon, from which Sargon, Sennacherib's father, had expelled him in the twelfth year of his reign. The battle appears to have been fought considerably to the north of that city. The result was that Sennacherib totally defeated Merodach Baladan, who fled to save his life, leaving behind him his chariots, *waggons* (?), horses, mares, *asses* (?) camels, and *riding horses with their trappings for war* (?). The victorious king then advanced to Babylon, where he plundered the palace, carrying off a vast treasure of gold, silver, vessels of gold and silver, precious stones, men and women servants, and a variety of objects which cannot yet be satisfactorily determined. No less than seventy-nine cities (or fortresses), all the castles of the Chaldæans, and eight hundred and twenty small towns (or villages), dependent upon them, were taken and spoiled by the Assyrian army, and the great wandering tribes " that dwelt around the cities of Mesopotamia," the Syrians (Arameans), and Chaldæans, &c. &c. were brought under

---

* Isaiah, xxxix. 1. and 2 Kings, xx. 12. where the name is written Berodach.

subjection. Sennacherib having made Belib *, one of his own officers, sovereign of the conquered provinces, proceeded to subdue the powerful tribes who border on the Euphrates and Tigris, and amongst them the Hagarenes and Nabathæans. From these wandering people he declares that he carried off to Assyria, probably colonising with them, as was the custom, new-built towns and villages, 208,000 men, women, and children, together with 7200 horses and mares, 11,063 *asses* (?), 5230 camels, 120,100 oxen, and 800,500 sheep. It is remarkable that the camels should bear so small a proportion to the oxen and asses in this enumeration of the spoil. Amongst the Bedouin tribes, who now inhabit the same country, the camels would be far more numerous.† It is interesting to find, that in those days, as at a later period, there was both a nomade and stationary population in Northern Arabia.

In the same year Sennacherib received a great tribute from the conquered Khararah, and subdued the people of Kherimmi, whom he declares to have been long rebellious (neither people can as yet be identified), rebuilding (? or consecrating) the city of the latter, and sacrificing on the occasion, for its dedication to the gods of Assyria, one ox, ten sheep, ten goats or lambs, and twenty other animals.‡

In the second year of his reign, Sennacherib appears to have turned his arms to the north of Nineveh, having reduced in his first year the southern country to obedience. By the help of Ashur, he says, he went to Bishi and Yasubirablai (both names of doubtful reading and not identified), who had long been rebellious to the kings his fathers. He took Beth Kilamzakh, their principal city, and carried away their men, small and great, horses, mares, *asses* (?), oxen, and sheep. The people of Bishi and Yasubirablai, who had fled from his servants, he brought down from the mountains and placed them under one of his eunuchs, the governor of the city of Arapkha. He made tablets, and *wrote on them the laws (or tribute) imposed upon the conquered, and set them up* in the city. He took permanent possession of the country of Illibi (Luristan?), and Ispa-

---

* Col. Rawlinson reads Bel-adon. This Belib is the Belibus of Ptolemy's Canon. The mention of his name led Dr. Hincks to determine the accession o Sennacherib to be in 703 B.C.

† Col. Rawlinson gives 11,180 head of cattle, 5230 camels, 1,020,100 sheep, and 800,300 goats. He has also pointed out that both Abydenus and Polyhistor mention this campaign against Babylon.

‡ It is to be remarked that he does not say he gave a new name to this city, as was generally the case; it may have been a holy city (compare 'Harem") and consequently escaped destruction.

bara* its king, after being defeated, fled, leaving the cities of
Marubishti and Akkuddu, the royal residences, with thirty-four
principal towns, and villages not to be counted, to be destroyed by
the Assyrians, who carried away a large amount of captives and
cattle. Beth-barrua, the city itself and its dependencies, Sennacherib separated from Illibi, and added to his immediate dominions.
The city of *Ilbinzash* (?) he appointed to be the chief city in this
district. He abolished its former name, called it Kar-Sanakhirba
(*i. e.* the city of Sennacherib), and placed in it a new people, an-
nexing it to the government of Kharkhar, which must have been
in the neighbourhood of Holwan, commanding the pass through
mount Zagros. After this campaign he received tribute to a great
amount from some Median nations, so distant, that his predecessors
" had not even heard mention of their names," and made them
obedient to his authority.

In the third year of his reign Sennacherib appears to have over-
ran with his armies the whole of Syria. He probably crossed the
Euphrates above Carchemish, at or near the ford of Thapsacus,
and marched to the sea-coast, over the northern spur of Mount
Lebanon. The Syrians are called by their familiar biblical name
of Hittites, the Khatti, or Khetta, by which they were also known
to the Egyptians. The first opposition he appears to have received
was from Luli (or Luliya), king of Sidon, who had withheld his
homage ; but who was soon compelled to fly from Tyre to Yavan in
the middle of the sea. Dr. Hincks identifies this country with the
island of Crete, or some part of the southern coast of Asia Minor,
and with the Yavan (יָוָן) of the Old Testament, the country of the
Ionians or Greeks, an identification which I believe to be correct.†
This very Phœnician king is mentioned by Josephus (quoting from
Menander), under the name of Elulæus, as warring with Shalmaneser, a predecessor of Sennacherib. He appears not to have been
completely subdued before this, but only to have paid homage or

* We learn from the Khorsabad inscriptions, that in the eleventh year of the
reign of Sargon, Dalta, the king of this country, died, leaving two sons, one of
whom was supported by the king of Susa, and the other by the Assyrian
monarch, who sent a large army, under seven generals, to his assistance, and
totally defeating the Susianians, placed Ispabara on the throne. Ispabara
appears afterwards to have thrown off the Assyrian yoke. (Dr. Hincks.) Col.
Rawlinson places Illibi in northern Media, and reads most of the names in
the text differently. (P. 20. of his Memoir.)

† Col. Rawlinson identifies the name, which he reads Yetnan, with the Rhinocolura of the Greeks, and places it in the south of Phœnicia, on the confines of
Egypt.

tribute to the Assyrian monarchs.* Sennacherib placed a person, whose name is doubtful (Col. Rawlinson reads it Tubaal), upon the throne of Luli, and appointed his annual tribute. All the kings of the sea-coast then submitted to him, except Zidkaha (compare Zedekiah) or Zidkabal, king of Ascalon. This chief was, however, soon subdued, and was sent, with his household and wealth, to Assyria, —— (name destroyed), the son of *Rukipti* (?), a former king, being placed on the throne in his stead. The cities dependent upon Ascalon, which had not been obedient to his authority, he captured and plundered. A passage of great importance which now occurs is unfortunately so much injured that it has not yet been satisfactorily restored. It appears to state that the *chief priests* (?) and people of Ekron (?) had dethroned their king Padiya, who was dependent upon Assyria, and had delivered him up to Hezekiah, king of Judæa.† The kings of Egypt sent an army, the main part of which is said to have belonged to the king of Milukhkha (Meroe, or Æthiopia), to Judæa, probably to help their Jewish allies. Sennacherib joined battle with the Egyptians, and totally defeated them near the city of Al . . . . ku, capturing the charioteers of the king of Milukhkha, and placing them in confinement. This battle between the armies of the Assyrians and Egyptians appears to be hinted at in Isaiah and in the Book of Kings.‡ Padiya having been brought back from Jerusalem was replaced by Sennacherib on his throne. " Hezekiah, king of Judah," says the Assyrian king, " who had not submitted to my authority, forty-six of his principal cities, and fortresses and villages depending upon them, of which I took no account, I cap-

---

* Joseph. l. ix. c. 14., and see Nineveh and its Remains, vol. ii. p. 400., where I had long before the deciphering of the inscriptions endeavoured to point out the representation of this event, in some bas-reliefs at Kouyunjik. This flight of Luliya, indeed, appears to be represented in plate No. 71. of the first series of the " Monuments of Nineveh."

† Col. Rawlinson reads the name of the king Haddiya. That of Ekron is very doubtful.

‡ Isaiah, xxxvii. 2 Kings, xix. 9. It is not stated that the armies of the two great antagonistic nations of the ancient world actually met in battle, but that Sennacherib " heard say concerning Tirhakah king of Ethiopia, He is coming forth to make war with thee." Herodotus, however, appears to have preserved the record of the battle in the celebrated story of the mice which gnawed the bowstrings and the thongs of the shields of the Assyrian soldiers during the night, and left them an easy prey to the Egyptians (lib. iii. s. 141.). This looks very much like a defeat sustained by the Egyptians, which the vanity of their priests had converted into this marvellous story. The fact, intimated in the inscriptions, of Tirhakah having not one but several Egyptian kings dependent upon him is new to history.

tured and carried away their spoil. I *shut up* (?) himself within Jerusalem, his capital city. The fortified towns, and the rest of his towns, which I spoiled, I severed from his country, and gave to the kings of Ascalon, Ekron, and Gaza, so as to make his country small. In addition to the former tribute imposed upon their countries, I added a tribute, the nature of which I fixed." The next passage is somewhat defaced, but the substance of it appears to be, that he took from Hezekiah the treasure he had collected in Jerusalem, 30 talents of gold and 800 talents of silver, the treasures of his palace, besides his sons and his daughters, and his male and female servants or slaves, and brought them all to Nineveh.* The city itself, however, he does not pretend to have taken.

There can be little doubt that the campaign against the cities of Palestine recorded in the inscriptions of Sennacherib at Kouyunjik, is that described in the Old Testament. The events agree with considerable accuracy. We are told in the Book of Kings, that the king of Assyria, in the fourteenth year of the reign of Hezekiah, "came up against all the fenced cities of Judah and took them,"† as he declares himself to have done in his annals. And, what is most important, and perhaps one of the most remarkable coincidences of historic testimony on record, the amount of the treasure in gold taken from Hezekiah, thirty talents, agrees in the two perfectly independent accounts.‡ Too much stress

---

* Col. Rawlinson gives a somewhat different version of this part of the inscription. He translates, "Because Hezekiah, king of Judæa, did not submit to my yoke, forty-six of his strong-fenced cities, and innumerable smaller towns which depended on them, I took and plundered; but I left to him Jerusalem, his capital city, and some of the inferior towns around it. . . . . And because Hezekiah still continued to refuse to pay me homage, I attacked and carried off the whole population, fixed and nomade, which dwelled around Jerusalem, with 30 talents of gold, and 800 talents of silver, the accumulated wealth of the nobles of Hezekiah's court, and of their daughters, with the officers of his palace, men slaves and women slaves. I returned to Nineveh, and I accounted their spoil for the tribute which he refused to pay me." He identifies Milukh-kha (or Mirukha) with Meroe or Æthiopia, and Al . . . ku, which he reads Al-lakis, with Lachish, the city besieged by Sennacherib, when he sent Rabshakeh to Hezekiah, and of which, I shall endeavour to show, we have elsewhere a more certain mention.

† 2 Kings, xviii. 13.; and compare Isaiah, xxxvi. 1. I may here observe that the names of Hezekiah and Judæa, with others mentioned in the text, occur in inscriptions on other bulls of Kouyunjik already published. (See British Museum Series, p. 61. l. 11.)

‡ "And the king of Assyria appointed unto Hezekiah, king of Judah, 300 talents of silver and 30 talents of gold." (2 Kings, xviii. 14.)

cannot be laid on this singular fact, as it tends to prove the general accuracy of the historical details contained in the Assyrian inscriptions. There is a difference of 500 talents, as it will be observed, in the amount of silver. It is probable that Hezekiah was much pressed by Sennacherib, and compelled to give him all the wealth that he could collect, as we find him actually taking the silver from the house of the Lord, as well as from his own treasury, and cutting off the gold from the doors and pillars of the temple, to satisfy the demands of the Assyrian king. The Bible may therefore only include the actual amount of money in the 300 talents of silver, whilst the Assyrian records comprise all the precious metal taken away. There are some chronological discrepancies which cannot at present be satisfactorily reconciled, and which I will not attempt to explain.* It is natural to suppose that Sennacherib would not perpetuate the memory of his own overthrow; and that, having been unsuccessful in an attempt upon Jerusalem, his army being visited by the plague described in Scripture, he should gloss over his defeat by describing the tribute he had previously received from Hezekiah as the general result of his campaign.

There is no reason to believe, from the biblical account, that Sennacherib was slain by his sons *immediately* after his return to Nineveh; on the contrary, the expression " he returned and dwelt at Nineveh," infers that he continued to reign for some time over Assyria. We have accordingly his further annals on the monuments he erected. In his fourth year he went southward, and subdued the country of Beth-Yakin, defeating Susubira, the Chaldæan, who dwelt in the city of Bittut on the river—(Agammi, according to Rawlinson). Further mention is made of Merodach Baladan. " This king, whom I had defeated in a former campaign, escaped from my principal servants, and fled to an island (name lost); his brothers, the seed of his father's house, whom he left behind him on the coast, with the rest of the men of his country from Beth-Yakin, near the *salt* (?) river (the Shat-el-Arab, or united waters of the Tigris and Euphrates), I carried away, and several of his towns I threw down, and burnt; Assurnad*immi* (? Assurnadin, according to Rawlinson), my son, I placed on the

---

* According to Dr. Hincks (Chronological Appendix to a Paper on the Assyrio-Babylonian Characters in vol. xxii. of the Transactions of the Royal Irish Academy), it is necessary to read the twenty-fifth for the fourteenth year of Hezekiah as the date of Sennacherib's invasion. The illness of Hezekiah, and the embassy of Merodach Baladan, he places eleven years earlier. Certainly the phrase " in those days" was used with great latitude.

throne of his kingdom." He appears then to have made a large government, of which Babylon was the chief place.*

In the fifth year he defeated the Tokkari, capturing their principal stronghold or Nipour (*detached hill fort?*), and others of their castles. He also attacked Maniyakh, king of *Okku* or *Wukku* (?), a country to which no previous Assyrian king had penetrated. This chief deserted his capital and fled to a distance. Sennacherib carried off the spoil of his palace and plundered his cities. This expedition seems to have been to the north of Assyria, in Armenia or Asia Minor.

In the following year Sennacherib again marched to the mouths of the Euphrates and Tigris, and attacked the two cities of Naghit and Naghit Dibeena. They appear to have stood on opposite sides of the great salt river, a name anciently given, it is conjectured, to the Shat-el-Arab, or united waters of the Euphrates and Tigris, which are affected by the tides of the Persian Gulf, and are, consequently, salt. Both cities belonged to the King of Elam (Elamti), or Nuvaki, the two names being used indifferently for the same country. The Assyrian king, in order to reach them, was compelled to build ships, and to employ the mariners of Tyre, Sidon, and Yavan, as navigators. He brought these vessels down the Tigris, and crossed on them to the Susianian side of the river, after having first, it would seem, taken the city Naghit which stood on the western bank. He offered precious sacrifices to a god (? Neptune, but name doubtful) on the bank of the salt river, and dedicated to him a ship of gold, and two other golden objects, the nature of which has not been determined. Mention is then made of his having captured Naghit Dibeena, together with three other cities, whose names cannot be well ascertained, and of his crossing the river *Ula* (? the Ulai of Daniel, the Eulæus of the Greeks, and the modern Karoon). Unfortunately the whole of the passage which contains the record of the expedition against these cities is much defaced, and has not yet been satisfactorily restored. It appears to give interesting details of the building of the ships on the Tigris, by the men of Tyre and Sidon and of the navigation of that river.

Such are the principal historical facts recorded on the bulls placed by Sennacherib in his palace at Nineveh. I have given them fully, in order that we may endeavour to identify the sculp-

---

* Dr. Hincks identifies the son of Sennacherib with the Aparanadius of Ptolemy's canon, whose reign began three years after that of Belibus. He supposes ϖ to be a corruption of σσ.

tured representations of these events on the walls of the chambers and halls of that magnificent building, described in the course of this work. Appended to the historical annals, and frequently embracing the whole of the shorter inscriptions on the colossi at the entrances, are very full and minute details of the form of the palace, the mode of its construction, and the materials employed, which will be alluded to when I come to a description of the architecture of the Assyrians.

As the name of Sennacherib, as well as those of many kings, countries and cities, are not written phonetically, that is, by letters having a certain alphabetic value, but by monograms, and the deciphering of them is a peculiar process, which may sometimes appear suspicious to those not acquainted with the subject, a few words of explanation may be acceptable to my readers. The greater number of Assyrian proper names with which we are acquainted, whether royal or not, appear to have been made up of the name, epithet, or title, of one of the national deities, and of a second word such as " slave of," " servant of," " beloved of," " protected by ; " like the " Theodosius," " Theodorus," &c. of the Greeks, and the " Abd-ullah," and " Abd-ur-Rahman," of Mohammedan nations. The names of the gods being commonly written with a monogram, the first step in deciphering is to know which God this particular sign denotes. Thus, in the name of Sennacherib, we have first the determinative of " god," to which no phonetic value is attached; whilst the second character denotes an Assyrian god, whose name was San. The first component part of the name of Essarhaddon, is the monogram for the god Assur. It is this fact which renders it so difficult to determine, with any degree of confidence, most of the Assyrian names, and which leads me to warn my readers that, with the exception of such as can with certainty be identified with well-known historic kings, as Sargon, Sennacherib, and Essarhaddon, the interpretation of all those which are found on the monuments of Nineveh, is liable to very considerable doubt. In speaking of them I shall, therefore, not use any of the readings which have been suggested by different writers.

Although no question can reasonably exist as to the identification of the king who built the palace of Kouyunjik with the Sennacherib of Scripture, it may still be desirable to place before my readers all the corroborative evidence connected with the subject. In so doing, however, I shall have to refer to discoveries made at a subsequent period, and which ought consequently to be described, if the order of the narrative be strictly preserved, in a subsequent part of this work. In the first place, it must be remembered that the Kouyunjik king was undoubtedly the son of the founder of the palace at Khorsabad. He is so called in the inscriptions behind the bulls in the S. W. palace at Nimroud, and in numerous detached inscriptions on bricks, and on other remains from those ruins and from Kouyunjik. Now the name of the Khorsabad king was generally admitted to be Sargon*, even before his relationship to the Kouyunjik king was known; although here again we are obliged to attach phonetic powers to characters used as monograms, which, when occurring as simple letters, appear to have totally different values.† Colonel Rawlinson states‡, that this king bears in other inscriptions the name of Shalmaneser, by which he was better known to the Jews.§ Dr. Hincks denies that the two names belong to the same person. It would appear, however, that there are events mentioned in the inscriptions of Khorsabad, which lead to the identification of its founder with the Shalmaneser of Scripture, and the ruins of the palace itself, were known even at the time of the Arab conquest by the name of "Sarghun."

Unfortunately the upper parts of nearly all the bas-reliefs at Kouyunjik having been destroyed the epigraphs are wanting; and we are unable, as yet, to identify with certainty the subjects represented with any known event in the reign of Sennacherib. There is, however, one remarkable exception.

During the latter part of my residence at Mosul a chamber was discovered in which the sculptures were in better preservation

---

* First, I believe, though on completely false premises, by M. Lowenstein.

† Col. Rawlinson reads the name "Sargina."

‡ Athenæum, Aug. 23. 1851.

§ Shalmaneser, who made war against Hoshea, and who is generally supposed to have carried away the ten tribes from Samaria, although the sacred historian does not distinctly say so (2 Kings, xvii.), is identified by general consent with Sargon, who sent his general against Ashdod (Isaiah, xx.). Dr. Hincks questioned this identification (Athenæum for Sept. 13. 1851), considering Shalmaneser as son of Sargon, and brother to Sennacherib. In his last paper, however (Trans. Royal Irish Acad. vol. xxii.), he has taken a different view. He considers Shalmaneser to be the predecessor of Sargon, who went up against Jerusalem in his last year, B.C. 722. "The king of Assyria," that is Sargon, took the city in his second year, B.C. 720. In either case, no monument whatever has yet been discovered bearing the name of this king. There is certainly nothing in Scripture to identify the two names as belonging to the same king, except that their general, in both instances, is called Tartan, which we now find from the inscriptions was merely the common title of the commander of the Assyrian armies.

than any before found at Kouyunjik.* Some of the slabs, indeed, were almost entire, though cracked and otherwise injured by fire; and the epigraph, which fortunately explained the event portrayed, was complete. These bas-reliefs represented the siege and capture by the Assyrians, of a city evidently of great extent and importance. It appears to have been defended by double walls, with battlements and towers, and by fortified outworks. The country around it was hilly and wooded, producing the fig and the vine. The whole power of the great king seems to have been called forth to take this stronghold. In no other sculptures were so many armed warriors seen drawn up in array before a besieged city. In the first rank were the kneeling archers, those in the second were bending forward, whilst those in the third discharged their arrows standing upright, and were mingled with spearmen and slingers; the whole forming a compact and organised phalanx. The reserve consisted of large bodies of horsemen and charioteers. Against the fortifications had been thrown up as many as ten banks or mounts, compactly built of stones, bricks, earth, and branches of trees, and seven battering-rams had already been rolled up to the walls. The besieged defended themselves with great determination. Spearmen, archers, and slingers thronged the battlements and towers, showering arrows, javelins, stones, and blazing torches upon the assailants. On the battering-rams were bowmen discharging their arrows, and men with large ladles pouring water upon the flaming brands, which, hurled from above, threatened to destroy the engines. Ladders, probably used for escalade, were falling from the walls upon the soldiers who mounted the inclined ways to the assault. Part of the city had, however, been taken. Beneath its walls were seen Assyrian warriors impaling their prisoners, and from the gateway of an advanced tower, or fort, issued a procession of captives, reaching to the presence of the king, who, gorgeously arrayed, received them seated on his throne. Amongst the spoil were furniture, arms, shields, chariots, vases of metal of various forms, camels, carts drawn by oxen, and laden with women and children, and many objects the nature of which cannot be determined. The vanquished people were distinguished from the conquerors by their dress, those who defended the battlements wore a pointed helmet, differing from that of the Assyrian warriors in having a fringed lappet falling over the ears. Some of the captives had a kind of turban with one end hanging down to the

* No. XXXVI.   Plan I.   38 feet by 18.

shoulder, not unlike that worn by the modern Arabs of the Hedjaz. Others had no head-dress, and short hair and beards. Their garments consisted either of a robe reaching to the ankles, or of a tunic scarcely falling lower than the thigh, and confined at the waist by a girdle. The latter appeared to be the dress of the fighting-men. The women wore long shirts, with an outer cloak thrown, like the veil of modern Eastern ladies, over the back of the head and falling to the feet.

Several prisoners were already in the hands of the torturers. Two were stretched naked on the ground to be flayed alive, others were being slain by the sword before the throne of the king. The haughty monarch was receiving the chiefs of the conquered nation, who crouched and knelt humbly before him. They were brought into the royal presence by the Tartan of the Assyrian forces, probably the Rabshakeh himself, followed by his principal officers. The general was clothed in embroidered robes, and wore on his head a fillet adorned with rosettes and long tasseled bands.

Sennacherib on his Throne before Lachish

The throne of the king stood upon an elevated platform, probably an artificial mound, in the hill country. Its arms and sides were supported by three rows of figures one above the other. The wood was richly carved, or encased in embossed metal, and the legs ended in pine-shaped ornaments, probably of bronze. The throne, indeed, appears to have resembled, in every respect, one discovered in the north-west palace at Nimroud, which I shall hereafter describe.* Over the high back was thrown an embroidered cloth, doubtless of some rare and beautiful material.

The royal feet rested upon a high footstool of elegant form, fashioned like the throne, and cased with embossed metal; the legs ending in lion's paws. Be-

* Chap. VIII.

hind the king were two attendant eunuchs raising fans above his head, and holding the embroidered napkins.

The monarch himself was attired in long loose robes richly ornamented, and edged with tassels and fringes. In his right hand he raised two arrows, and his left rested upon a bow ; an attitude, probably denoting triumph over his enemies, and in which he is usually portrayed when receiving prisoners after a victory.

Behind the king was the royal tent or pavilion * : and beneath him were his led horses, and an attendant on foot carrying the parasol, the emblem of royalty. His two chariots with their charioteers, were waiting for him. One had a peculiar semicircular ornament of considerable size, rising from the pole between the horses, and spreading over their heads. It may originally have contained the figure of a deity, or some mythic symbol. It was attached to the chariot by that singular contrivance joined to the yoke and represented in the early sculptures of Nimroud, the use and nature of which I am still unable to explain.† This part of the chariot was richly adorned with figures and ornamental designs, and appeared to be supported by a prop resting on the pole. The trappings of the horses were handsomely decorated, and an embroidered cloth, hung with tassels, fell on their chests. Two quivers, holding a bow, a hatchet, and arrows, were fixed to the side of the chariot.

This fine series of bas-reliefs ‡, occupying thirteen slabs, was finished by the ground-plan of a castle, or of a fortified camp containing tents and houses. Within the walls was also seen a fire-altar with two beardless priests, wearing high conical caps, standing before it. In front of the altar, on which burned the sacred flame, was a table bearing various sacrificial objects, and beyond it two sacred chariots, such as accompanied the Persian kings in their wars.§ The horses had been taken out, and the yokes rested upon stands. Each chariot carried a lofty pole surmounted by a globe, and long tassels or streamers ; similar standards were introduced into scenes representing sacrifices ‖ in the sculptures of Khorsabad.

* I presume this to be a tent, or moveable dwelling-place. It is evidently supported by ropes. Above it is an inscription declaring that it is " the *tent* (?) (the word seems to read *sarata*) of Sennacherib, king of Assyria."

† It has been suggested to me that it may have been a case in which to place the bow ; but the bow and arrows are contained in the quiver suspended to the side of the chariot.

‡ For detailed drawings, see 2nd series of the Monuments of Nineveh, Plates 20. to 24.

§ Xenophon Cyrop. lvii. c. 3.   Quintus Curtius, liii. c. 3.

‖ Botta's Monumens de Ninive, Plate 146.

Above the head of the king was the following inscription,

which may be translated, " Sennacherib, the mighty king, king of
the country of Assyria, sitting on the throne of judgment, before
(or at the entrance of) the city of Lachish (Lakhisha). I give
permission for its slaughter."

Here, therefore, was the actual picture of the taking of Lachish,
the city, as we know from the Bible, besieged by Sennacherib,
when he sent his generals to demand tribute of Hezekiah, and
which he had captured before their return *; evidence of the most

Jewish Captives from Lachish (Kouyunjik)

remarkable character to confirm the interpretation of the inscriptions, and to identify the king who caused them to be engraved with
the Sennacherib of Scripture. This highly interesting series of
bas-reliefs contained, moreover, an undoubted representation of a
king, a city, and a people, with whose names we are acquainted,
and of an event described in Holy Writ. They furnish us, there-

* 2 Kings, xviii. 14. Isaiah xxxvi. 2. From 2 Kings, xix. 8., and Isaiah, xxxvii.
8., we may infer that the city soon yielded.

fore, with illustrations of the Bible of very great importance * The captives were undoubtedly Jews, their physiognomy was strikingly indicated in the sculptures, but they had been stripped of their ornaments and their fine raiment, and were left barefooted and half-clothed. From the women, too, had been removed " the splendor of the foot ornaments and the caps of network, and the crescents ; the ear-pendents, and the bracelets, and the thin veils ; the head-dress, and the ornaments of the legs and the girdles, and the perfume-boxes and the amulets; the rings and the jewels of the nose; the embroidered robes and the tunics, and the cloaks and the satchels; the transparent garments, and the fine linen vests, and the turbans and the mantles, "for they wore instead of a girdle, a rope ; and instead of a stomacher, a girdling of sackcloth." †

Other corroborative evidence as to the identity of the king who built the palace of Kouyunjik with Sennacherib, is scarcely less remarkable. In a chamber, or passage, in the south-west corner of this edifice ‡, were found a large number of pieces of fine clay bearing the impressions of seals §, which, there is no doubt, had been affixed, like modern official seals of wax, to documents written on leather, papyrus, or parchment. Such documents, with seals in clay still attached, have been discovered in Egypt, and specimens are preserved in the British Museum. The writings themselves

---

* Col. Rawlinson has, I am aware, denied that this is the Lachish mentioned in Scripture, which he identifies with the All...ku of the bull inscriptions, and places on the sea-coast between Gaza and Rhinocolura. (Outlines of Assyrian History, p. xxxvi.) But I believe this theory to be untenable, and I am supported in this view of the subject by Dr. Hincks, who also rejects Col. Rawlinson's reading of Lubana (Libnah). Lachish is mentioned amongst "the uttermost cities of the tribe of Judah." (Joshua, xv. 39.) From verse 21 to 32 we have one category of twenty-nine cities "toward the coast of Edom southward." The next category appears to extend to verse 46, and includes cities in the valley, amongst which is Lachish. We then come to Ashdod and the sea. It was therefore certainly situated in the hill country. (See also Robinson's Biblical Researches in Palestine, vol. ii. p. 388.)

† Isaiah, iii. 18—24. &c. (See translation by the Rev. J. Jones.) This description of the various articles of dress worn by the Jewish women is exceedingly interesting. Most of the ornaments enumerated, probably indeed the whole of them, if we were acquainted with the exact meaning of the Hebrew words, are still to be traced in the costumes of Eastern women inhabiting the same country. Many appear to be mentioned in the Assyrian inscriptions amongst objects of tribute and of spoil brought to the king. See also Ezekiel xvi. 10—14. for an account of the dress of the Jewish women.

‡ No. LXI. Plan I.

§ Resembling the γῆ σημαντρίς (the sealing earth) of the Greeks.

had been consumed by the fire which destroyed the building or
had perished from decay.   In the stamped clay, however, may still

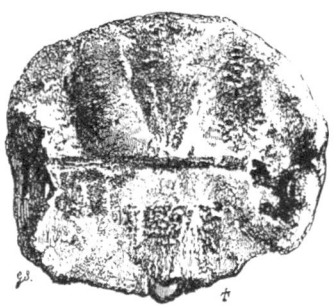

Impression of a Seal on Clay.   Back of the same seal, showing the Marks of
the String and the Fingers.

be seen the holes for the string, or strips of skin, by which the seal
was fastened; in some instances the ashes of the string itself remain *, with the marks of the fingers and thumb.

The greater part of these seals are Assyrian, but with them are
others bearing Egyptian, Phœnician, and doubtful symbols and characters.   Sometimes the same seal is impressed more than once
on the same piece of clay.   The Assyrian devices are of various
kinds; the most common is that of a king plunging a dagger into
the body of a rampant lion.   This appears to have been the royal,
and, indeed, the national, seal or signet.   It is frequently encircled by a short inscription, which has not yet been deciphered,
or by a simple guilloche border.   The same group, emblematic
of the superior power and wisdom of the king, as well as of his
sacred character, is found on Assyrian cylinders, gems, and monuments.   From the Assyrians it was adopted by the Persians,
and appears upon the walls of Persepolis and on the coins of
Darius.

Other devices found among these impressions of seals are : — 1.
A king, attended by a priest, in act of adoration before a deity
standing on a lion, and surrounded by seven stars : above the
god's head, on one seal, is a scorpion.   2. The king, followed by
an attendant bearing a parasol, and preceded by a rampant horse.
3. A god, or the king, probably the former, rising from a crescent.
There appears to be a fish in front of the figure.   4. The king,
with an eunuch or priest before him; a flower, or ornamented

* M. Botta also found, at Khorsabad, the ashes of string in lumps of clay
impressed with a seal, without being aware of their origin.

staff, between them. 5. A scorpion, surrounded by a guilloche border (a device of very frequent occurrence, and probably astronomical). 6. A priest worshipping before a god, encircled by stars. 7. A priest worshipping before a god. Behind him are a bull, and the sacred astronomical emblems. 8. An ear of corn, surrounded by a fancy border. 9. An object resembling a dagger, with flowers attached to the handle; perhaps a sacrificial knife. 10. The head of a bull and a trident, two sacred symbols of frequent occurrence on Assyrian monuments. 11. A crescent in the midst of a many-rayed star. 12. Several rudely cut seals, representing priests and various sacred animals, stars*, &c.

The seals most remarkable for beauty of design and skilful

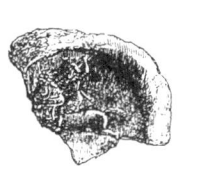

execution represent horsemen, one at full speed raising a spear, the other hunting a stag. The impressions show that they were little inferior to Greek intaglios. No Assyrian or Babylonian relics yet discovered, equal them in delicacy of workmanship, and the best examples of the art of en-

Assyrian Seals

graving on gems, — an art which appears to have reached great perfection amongst the Assyrians, — are unknown to us, except through these impressions.

There are three seals apparently Phœnician; two of them bear-

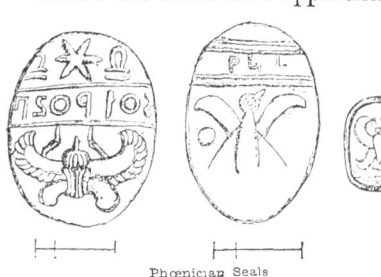

ing Phœnician characters†, for which I cannot suggest any interpretation. A few have doubtful symbols upon them, which I will not attempt to explain; perhaps hieroglyphical signs.

Of the purely Egyptian seals there are four. One

Phœnician Seals

has two cartouches placed on the symbol of gold, and each surmounted by a tall plume; they probably contained the prænomen and name of a king, but not the slightest trace remains of the hieroglyphs. The impression is concave, having been made from

---

* For engravings of these seals, see 2nd series of Monuments of Nineveh, Plate 69.

† It is, however, possible that these characters may belong to some other Semitic nation, as a cursive alphabet, having a close resemblance to the Phœnician, was used from Tadmor to Babylon.

a convex surface : the back of some of the Egyptian ovals, the
rudest form of the scarabæus, are of this shape. On the second
seal is the figure of the Egyptian god Harpocrates, seated on a
lotus flower, with his finger placed upon his mouth; an attitude
in which he is represented
on an ivory from Nimroud.
The hieroglyph before him
does not appear to be
Egyptian.

Egyptian Seals.

But the most remark-
able and important of the
Egyptian seals are two
impressions of a royal signet, which, though imperfect, retain
the cartouche, with the name of the king, so as to be perfectly
legible. It is one well known to Egyptian scholars, as that
of the second Sabaco the Æthiopian, of the twenty-fifth dynasty.
On the same piece of clay is impressed an Assyrian seal, with a
device representing a priest ministering before the king, probably
a royal signet.

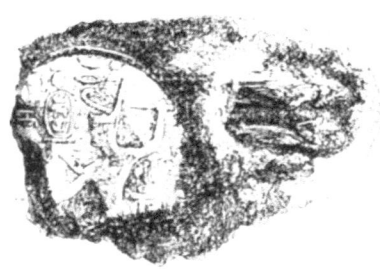

Impressions of the Signets of the Kings of Assyria
and Egypt. (Original Size.)

Part of Cartouche of Sabaco, enlarged from
the Impression of his Signet.

There can be no doubt whatever as to the identity of the cartouche.* Sabaco reigned in Egypt at the end of the seventh cen-

---

* I am indebted to Mr. Birch for the following remarks upon this seal : —
"The most important of the numerous seals discovered at Kouyunjik is one
which has received two impressions — an Assyrian, representing a personage in
adoration before a deity ; and a second, with the representation and name of the
Egyptian monarch, Sabaco, of the twenty-fifth dynasty of Æthiopians, and
evidently impressed from a royal Egyptian seal. Similar impressions are by no
means unknown, and a few examples have reached the present time. Not to
instance the clay seals found attached to the rolls of papyrus containing letters

tury before Christ, the exact time at which Sennacherib came to the throne. He is probably the So mentioned in the second book of Kings (xvii. 4.) as having received ambassadors

written in the time of the Ptolemies and Romans, there are in the British Museum seals bearing the name of Shashank or Shishak (No. 5585.) of Amasis II. of the twenty-sixth dynasty (No. 5584.) and of Nafuarut or Nepherophis, of the twenty-ninth dynasty (No. 5585.). Such seals were, therefore, affixed by the Egyptians to public documents, and it was in accordance with this principle, common to the two monarchies, that the seal of the Egyptian king has been found in Assyria. It appears to have been impressed from an oval, in all probability the bezel of a metallic finger ring, like the celebrated seal of Cheops ; in this case an oval, two inches in length by one inch wide. The king Sabaco is represented upon the left in an action very commonly seen in the historical monuments of Egypt, wearing the red cap *teshr*. He bends down, seizing with his left hand the hair of the head of an enemy, whom he is about to smite with a kind of mace or axe in his right, having slung his bow at his side. Above and before him are hieroglyphs, expressing *Netr nfr nb ar cht Shabaka*, 'the perfect God, the Lord who produces things, Shabaka (or Sabaco).' Behind is an expression of constant occurrence in Egyptian texts : *sha (s)anch-ha f*, 'life follows his head.' Although no figure of any deity is seen, the hieroglyphs at the left edge show that the king was performing this action before one — *ma, na nak*, 'I have given to thee,' which must have been followed by some such expression as 'a perfect life,' 'all enemies or countries under thy sandals.' It is impossible to determine which god of the Pantheon was there, probably Amon-Ra, or the Theban Jupiter. These seals, therefore, assume a most important character as to the synchronism of the two monarchies. There can, indeed, be no doubt that the Shabak found upon them is the usual king of the inscriptions ; and it is owing alone to the confusion of Herodotus and Diodorus that the difficulty of identifying the true chronological position has occurred. The twenty-fifth dynasty of Manetho, according to all three versions, consisted of three Æthiopic kings, the seat of whose empire was originally at Gebel Barkal, or Napata, and who subsequently conquered the whole of Egypt. The first monarch of this line was called Sabaco by the Greek writers ; the second Sebechos, or Seuechos, his son ; the third was Tarkos or Taracus. Now, corresponding to Sabacon and Seuechos are two kings, or at least two prænomens, each with the name of Shabak : one reads *Ra-nefer-kar*, the other *Ra-tat-karu*, although the correctness of this last prænomen is denied, and it is asserted that only one king is found on the monuments. Even the existence of the first Shabak or Sabacon is contested, and the eight or twelve years of his reign credited to his successor ; and it is remarkable to find that in two versions of Manetho each reigned twelve years. Still the non-appearance of the first Shabak on the monuments of Egypt would be intelligible, owing to the trouble he may have had to establish his sway, although then it would be probable that he should be found at Napata, his Æthiopian capital. As Rosellini, however, gives so distinctly the second prænomen (M.R. cli. 5.), it is difficult to conceive that it does not exist. In the other scenes at Karnak, Shabak, wearing the upper and lower crown, showing his rule over the Delta, is seen embraced by Athor and Amen-t, or T-Amen (Rosell. M.R. cli. 2 and 3.), or else wearing a plain head-dress, he is received by Amen and Mut ; but as he is unaccompanied by his prænomen, it is uncertain whether Shabak I.

from Hoshea, the king of Israel, who, by entering into a league
with the Egyptians, called down the vengeance of Shalmaneser,
whose tributary he was, which led to the first great captivity of

or Shabak II. is intended. In the legends, Shabak II. is said to be 'crowned
on the throne of Tum (Tomos), like the sun for ever,' from which it is evident
that Sabaco claimed to be at that time king of Upper and Lower Egypt. The
hypothesis originally proposed by Marsham (1 Chron. Com. p. 457.), and subsequently adopted by others is, that Sabaco is the king Sua or So, mentioned in
Kings, xvii. 4., to whom Hoshea, in the sixth year of his reign, sent an embassy.
' Against him came up Shalmaneser king of Assyria ; and Hoshea became his
servant, and gave him presents. And the king of Assyria found conspiracy in
Hoshea : for he had sent messengers to So king of Egypt, and brought no
present to the king of Assyria, as he had done year by year: therefore the king
of Assyria shut him up, and bound him in prison.' According to some chrono-
logers, this was B.C. 723—722. (Winer, Bibl. Real-Worterbuch, ii. s. 876. Bd. i.
730 f.) ; according, however, to De Vignolles, 721—720. Of the later chrono-
logists, Rosellini places Sabaco I. B.C. 719., and Sabaco II. B.C. 707.; Sir Gardner
Wilkinson, B.C. 778—728. If Sabaco be really So, the reckoning of Rosellini
and Böckh (Manetho, s. 393.), B.C. 711., for Sabaco II. is nearest the truth.
The name of So is written נוא, סוא, Sva or Sia. The great difficulty is the
dreadful confusion of the period. The duration of the Æthiopian dynasty,
according to Africanus and Eusebius, is,

				Years.	Years.
Sabacon	-	-	-	8 (Africanus)	12 (Eusebius)
Seuechos	-	-	-	14 „	12 „
Taracus	-	-	-	18 „ .	20 „
Total -	-	-	-	40	44

Herodotus (ii. 152.), in his usual confusion, places Sabaco, (who, he says,
reigned after Anysis, a blind man, who fled to the island of Elbo in the marshes,)
after Mycerinus, of the fourth dynasty, and states that he reigned fifty years,
more than the whole time of the dynasty. Diodorus placed Sabaco after Boc-
choris, whom, he declares, he burnt alive. This might be the deed of Sabaco I.,
while the burning of Nechao I. may have been the act of Sabaco II. Hence,
M. Bunsen (Aegyptens Stelle, iii. 137, 138.) and Lepsius have adopted the hypothesis that the twenty-fifth and twenty-sixth dynasties were contemporaneous,
and that the capital of the Æthiopian dynasty was at Napata, or Mt. Barkal,
whence, from time to time, the Æthiopians successfully invaded Egypt, or the
hypothesis that Amenartas, the Æthiopian, was not expelled when the Saites
commenced their reign. (M. De Rougé, Exam. ii. p. 66.)

XXIV.
Anysis, in the Delta.
XXV.
Sabaco (Thebes).
Sebichus.
Amenartas

XXVI.
Stephinates.
Nechepsos.
The Dodekarchy (League of No-
marchs).
Psammetichus I. (M. Maury, Rev.
Arch. 1851, p. 277.)

The great interest attached to the Kouyunjik seals depends upon having

the people of Samaria. Shalmaneser we know to have been an
immediate predecessor of Sennacherib, and Tirakhah, the Egyptian
king, who was defeated by the Assyrians near Lachish, was the
immediate successor of Sabaco II.

It would seem that a peace having been concluded between the
Egyptians and one of the Assyrian monarchs, probably Sennacherib, the royal signets of the two kings, thus found together, were
attached to the treaty, which was deposited amongst the archives
of the kingdom. Whilst the document itself, written upon parchment or papyrus, has completely perished, this singular proof of
the alliance, if not actual meeting, of the two monarchs is still preserved amidst the remains of the state papers of the Assyrian
empire ; furnishing one of the most remarkable instances of con-
firmatory evidence on record *, whether we regard it as verifying
the correctness of the interpretation of the cuneiform character, or
as an illustration of Scripture history.

Little doubt, I trust, can now exist in the minds of my readers
as to the identification of the builder of the palace of Kouyunjik, with the Sennacherib of Scripture. Had the name stood
alone, we might reasonably have questioned the correctness of
the reading, especially as the signs or monograms, with which it
is written, are admitted to have no phonetic power. But when
characters, whose alphabetic values have been determined from
a perfectly distinct source, such as the Babylonian column of the
trilingual inscriptions, furnish us with names in the records attributed to Sennacherib, written almost identically as in the Hebrew
version of the Bible, such as Hezekiah, Jerusalem, Judah, Sidon,
and others, and all occurring in one and the same paragraph, their
reading, moreover, confirmed by synchronisms, and illustrated by
sculptured representations of the events, the identification must be
admitted to be complete.

the precise date of this king, as they were probably affixed to a treaty with
Assyria, or some neighbouring nation.   There can be no doubt as to the
name of Sabaco. Herodotus (ii. 139.) writes ΣΑΒΑΚΩΣ ; Diodorus (i. 59.)
ΣΑΒΑΚΩΝ, Africanus Sabakôn, for the first Sabach, and Sebechos or Senechos
(ΣΕΒΗΧΩΣ) for the second. The Armenian version reads Sabbakón, for the
name of the first king (M. Böckh, Manetho, 326.). Some MSS. of the Sep-
tuagint have ΣΗΓΩΡ (Segoor). (Cf. Winer, l. c. ; Gesenius, Com. in Test. i.
696.) It is indeed highly probable, that this is the monarch mentioned in
the Book of Kings as Sua or So, and that his seal was affixed to some treaty
between Assyria and Egypt."

* The impressions of the signets of the Egyptian and Assyrian kings, besides
a large collection of seals found in Kouyunjik, are now in the British Museum.

The palace of Khorsabad, as I have already observed, was built by the father of Sennacherib. The edifice in the south-west corner of Nimroud was raised by the son, as we learn from the inscription on the back of the bulls discovered in that building.* The name of the king is admitted to be Essarhaddon, and there are events, as it will hereafter be seen, mentioned in his records, which further tend to identify him with the Essarhaddon of Scripture, who, after the murder of his father Sennacherib, succeeded to the throne.

I may mention in conclusion, as connected with the bulls forming the grand entrance, that in the rubbish at the foot of one of them were found four cylinders and several beads, with a scorpion in lapis lazuli, all apparently once strung together. On one cylinder of translucent green felspar, called amazon stone, which I believe to have been the signet, or amulet, of Sennacherib himself, is engraved the king standing in an arched frame as on the rock tablets at Bavian, and at the Nahr-el-Kelb in Syria. He holds in one hand the sacrificial mace, and raises the other in the act of adoration before the winged figure in a circle, here represented as a triad with three heads. This mode of portraying this emblem is very rare on Assyrian relics, and is highly interesting, as confirming the conjecture that the mythic human figure, with the wings and tail of a bird, inclosed in a circle,

Royal Cylinder of Sennacherib.

was the symbol of the triune god, the supreme deity of the Assyrians, and of the Persians, their successors, in the empire of the East.* In front of the king is an eunuch, and the sacred tree, whose flowers are, in this instance, in the form of an acorn. A mountain goat, standing upon a flower resembling the lotus, occupies the rest of the cylinder. The intaglio of this beautiful gem

* The relationship between the various Assyrian kings whose names are found on the monuments, was discovered by me during the first excavations, and published in my Nineveh and its Remains, vol. ii. 2nd part, chap. 1. Colonel Rawlinson in his first memoir declares, that I had been too hasty in attributing the south-west palace to the son of Sennacherib, but he appears since to have adopted the same opinion. (Outlines of Assyrian History, p. 40.)

† M. Lajard had conjectured that the component parts of this representation of the triune deity were a circle or crown to denote time without bounds, or

is not deep but sharp and distinct, and the details are so minute, that a magnifying glass is almost required to perceive them.

On a smaller cylinder, in the same green felspar *, is a cuneiform inscription, which has not yet been deciphered, but which does not appear to contain any royal name. On two cylinders of onyx, also found at Kouyunjik, and now in the British Museum, are, however, the name and titles of Sennacherib.

eternity, the image of Baal the supreme god, and the wings and tail of a dove, to typify the association of Mylitta, the Assyrian Venus. (Nineveh and its Remains, vol. ii. p. 449. note.)

    * A cylinder, not yet engraved or pierced, and several beads, are in the same material. Part of another cylinder appears to be of a kind of vitreous composition. I shall, hereafter, describe the nature and use of these relics, which are so frequently found in Assyrian and Babylonian ruins.

Piece of clay with impressions of seals.

Vaulted Drain beneath the North-west Palace at Nimroud

CHAP. VII.

ROAD OPENED FOR REMOVAL OF WINGED LIONS. — DISCOVERY OF VAULTED DRAIN — OF OTHER ARCHES — OF PAINTED BRICKS. — ATTACK OF THE TAI ON THE VILLAGE OF NIMROUD. — VISIT TO THE HOWAR. — DESCRIPTION OF THE ENCAMPMENT OF THE TAI. — THE PLAIN OF SHOMAMOK. — SHEIKH FARAS. — WALI BEY. — RETURN TO NIMROUD.

THE gigantic human-headed lions, first discovered in the north-west palace at Nimroud*, were still standing in their original position. Having been carefully covered up with earth previous to my departure in 1848, they had been preserved from exposure to the effects of the weather, and to wanton injury on the part of the Arabs. The Trustees of the British Museum wishing to add these fine sculptures to the national collection I was directed

* Nineveh and its Remains, vol. i. p. 65.

to remove them entire. A road through the ruins, for their transport to the edge of the mound, was in the first place necessary,
and it was commenced early in December. They would thus be
ready for embarkation as soon as the waters of the river were
sufficiently high to bear a raft so heavily laden, over the rapids and
shallows between Nimroud and Baghdad. This road was dug to
the level of the pavement or artificial platform, and was not finished
till the end of February, as a large mass of earth and rubbish
had to be taken away to the depth of fifteen or twenty feet.
During the progress of the work we found some carved fragments
of ivory similar to those already placed in the British Museum;
and two massive sockets in bronze, in which turned the hinges

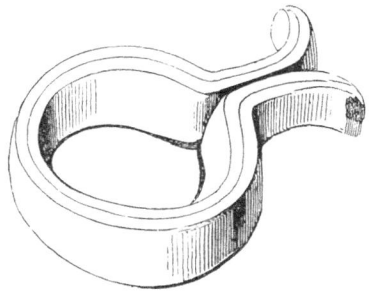

of a gate of the palace. No
remains of the door-posts, or
other parts of the gate, were
discovered in the ruins, and it
is uncertain whether these rings
were fixed in stone or wood.*

In the south-eastern corner of
the mound tunnels carried beneath the ruined edifice, which
is of the seventh century B. C.,
showed the remains of an earlier

Bronze Socket of the Palace Gate (Nimroud).

building. A *vaulted* drain, about five feet in width, was also
discovered. The arch was turned with large kiln-burnt bricks,
and rested upon side walls of the same material. The bricks
being square, and not expressly made for vaulting, a space
was left above the centre of the arch, which was filled up by bricks
laid longitudinally.

Although this may not be a perfect arch, we have seen from the
vaulted chamber discovered in the very centre of the high mound
at the north-west corner, that the Assyrians were well acquainted
at an early period with its true principle. Other examples were
not wanting in the ruins. The earth falling away from the sides
of the deep trench opened in the north-west palace for the removal
of the bull and lion during the former excavations, left uncovered
the entrance to a vaulted drain or passage built of sun-dried bricks.
Beneath was a small watercourse, inclosed by square pieces of

* The sockets, which are now in the British Museum, weigh 6lb. 3¾oz.; the
diameter of the ring is about five inches. The hinges and frames of the brass
gates at Babylon were also of brass (Herod. i. 178.).

alabaster.* A third arch, equally perfect in character, was found
beneath the ruins of the south-east edifice. A tunnel had been

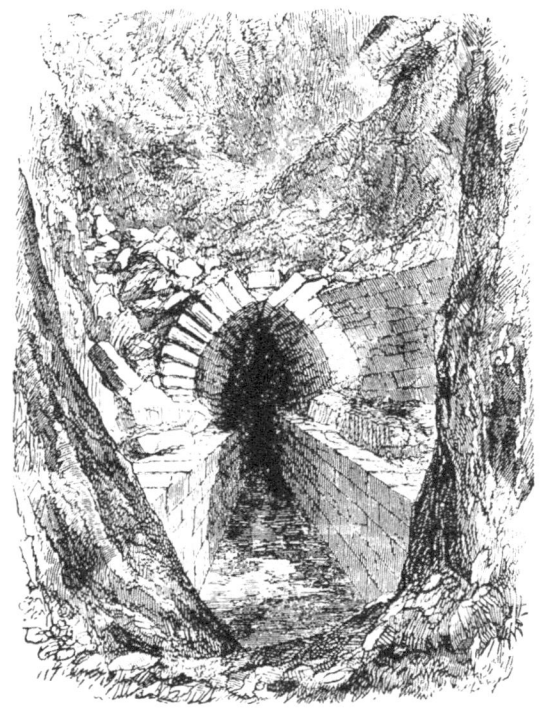

Vaulted Drain beneath South-east Palace (Nimroud).

opened almost on a level with the plain, and carried far into the
southern face of the mound, but without the discovery of any other
remains of building than this solitary brick arch. This part of the
artificial elevation or platform appears to consist entirely of earth,
heaped up without any attempt at regular construction. It contained no relics except a few rude vessels, or vases, in the coarsest
clay.

In the south-east corner of the quadrangle, formed by the low
mounds marking the walls once surrounding this quarter of the
city of Nineveh, or the park attached to the royal residence, the
level of the soil is considerably higher than in any other part of
the inclosed space. This sudden inequality evidently indicates the

* See woodcut at the head of this chapter. This drain was beneath chambers
S. and T. of the north-west palace. (See Plan III. Nineveh and its Remains,
vol. i. p. 62.)

site of some ancient edifice. Connected with it, rising abruptly, and almost perpendicularly, from the plain, and forming one of

Perfect Arch beneath South-east Edifice (Nimroud).

the corners of the walls, is a lofty, irregular mound, which is known to the Arabs by the name of the Tel of Athur, the Lieutenant of Nimroud.* Tunnels and trenches opened in it showed nothing but earth, unmingled even with bricks or fragments of stone. Remains of walls and a pavement of baked bricks were, however, discovered in the lower part of the platform. The bricks had evidently been taken from some other building, for upon them were traces of colored figures and patterns, of the same character as those on the sculptured walls of the palaces. Their painted faces were placed downwards, as if purposely to conceal them, and the designs upon them were in most instances injured or destroyed. A few fragments were collected,

* " Out of that land went forth Asshur, and builded Nineveh." (Gen. x. 11.)

and are now in the British Museum. The colors have faded, but were probably once as bright as the enamels of Khorsabad.* The outlines are white, and the ground a pale blue and olive green. The only other color used is a dull yellow. The most interesting specimens are †,

1. Four captives tied together by their necks, the end of the rope being held by the foremost prisoner, whose hands are free, whilst the others have their arms bound behind. They probably formed part of a line of captives led by an Assyrian warrior. They are beardless, and have bald heads, to which is attached a single feather.‡ Two of them have white cloths round their loins, the others long white shirts open in front, like the shirt of the modern Arab. The figures on this fragment are yellow on a blue ground.

2. Similar captives followed by an Assyrian soldier. The armour of the warrior is that of the later period, the scales and greaves are painted blue and yellow, and the tunic blue. The ground blue.

3. Parts of two horses, of a man holding a dagger, and of an Assyrian warrior. The horses are blue. The man appears to have been wounded or slain in battle, and is naked, with the exception of a twisted blue cloth round the loins. Ground an olive green.

4. Fragment, with Assyrian warriors on horses. Horses yellow, with blue trappings. Ground olive green.

5. Part of a chariot and horse, yellow on a blue ground.

6. A man, with a white cloth round his loins, pierced by two arrows. A fish, blue, with the scales marked in white; and part of a horse's head, yellow. Ground yellow.

* The colors on the Nineveh bricks have not yet been fully examined, but they appear to be precisely the same as those on the Babylonian, which have been carefully analyzed by Sir Henry De la Beche and Dr. Percy. The yellow is an antimoniate of lead, from which tin has also been extracted, called Naples yellow, supposed to be comparatively a modern discovery, though also used by the Egyptians. The white is an enamel or glaze of oxide of tin, an invention attributed to the Arabs of Northern Africa in the eighth or ninth century. The blue glaze is a copper, contains no cobalt, but some lead; a curious fact, as this mineral was not added as a coloring matter, but to facilitate the fusion of the glaze, to which use, it was believed, lead had only been turned in comparatively modern times. The red is a sub-oxide of copper.

† For facsimiles of these colored fragments, see 2nd series of Monuments of Nineveh, Plates 53, 54, 55.

‡ On Egyptian monuments captives are portrayed with similar feathers attached to their heads; but they appear to be of a negro race, whilst those on the Nimroud bricks bear no traces of negro color or physiognomy. (Wilkinson's Ancient Egyptians, vol. i. plate, p. 385.)

7. Part of a walled tower, or fort, with square battlements; white, on a blue ground.

8. Fragment of a very spirited design representing a chariot and horses passing over a naked figure, pierced through the neck by an arrow. Under this group are the heads, and parts of the shields, of two Assyrian warriors. The wounded man wears a fillet round his head, to which is attached a feather. The horses are blue, and their trappings white; the wheels of the chariot, yellow. The shields of the warriors are blue, edged by a band of alternate squares of blue and yellow; their helmets are yellow, but the faces appear to be merely outlined in white on the olive green ground.

9. The lower part of an Assyrian warrior, his armour and greaves blue, yellow, and white. The naked hand is of a pale brown color. Ground olive green.

10. A castle, with angular battlements; white, with yellow bands on a blue ground. A square door is painted blue.

All these fragments evidently belong to the same period, and probably to the same general subject, the conquest of some distant nation by the Assyrians. It is evident, from the costume of the warriors, and the form of the chariots, that they are of the later epoch, and without attempting to fix their exact date, I should conjecture that they had been taken from the same building as the detached bas-reliefs in the south-west palace, and that consequently they may be attributed to the same king.* The outlines are spirited, in character and treatment resembling the sculptures.

A fragment of painted brick, found in the ruins of the north-west palace, is undoubtedly of a different, and of an earlier, period.† The outline is in black, and not in white. The figures, of which the heads have been destroyed, wear the same dress as the tribute-bearers bringing the monkey and ornaments, on the exterior walls of the same building. ‡ The upper robe is blue, the under yellow, and the fringes white. The ground is yellow.

But the most perfect and interesting specimen of painting is that on a brick, 12 inches by 9, discovered in the centre of the mound of Nimroud, and now in the British Museum. It represents the king followed by his attendant eunuch, receiving his general or vizir, a group very similar to those seen in the sculptures from the north-west palace. Above his head is a kind of fringed pavilion,

---

* That is, as will be hereafter shown, to Pul, or Tiglath Pileser.
† No. 6. Plate 53. 2nd series of Monuments of Nineveh.
‡ First series of Monuments of Nineveh, Plate 40.

and part of an inscription, which appears to have contained his name; beneath him is the Assyrian guilloche border.* The outline is in black upon a pale yellow ground, the colors having probably faded. From the costume of the king I believe him to be either the builder of the north-west or centre palace. This is an unique specimen of an entire Assyrian painting.

During the greater part of the month of December I resided at Nimroud. One morning, I was suddenly disturbed by the reports of firearms, mingled with the shouts of men and the shrieks of women. Issuing immediately from the house, I found the open space behind it a scene of wild excitement and confusion. Horsemen, galloping in all directions and singing their war song, were driving before them with their long spears the cattle and sheep of the inhabitants of the village. The men were firing at the invaders; the women, armed with tent poles and pitchforks, and filling the air with their shrill screams, were trying to rescue the animals. The horsemen of the Arab tribe of Tai had taken advantage of a thick mist hanging over the Jaif, to cross the Zab early in the morning, and to fall upon us before we were aware of their approach. No time was to be lost to prevent bloodshed, and all its disagreeable consequences. A horse was soon ready, and I rode towards the one who appeared to be the chief of the attacking party. Although his features were concealed by the *keffieh* closely drawn over the lower part of his face, after the Bedouin fashion in war, he had been recognised as Saleh, the brother of the Howar, the Sheikh of the Tai. He saluted me as I drew near, and we rode along side by side, whilst his followers were driving before them the cattle of the villagers. Directing Hormuzd to keep back the Shemutti, I asked the chief to restore the plundered property. Fortunately, hitherto only one man of the attacking party had been seriously wounded. The expedition was chiefly directed against the Jebours, who some days before had carried off a large number of the camels of the Tai. I promised to do my best to recover them. At length Saleh, for my sake, as he said, consented to restore all that had been taken, and the inhabitants of Nimroud were called upon to claim each his own property. As we approached the ruins, for the discussion had been carried on as we rode from the village, my Jebour workmen, who had by this time heard of the affray, were preparing to meet the enemy. Some had ascended to the top of the high conical mound, where they had collected stones and bricks ready to hurl against the Tai should

---

* Plate 55. 2nd series of Monuments of Nineveh.

they attempt to follow them. Thus probably assembled on this very mound, which Xenophon calls a pyramid, the people of Larissa when the ten thousand Greeks approached their ruined city.* Others advanced towards us, stripped to their waists, brandishing their swords and short spears in defiance, and shouting their war-cry. It was with difficulty that, with the assistance of Hormuzd, I was able to check this display of valour, and prevent them from renewing the engagement. The men and women of the village were still following the retreating horsemen, clamoring for various articles, such as cloaks and handkerchiefs, not yet restored. In the midst of the crowd of wranglers, a hare suddenly sprang from her form and darted over the plain. My greyhounds, who had followed me from the house, immediately pursued her. This was too much for the Arabs; their love of the chase overcame even their propensity for appropriating other people's property ; cattle, cloaks, swords, and *keffiehs* were abandoned to their respective claimants, and the whole band of marauders joined wildly in the pursuit. Before we had reached the game we were far distant from Nimroud. I seized the opportunity to conclude the truce, and Saleh with his followers rode slowly back towards the ford of the Zab to seek his brother's tents. I promised to visit the Howar in two or three days, and we parted with mutual assurances of friendship.

Accordingly, two days afterwards, I started with Hormuzd, Schloss, and a party of Abou-Salman horsemen, for the tents of the Tai. We took the road by an ancient Chaldæan monastery, called Kuther Elias, and in three hours reached the Zab. The waters, however, were so much swollen by recent rains, that the fords were impassable, and having vainly attempted to find some means of crossing the river, we were obliged to retrace our steps.

I spent Christmas-day at Nimroud, and on the 28th renewed the attempt to visit the Howar. Schloss again accompanied me, Mr. Rolland (a traveller, who had recently joined us), Hormuzd, and Awad being of the party. Leaving the Kuther Elias to the left, we passed the ruined village of Kini-Haremi, taking the direct track to the Zab. The river, winding through a rich alluvial plain, divides itself into four branches, before entering a range of low conglomerate hills, between which it sweeps in its narrowed bed with great velocity. The four channels are each fordable, except during floods, and the Arabs generally cross at this spot. The water reached above the bellies of our horses, but we found no difficulty in stemming the current. The islands and the banks were clothed

* Anab. l. iii. c. 4.

with trees and brushwood.   In the mud and sand near the jungle
were innumerable deep, sharp prints of the hoof of the wild boar.
About two miles above the ford, on the opposite side, rose a
large, table-shaped mound, called Abou-Sheetha.   We rode to it,
and I carefully examined its surface and the deep rain-worn
ravines down its sides, but there were no remains of building;
and although fragments of brick and pottery were scattered over
it, I could see no traces upon them of cuneiform characters; yet
the mound was precisely of that form which would lead to the conjecture that it covered an edifice of considerable extent.   Awad,
however, subsequently excavated in it without finding any ruins
of the Assyrian period.   A few urns and vases were the only
objects discovered.

The tents of the Howar were still higher up the Zab.   Sending
a horseman to apprise the chief of our approach, we rode leisurely
towards them.   Near Abou-Sheetha is a small village named
Kaaitli, inhabited by sedentary Arabs, who pay tribute to the
Sheikh.   A few tents of the Tai were scattered around it.   As we
passed by, the women came out with their children, and pointing
to me exclaimed, " Look, look ! this is the Beg who is come from
the other end of the world to dig up the bones of our grandfathers
and grandmothers!" a sacrilege which they seemed inclined to
resent.   Saleh, at the head of fifty or sixty horsemen, met us
beyond the village, and conducted us to the encampment of his
brother.

The tents were pitched in long, parallel lines.   That of the chief
held the foremost place, and was distinguished by its size, the upright
spears tufted with ostrich feathers at its entrance, and the many high-
bred mares tethered before it.   As we approached, a tall, commanding
figure, of erect and noble carriage, issued from beneath the black
canvass, and advanced to receive me.   I had never seen amongst
the Arabs a man of such lofty stature.   His features were regular
and handsome, but his beard, having been fresh dyed with hennah
alone *, was of a bright brick-red hue, ill suited to the gravity and
dignity of his countenance.   His head was encircled by a rich
cashmere shawl, one end falling over his shoulder, as is the custom
amongst the Arabs of the Hedjaz.   He wore a crimson satin robe
and a black cloak, elegantly embroidered down the back, and on
one of the wide sleeves with gold thread and many-colored silks.
This was Sheikh Howar, and behind him stood a crowd of fol-

* In order to dye the hair black, a preparation of indigo should be used
after the hennah.

lowers and adherents, many of whom had the features and stature which marked the family of the chief.

As I dismounted, the Sheikh advanced to embrace me, and when his arms were round my neck my head scarcely reached to his shoulder. He led me into that part of the tent which is set aside for guests. It had been prepared for my reception, and was not ill furnished with cushions of silk and soft Kurdish carpets. The tent itself was more capacious than those usually found amongst Arabs. The black goat-hair canvass alone was the load of three camels *, and was supported by six poles down the centre, with the same number on either side. Around a bright fire was an array of highly burnished metal coffee-pots, the largest containing several quarts, and the smallest scarcely big enough to fill the diminutive cup reserved for the solitary stranger. Several noble falcons, in their gay hoods and tresses, were perched here and there on their stands. The Howar seated himself by my side, and the head men of his tribe, who had assembled on the occasion, formed a wide circle in front of us; Saleh, his brother, standing without, and receiving the commands of the Sheikh.

Coffee was, of course, the first business. It was highly spiced, as drank by the Bedouins. The Howar, after some general conversation, spoke of the politics of the Tai, and their differences with the Turkish government. The same ruinous system which has turned some of the richest districts of Asia into a desert, and has driven every Arab clan into open rebellion against the Sultan, had been pursued towards himself and his tribe. He was its acknowledged hereditary chief, and enjoyed all the influence such a position can confer. For years he had collected and paid the appointed tribute to the Turkish authorities. Fresh claims had, however, been put forward: the governors of Arbil, in whose district the Tai pastured their flocks, were to be bribed ; the Pashas of Baghdad required presents, and the tribute itself was gradually increased. At length the Howar could no longer satisfy the growing demands upon him. One of the same family was soon found who promised to be more yielding to the insatiable avarice of the Osmanlis, and, in consideration of a handsome bribe, Faras, his cousin, was named Sheikh of the tribe. The new chief had his own followers, the support of the government gave him a certain authority, and the Tai were now divided into two parties. The Pasha of Baghdad and the governor of Arbil profited by their

* The canvass of such tents is divided into strips, which, packed separately on the camels during a march, are easily united again by coarse thread, or by small wooden pins.

dissensions, received bribes from both, and from others who aimed
at the sheikhship, and the country had rapidly been reduced to a
state of anarchy. The Arabs, having no one responsible chief,
took, of course, to plundering. The villages on the Mosul side of
the Zab, as well as in the populous district of Arbil, were laid
waste. The Kurds, who came down into the plains during the
winter, were encouraged to follow the example of the Tai, and,
from the rapaciousness and misconduct of one or two officers of the
Turkish government, evils had ensued whose consequences will be
felt for years, and which will end in adding another rich district to
the desert. Such is the history of almost every tribe in Turkey,
and such the causes of the desolation that has spread over her
finest provinces.

The Tai, now reduced to two comparatively small branches,
one under the Howar, the other residing in the desert of Nisibin,
watered by the eastern branch of the Khabour, is a remnant of
one of the most ancient and renowned tribes of Arabia. The
Howar himself traces his descent from Hatem, a sheikh of the
tribe who lived in the seventh century, and who, as the imper-
sonation of all the virtues of Bedouin life, is the theme to this
day of the Arab muse. His hospitality, his generosity, his
courage, and his skill as a horseman were alike unequalled, and
there is no name more honored amongst the wild inhabitants of
the desert than that of Hatem Tai. The Howar is proud of his
heroic ancestor, and the Bedouins acknowledge and respect his
descent.*

We dined with the Sheikh and sat until the night was far spent,
listening to tales of Arab life, and to the traditions of his tribe.

On the following morning the tents were struck at sunrise, and
the chief moved with his followers to new pastures. The crowd
of camels, flocks, cattle, laden beasts of burden, horsemen, footmen,
women and children darkened the plain for some miles. We
passed through the midst of them with the Sheikh, and leaving
him to fix the spot for his encampment, we turned from the river
and rode inland towards the tents of his rival, Faras. Saleh, with
a few horsemen, accompanied me, but Schloss declared that it was

---

* The reader may remember a well-known anecdote of this celebrated
Sheikh, still current in the desert. He was the owner of a matchless mare
whose fame had even reached the Greek Emperor. Ambassadors were sent
from Constantinople to ask the animal of the chief, and to offer any amount of
gold in return. When they announced, after dining, the object of their
embassy, it was found, that the tribe suffering from a grievous famine, and
having nothing to offer to their guests, the generous Hatem had slain his own
priceless mare to entertain them.

against all the rules of Arab etiquette for a stranger, like myself, to take undue advantage of the rights of hospitality by introducing an enemy under my protection into an encampment. There was a feud between the two chiefs, blood had actually been spilt, and if Saleh entered the dwellings of his rivals, disagreeable consequences might ensue, although my presence and the fact of his having eaten bread with me would save him from actual danger. However, one of my objects was to bring about a reconciliation between the two chiefs, and as Saleh had consented to run the risk of accompanying me, I persevered in my determination. Schloss was not to be persuaded, he hung behind, sulked, and finally turning the head of his mare, rode back with his companions to the river. I took no notice of his departure, anticipating his speedy return. He recovered from his ill humor, and joined us again late in the evening.

The plain, bounded by the Tigris, the great and lesser Zab, and the Kurdish hills, is renowned for its fertility. It is the granary of Baghdad, and it is a common saying amongst the Arabs, " that if there were a famine over the rest of the earth, Shomamok (for so the principal part of the plain is called) would still have its harvest." This district belongs chiefly to the Tai Arabs, who wander from pasture to pasture, and leave the cultivation of the soil to small sedentary tribes of Arabs, Turcomans, and Kurds, who dwell in villages, and pay an annual tribute in money or in kind.

As we rode along we passed many peasants industriously driving the plough through the rich soil. Large flocks of gazelles grazed in the cultivated patches, scarcely fearing the husbandman, though speedily bounding away over the plain as horsemen approached. Artificial mounds rose on all sides of us, and near one of the largest, called Abou-Jerdeh, we found the black tents of Sheikh Faras. The rain began to fall in torrents before we reached the encampment. The chief had ridden out to a neighbouring village to make arrangements for our better protection against the weather. He soon returned urging his mare to the top of her speed. In person he was a strange contrast to the elder member of his family. He was short, squat, and fat, and his coarse features were buried in a frame of hair dyed bright red. He was, however, profuse in assurances of friendship, talked incessantly, agreed to all I proposed with regard to a reconciliation with the other branch of the tribe, and received Saleh with every outward sign of cordiality. His son had more of the dignity of his race, but the expression of his countenance was forbidding and sinister. The two young men, as they sat, cast looks of defiance at each other, and I had some difficulty in restraining Saleh from breaking out in invectives, which probably would have ended in an appeal to the sword.

As the rain increased in violence, and the tent offered but an imperfect shelter, we moved to the village, where a house had been prepared for us by its honest, kind-hearted Turcoman chief, Wali Bey. With unaffected hospitality he insisted that we should become his guests, and had already slain the sheep for our entertainment. I have met few men who exceed, in honesty and fidelity, the descendants of the pure Turcoman race, scattered over Asia Minor and the districts watered by the Tigris.

On the following morning, Wali Bey having first provided an ample breakfast, in which all the luxuries of the village were set before us, we again visited the tents of the Howar. After obtaining his protection for Awad, who was to return in a few days with a party of workmen, to explore the mounds of Sho-mamok, and settling the terms of reconciliation between himself and Faras, we followed the baggage, which had been sent before us to the ford. On reaching the Zab, we found it rising rapidly from the rains of the previous day. Our servants had already crossed, but the river was now impassable. We sought a ford higher up, and above the junction of the Ghazir. Having struggled in vain against the swollen stream, we were compelled to give up the attempt. Nothing remained but to seek the ferry on the high road, between Arbil and Mosul. We did not reach the small village, where a raft is kept for the use of travellers and caravans, until nearly four o'clock in the afternoon, and it was sunset before we had crossed the river.

We hurried along the direct track to Nimroud, hoping to cross the Ghazir before night-fall. But fresh difficulties awaited us. That small river, collecting the torrents of the Missouri hills, had overflown its bed, and its waters were rushing tumultuously onwards, with a breadth of stream almost equalling the Tigris. We rode along its banks, hoping to find an encampment where we could pass the night. At length, in the twilight, we spied some Arabs, who immediately took refuge behind the walls of a ruined village, and believing us to be marauders from the desert, prepared to defend themselves and their cattle. Directing the rest of the party to stop, I rode forward with the Bairakdar, and was in time to prevent a discharge of fire-arms pointed against us. The Arabs were of the tribe of Haddedeen, who having crossed the Ghazir, with their buffaloes, had been unable to regain their tents on the opposite side by the sudden swelling of the stream.

The nearest inhabited village was Tel Aswad, or Kara Tuppeh, still far distant. As we rode towards it in the dusk, one or two wolves lazily stole from the brushwood, and jackals and other beasts of prey occasionally crossed our path. We found the Kiayah seated with some travellers round a blazing fire. The miserable hut was soon cleared of its occupants, and we prepared to pass the night as we best could.

Towards dawn the Kiayah brought us word that the Ghazir had subsided sufficiently to allow us to ford. We started under his guidance, and found that the stream, although divided into three branches, reached in some places almost to the backs of the horses. Safe over, we struck across the country towards Nimroud, and reached the ruins as a thick morning mist was gradually withdrawn from the lofty mound.

During our absence, a new chamber had been opened in the north-west palace, to the south of the great centre hall. The walls were of plain, sun-dried brick, and there were no remains of sculptured slabs, but in the earth and rubbish which had filled it, were discovered some of the most interesting relics obtained from the ruins of Assyria. A description of its contents alone will occupy a chapter.

Arab Tent.

Excavated Chamber in which the Bronzes were discovered (Nimroud)

CHAP. VIII.

CONTENTS OF NEWLY DISCOVERED CHAMBER. — A WELL. — LARGE COPPER CALDRONS. — BELLS, RINGS, AND OTHER OBJECTS IN METAL. — TRIPODS.— CALDRONS AND LARGE VESSELS. — BRONZE BOWLS, CUPS, AND DISHES. — DESCRIPTION OF THE EMBOSSINGS UPON THEM. — ARMS AND ARMOUR. — SHIELDS. — IRON INSTRUMENTS. — IVORY REMAINS. — BRONZE CUBES INLAID WITH GOLD. — GLASS BOWLS. — LENS. — THE ROYAL THRONE.

THE newly discovered chamber was part of the north-west palace, and adjoined a room previously explored.* Its only entrance was to the west, and almost on the edge of the mound. It must, consequently, have opened upon a gallery or terrace running along the river front of the building. The walls were of sun-dried brick, panelled round the bottom with large burnt bricks, about three feet high, placed one against the other. They were coated with bitumen, and, like those forming the pavement, were inscribed with

---

* It was parallel to, and to the south of, the chamber marked A A, in the plan of the north-west palace. (Nineveh and its Remains, vol. i. Plan III.)

the name and usual titles of the royal founder of the building. In one corner, and partly in a kind of recess, was a well, the mouth of which was formed by brickwork about three feet high. Its sides were also bricked down to the conglomerate rock, and holes had been left at regular intervals for descent. When first discovered it was choked with earth. The workmen emptied it until they came, at the depth of nearly sixty feet, to brackish water.*

The first objects found in this chamber were two plain copper vessels or caldrons, about 2½ feet in diameter, and 3 feet deep, resting upon a stand of brickwork, with their mouths closed by large tiles. Near them was a copper jar, which fell to pieces almost as soon as uncovered. Several vases of the same metal, though smaller in size, had been dug out of other parts of the ruins; but they were empty, whilst those I am describing were filled with curious relics. I first took out a number of small bronze

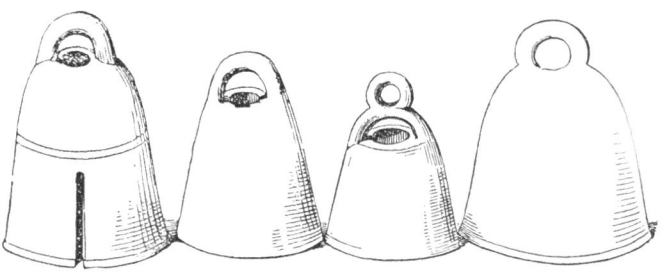

Bronze Bells found in a Caldron (Nimroud).

bells† with iron tongues, and various small copper ornaments, some suspended to wires. With them were a quantity of tapering bronze rods, bent into a hook, and ending in a kind of lip. Beneath were several bronze cups and dishes, which I succeeded in removing entire. Scattered in the earth amongst these objects were several hundred studs and buttons in mother of pearl and ivory, with many small rosettes in metal.

All the objects contained in these caldrons, with the exception of

---

* Few wells in the plains bordering on the Tigris yield sweet water.

† The caldrons contained about eighty bells. The largest are 3¼ inches high, and 2¼ inches in diameter, the smallest 1¾ inch high, and 1¼ inch in diameter. With the rest of the relics they are now in the British Museum.

the cups and dishes, were probably ornaments of horse and chariot furniture. The accompanying woodcut from a bas-relief at Kouyun-

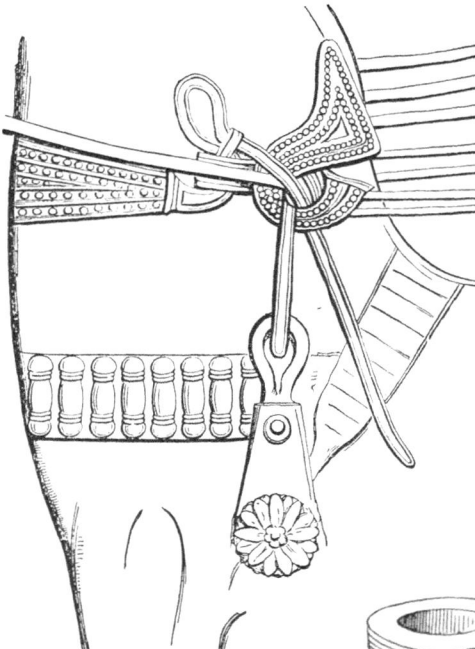

jik, will show the way in which the studs of ivory and mother of pearl, and the rosettes or stars of metal, were probably used. The horses of the Assyrian cavalry, as well as those harnessed to chariots, are continually represented in the sculptures with bells round their necks, and in the Bible we find allusion to this custom.* The use of the metal hooks cannot be so satisfactorily traced; they probably belonged to

Horse Trappings, from a Bas-relief at Kouyunjik, showing probable Use of Ivory Studs and Metal Rosettes

some part of the chariot, or the horse trappings.

Beneath the caldrons were heaped lions' and bulls' feet of bronze; and the remains of iron rings and bars, probably parts of tripods, or stands, for supporting vessels and bowls†;

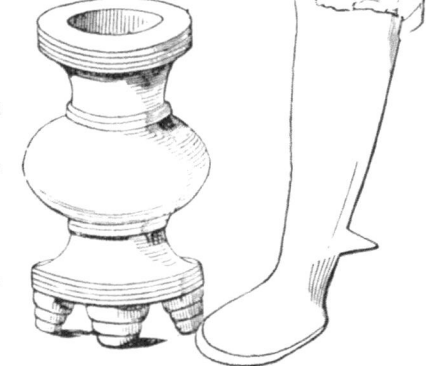

Feet of Tripods in Bronze and Iron.

* Zech. xiv. 20.

† Tripod-stands, consisting of a circular ring raised upon feet, to hold jars and vases, are frequently represented in the bas-reliefs. (See particularly Botta's large work, plate 141.) The ring was of iron, bound in some places with copper, and the feet partly of iron and partly of bronze ingeniously cast over it

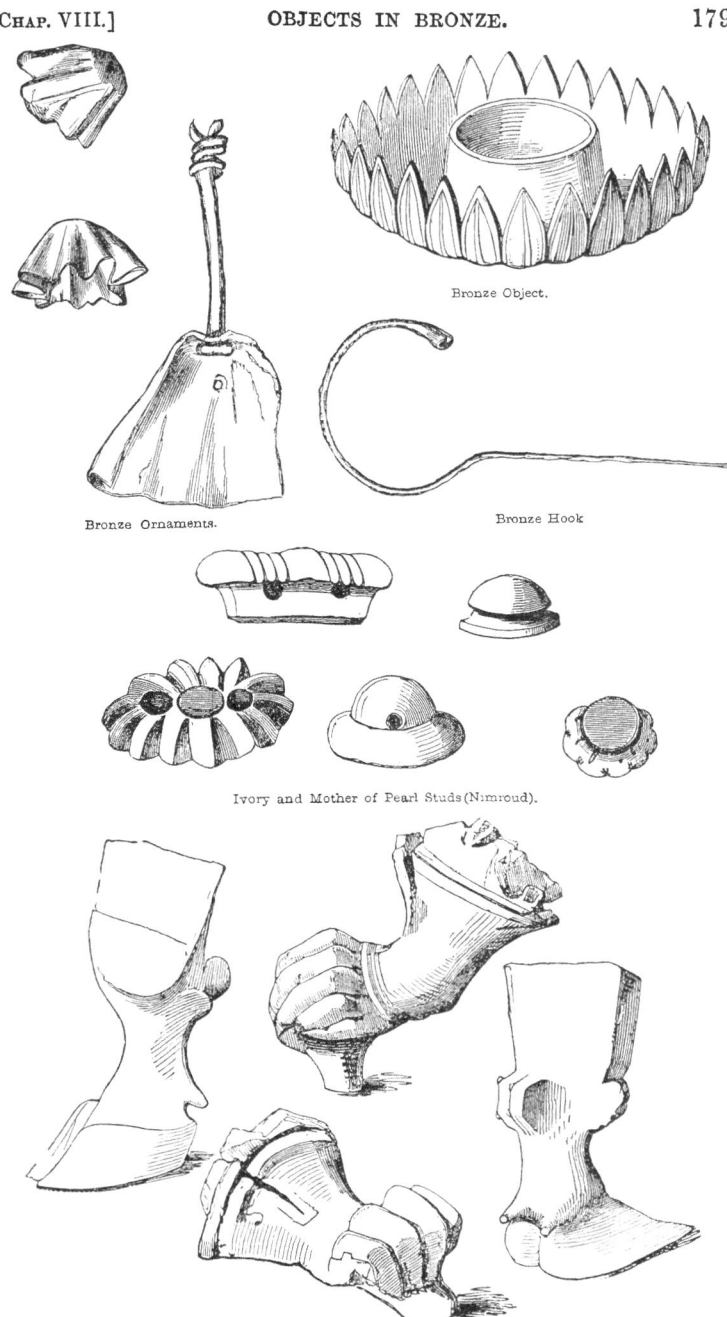

Bronze Object.

Bronze Ornaments.

Bronze Hook

Ivory and Mother of Pearl Studs (Nimroud).

Feet of Tripods in Bronze and Iron.

which, as the iron had rusted away, had fallen to pieces, leaving such parts entire as were in the more durable metal.

Two other caldrons, found further within the chamber, contained, besides several plates and dishes, four crown shaped bronze ornaments, perhaps belonging to a throne or couch * ; two long ornamented bands of copper, rounded at both ends, apparently belts, such as were worn by warriors in armour † ; a grotesque head in bronze, probably the top of a mace; a metal wine-strainer of elegant shape; various metal vessels of peculiar form, and a bronze ornament, probably the handle of a dish or vase.

Eight more caldrons and jars were found in other parts of the chamber. One contained ashes and bones, the rest were empty.‡ Some of the larger vessels were crushed almost flat, probably by the falling in of the upper part of the building.

With the caldrons were discovered two circular flat vessels, nearly six feet in diameter, and about two feet deep, which I can only compare with the brazen sea that stood in the temple of Solomon. §

Caldrons are frequently represented as part of the spoil and tribute, in the sculptures of Nimroud and Kouyunjik. ‖ They were so much valued by the ancients that, it appears from the Homeric poems, they were given as prizes at public games, and were considered amongst the most precious objects that could be carried away from a captured city. They were frequently embossed with flowers and other ornaments. Homer declares one so adorned to be worth an ox.¶

---

* If, however, they were part of a throne, it is difficult to account for their being found detached in the caldron. They measured 6 inches in diameter, and 2 inches in depth.

† Resembling those of the eunuch warriors in Plate 28. of the 1st series of the Monuments of Nineveh.

‡ One of the jars was 4 feet 11 inches high. Two of the caldrons with handles on each side were 2 feet 5 inches in diameter, and 1 foot 6 inches deep.

§ 2 Chron. iv. 2. The dimensions, however, of this vessel were far greater. It is singular that in some of the bas-reliefs large metal caldrons supported on brazen oxen are represented.

‖ See particularly Monuments of Nineveh, 1st series, Plate 24., and 2nd series, Plate 35., and on the black obelisk. They were carried away by the Babylonians from Jerusalem. Jerem. lii. 18.

¶ They were dedicated to the gods in temples. Colœus dedicated a large vessel of brass, *adorned with griffins*, to Heré. Herod. iv. 152.

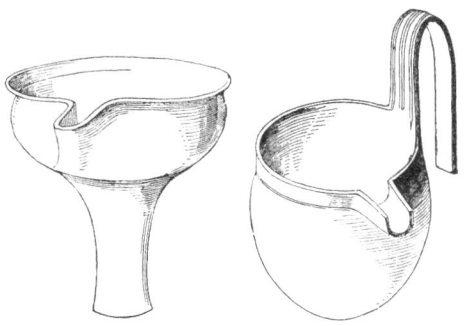

Bronze Vessels, taken from the Interior of a Caldron

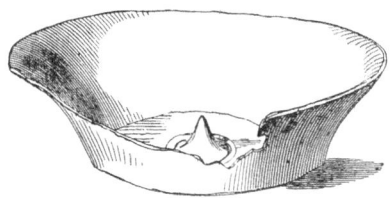

Bronze Vessel, taken from the Interior of a Caldron.

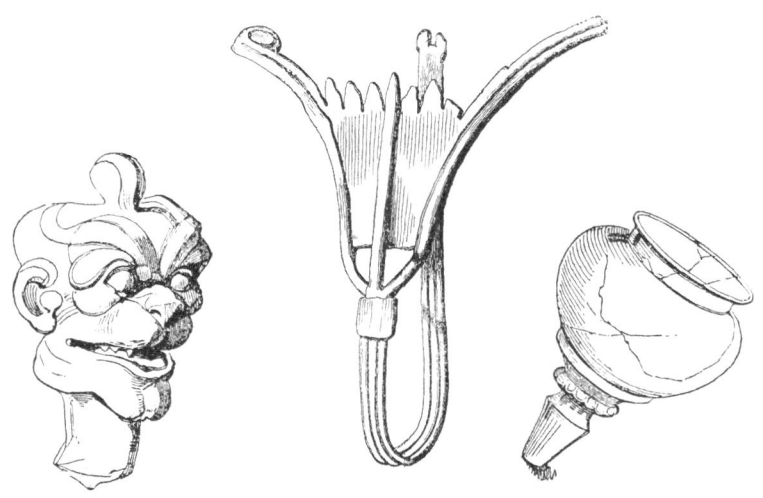

Bronze Head of a Mace.   Bronze Handle of a Dish or Vase.   Bronze Wine Strainer.

N 3

Behind the caldrons was a heap of curious and interesting objects. In one place were piled without order, one above the other, bronze cups, bowls, and dishes of various sizes and shapes. The upper vessels having been most exposed to damp, the metal had been eaten away by rust, and was crumbling into fragments, or into a green powder. As they were cleared away, more perfect specimens were taken out, until, near the pavement of the chamber, some were found almost entire. Many of the bowls and plates fitted so closely, one within the other, that they have only been detached in England. It required the greatest care and patience to separate them from the tenacious soil in which they were embedded.

Although a green crystaline deposit, arising from the decomposition of the metal, encrusted all the vessels, I could distinguish upon many of them traces of embossed and engraved ornaments. Since they have been in England they have been carefully and skilfully cleaned by Mr. Doubleday, of the British Museum*, and the very beautiful and elaborate designs upon them brought to light.†

The bronze objects thus discovered may be classed under four heads — dishes with handles, plates, deep bowls, and cups. Some are plain, others have a simple rosette, scarab, or star in the centre, and many are most elaborately ornamented with the figures of men and animals, and with elegant fancy designs, either embossed or incised. Although the style, like that of the ivories from the same palace, and now in the British Museum, is frequently Egyptian in character, yet the execution and treatment, as well as the subjects, are peculiarly Assyrian. The inside, and not the outside, of these vessels is ornamented. The embossed figures have

---

* I seize this opportunity of expressing my thanks to that gentleman, for the kind assistance and valuable information I have received from him during my connection with the British Museum, and of bearing testimony to the judgment and skill he has displayed as well in the disembarkation and removal of the great sculptures, as in the cleaning and repairing of the most minute and delicate objects confided to his care.

† Engravings of the most interesting of these vessels will be found in the 2nd series of my Monuments of Nineveh. They have been chiefly executed from the admirable drawings of Mr. Prentice, to whom I am indebted for the very accurate representations of the ivories, published in my former work. The Trustees of the British Museum have judiciously employed that gentleman to make exact copies of these interesting relics, which, it is feared, will ere long be utterly destroyed by a process of natural decomposition in the metal, that no ingenuity can completely arrest.

been raised in the metal by a blunt instrument, three or four strokes of which in many instances very ingeniously produce the image of an animal.* Even those ornaments which are not embossed but incised, appear to have been formed by a similar process, except that the punch was applied on the inside. The tool of the graver has been sparingly used.

The most interesting dishes in the collection brought to England are :—

No. 1., with moving circular handle (the handle wanting), se-

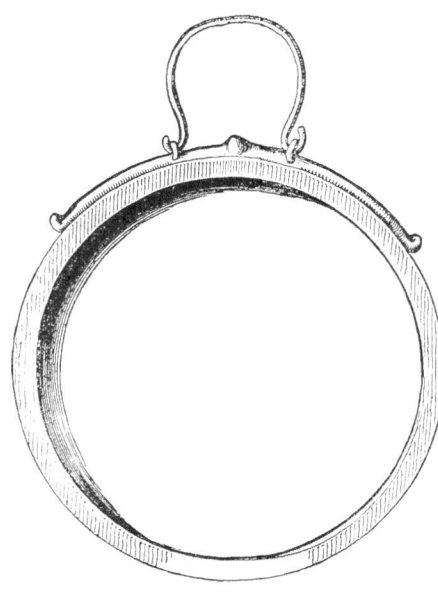

cured by three bosses; diameter 10¾ inches, depth 2¼ inches; divided into two friezes surrounding a circular medallion containing a male deity *with bull's ears* (?) and hair in ample curls†, wearing bracelets and a necklace of an Egyptian character, and a short tunic; the arms crossed, and the hands held by two *Egyptians* (?), who place their other hands on the head of the centre figure. The inner frieze contains horsemen draped as Egyptians, galloping round in pairs; the outer, figures, also wearing the Egyptian "*shenti*" or tunic,

Bronze Dish from Nimroud

hunting lions on horseback, on foot, and in chariots. The hair of these figures is dressed after a fashion, which prevailed in Egypt from the ninth to the eighth century B.C. Each frieze is separated by a band of guilloche ornament.‡

No. 2., diameter 10½ inches, having a low rim, partly destroyed; ornamented with an embossed rosette of elegant shape, surrounded

---

* The embossing appears to have been produced by a process still practised by silversmiths. The metal was laid upon a bed of mixed clay and bitumen, and then punched from the outside.

† The Egyptian goddess Athor is represented with similar ears and hair.

‡ Monuments of Nineveh, 2nd Series. Plate 65.

by three friezes of animals in high relief, divided by a guilloche
band.   The outer frieze contains twelve walking bulls, designed

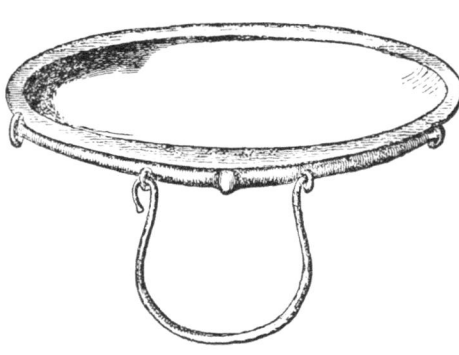

Bronze Dish, from Nimroud.

with considerable spirit;
between each is a dwarf
shrub or tree.   The se-
cond frieze has a bull, a
winged griffin, an ibex,
and a gazelle, walking one
behind the other, and the
same animals seized by
leopards or lions, in all
fourteen figures.   The
inner frieze contain twelve
gazelles.   The handle is
formed by a plain movable
ring.*   The ornaments

on this dish, as well as the design, are of an Assyrian character.
The bull, the wild-goat, and the griffin are the animals, evidently
of a sacred character, which occur so frequently in the sculptures
of Nimroud.   The lion, or leopard, devouring the bull and gazelle,
is a well-known symbol of Assyrian origin, afterwards adopted by
other Eastern nations, and may typify, according to the fancy of
the reader, either the subjection of a primitive race by the Assyrian
tribes, or an astronomical phenomenon.

No. 3., diameter 10¾ inches, and 1½ inch deep, with a raised star
in the centre; the handle formed by two rings, working in sockets
fastened to a rim, running about one third round the margin, and
secured by five nails or bosses; four bands of embossed ornaments
in low relief round the centre, the outer band consisting of alternate standing bulls and crouching lions, Assyrian in character and
treatment; the others, of an elegant pattern, slightly varied from the
usual Assyrian border by the introduction of a fanlike flower in the
place of the tulip.†

Other dishes were found still better preserved than those just
described, but perfectly plain, or having only a star, more or
less elaborate, embossed or engraved in the centre.   Many frag-

---

* Monuments of Nineveh, 2nd series.   Plate 60.

† Id. A. Plate 57.   I have called this flower, the lotus of the Egyptian sculptures, a tulip, as it somewhat resembles a bright scarlet tulip which abounds in
early spring on the Assyrian plains, and may have suggested this elegant ornament.   It has no resemblance whatever to the honeysuckle, by which name it
is commonly known, when used in Greek architecture.

ments were also discovered with elegant handles, some formed by
the figures of rams and bulls.

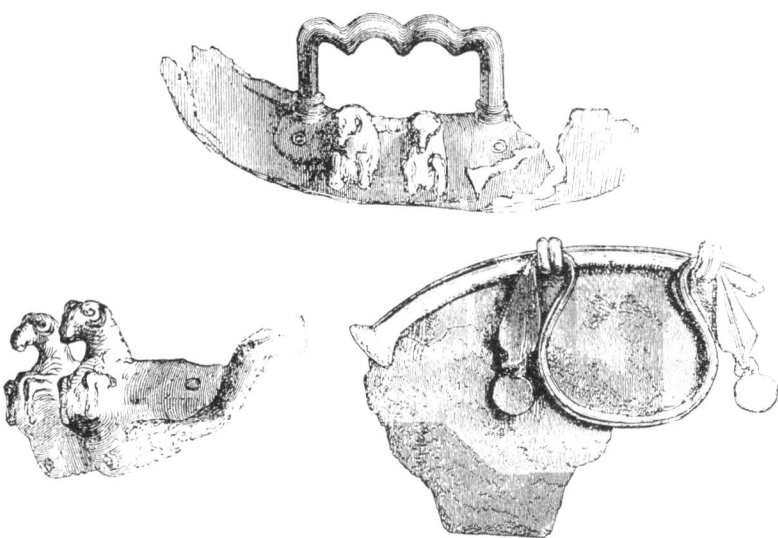

Handles of Bronze Dishes from Nimroud.

Of the plates the most remarkable are: —

No. 1., shallow, and 8¾ inches in diameter, the centre slightly
raised and incised with a star and five bands of tulip-shaped
ornaments; the rest occupied by four groups, each consisting
of two winged hawk-headed sphinxes, wearing the "pshent,"
or crown of the upper and lower country of Egypt; one paw
raised, and resting upon the head of a man kneeling on one knee,
and lifting his hands in the act of adoration. Between the
sphinxes, on a column in the form of a papyrus-sceptre, is the
bust of a figure wearing on his head the sun's disc, with the uræi
serpents, a collar round the neck, and four feathers; above are two
winged globes with the asps, and a row of birds. Each group is
inclosed by two columns with capitals in the form of the Assyrian
tulip ornament, and is separated from that adjoining by a scarab
with out-spread wings, raising the globe with its fore feet, and
resting with its hind on a papyrus-sceptre pillar.* This plate
is in good preservation, having been found at the very bottom of

* Monuments of Nineveh, 2nd series. Plate 63.

a heap of similar relics. Part of the bronze was still bright, and of a golden color; hence the report spread at the time of the discovery, that an immense treasure in vessels of gold had been dug up at Nimroud. The emblems are evidently derived from familiar objects in Egyptian mythology, which may have been applied by the Assyrians to other ideas. The workmanship, although not purely Egyptian, appears to be more so than that of any other specimen in the collection, except a fragment very closely resem-

Bronze Cup 6¼ in. diameter, and 1⅜ in. deep   Engraved Scarab in Centre of same Cup.

bling this plate.* A scarab, apparently more of a Phœnician than of an Egyptian form, occurs as an ornament on many of these bronzes; as in the centre of a well-preserved bowl otherwise plain, and on a dish.

No. 2., depth, 1¾ in.; diameter, 9⅓ in., with a broad, raised rim, like that of a soup plate, embossed with figures of greyhounds pursuing a hare. The centre contains a frieze in high relief, representing combats between men and lions, and a smaller border of gazelles, between guilloche bands, encircling an embossed star.† In this very fine specimen, although the costumes of the figures are Egyptian in character, the treatment and design are Assyrian.

No. 3., shallow; 9½ inches diameter; an oval in the centre, covered with dotted lozenges, and set with nine silver bosses, probably intended to represent a lake or valley, surrounded by four groups of hills, each with three crests in high relief, on which are incised in outline trees and stags, wild goats, bears, and leopards. On the sides of the hills, in relief, are similar figures of animals. The outer rim is incised with trees and deer.‡ The workmanship of this specimen is Assyrian, and very minute and curious. The sub-

* Monuments of Nineveh, 2nd series. Plate 68.
† Id. Plate 64.
‡ Id. Plate 66.

ject may represent an Assyrian paradise, or park, in a mountainous district.

No. 4., diameter, 7¼ inches, the centre raised, and containing an eight-rayed star, with smaller stars between each ray, encircled by a guilloche band. The remainder of the plate is divided into eight compartments, by eight double-faced figures of Egyptian character in high relief; between each figure are five rows of animals, inclosed by guilloche bands; the first three consisting of stags and hinds, the fourth of lions, and the fifth of hares, each compartment containing thirteen figures. A very beautiful specimen, unfortunately much injured.*

No. 5., diameter, 8¾ inches; depth, 1¼ inch. The embossings and ornaments on this plate are of an Egyptian character. The centre consists of four heads of the cow-eared goddess Athor (?), forming, with lines of bosses, an eight-rayed star, surrounded by hills, indicated as in plate No. 3., but filled in with rosettes and other ornaments. Between the hills are incised animals and trees. A border of figures, almost purely Egyptian, but unfortunately only in part preserved, encircles the plate ; the first remaining group is that of a man seated on a throne, beneath an ornamented arch, with the Egyptian Baal, represented as on the coins of Cossura, standing full face , to the right of this figure is a square ornament with pendants (resembling a sealed document), and beneath it the crux ansata or Egyptian symbol of life. The next group is that of a warrior in Egyptian attire, holding a mace in his right hand, and in his left a bow and arrow, with the hair of a captive of smaller proportions, who crouches before him. At his side is a tame lion, recalling to mind the pictures on Egyptian monuments of Rameses II., accompanied by a lion during his campaigns. A goddess, wearing a long Egyptian tunic, presents a falchion with her right hand to this warrior, and holds a sceptre in her left. Between these figures are two hieroglyphs, an ox's head and an ibis or an heron. Over the goddess is a square tablet for her name. The next group represents the Egyptian Baal (?), with a lion's skin round his body, and plumes on his head, having on each side an Egyptian figure wearing the " shent," or short tunic, carrying a bow, and plucking the plumes from the head of the god, perhaps symbolical of the victory of Horus over Typhon. This group is followed by a female figure, draped in the Assyrian fashion, but wearing on her head the triple crown of the Egyptian god Pnebta, holding in one hand a sword, and in the other a

* Monuments of Nineveh, 2nd series. A. Plate 61.

bow (?), and having on each side men, also dressed in the Assyrian costume, pouring out libations to her from a jug or chalice: the Egyptian symbol of life occurs likewise in this place. The Egyptian god Amon, bearing a bird in one hand and a falchion in the other, with female figures similar to that last described, appears to form the next group; but unfortunately this part of the plate has been nearly destroyed: the whole border, however, appears to have represented a mixture of religious and historical scenes.*

No. 6., diameter, 6 in.; depth, 1½ in.; a projecting rim, ornamented with figures of vultures with outspread wings; an embossed rosette, encircled by two rows of fan-shaped flowers and guilloche bands, occupies a raised centre, which is surrounded by a frieze, consisting of groups of two vultures devouring a hare. A highly finished and very beautiful specimen. On the back of this plate are five letters, either in the Phœnician or Assyrian cursive character.†

Nos. 7. and 8.; covered with groups of small stags, surrounding an elaborate star, one plate containing above 600 figures; the animals are formed by three blows from a blunt instrument or punch. These plates are ornamented with small bosses of silver and gold let into the copper.‡

No. 9., diameter, 7⅝ inches; depth, 1½ inch, of fine workmanship; the centre formed by an incised star, surrounded by guilloche and tulip bands. Four groups on the sides representing a lion, lurking amongst papyri or reeds, and about to spring on a bull.

No. 10., diameter 7½ inches. In the centre a winged scarab raising the disc of the sun, surrounded by guilloche and tulip bands, and by a double frieze, the inner consisting of trees, deer, winged uræi, sphinxes, and papyrus plants: the outer, of winged scarabs, flying serpents, deer, and trees, all incised.

The plates above described are the most interesting specimens brought to this country: there are others, indeed, scarcely less remarkable for beauty of workmanship §, or, when plain or ornamented with a simple star in the centre, for elegance of form. Of the seventeen deep bowls discovered, only three have embossings, sufficiently well preserved, to be described; the

---

* Monuments of Nineveh, 2nd series. B. Plate 61.   † Id. B. Plate 62.
‡ Id. E. Plate 57. and C. Plate 59.
§ I may instance in particular a fragment covered with a very elegant and classic design. Monuments of Nineveh, 2nd series. Plate 62., and see Plates 57, 58, 59. of same work.

greater part appear to be perfectly plain. The most remark-
able is $8\frac{1}{2}$ inches in diameter, and $3\frac{3}{4}$ inches deep, and has at the
bottom, in the centre, an embossed star, surrounded by a rosette,
and on the sides a hunting scene in bold relief. From a chariot,
drawn by two horses, and driven by a charioteer, a warrior turning
back shoots an arrow at a lion, which is already wounded; whilst a
second huntsman in armour, above whose head hovers a hawk,
pierces the animal from behind with a spear. These figures are
followed by a sphinx, wearing the Egyptian head-dress " pshent "
and a collar, on which is the bust of a winged, ram-headed god.
Two trees, with flowers or leaves in the shape of the usual Assyrian tulip ornament, are introduced into the group.

A second, $7\frac{1}{2}$ inches in diameter, and $3\frac{3}{4}$ inches deep, has in the
centre a medallion similar to that in the one last described, and

Embossed Figures on the Bronze Pedestal of a figure from Polledrara, in the British Museum.

on the sides, in very high relief, two lions and two sphinxes

Embossed Figure on the Bronze Pedestal of a Figure
from Polledrara.

of Egyptian character, wearing
a collar, feathers, and housings,
and a head-dress formed by a disc
with two uræi. Both bowls are
remarkable for the boldness of
the relief and the archaic treatment of the figures, in this respect resembling the ivories previously discovered at Nimroud.
They forcibly call to mind the
early remains of Greece, and especially the metal work, and painted
pottery found in very ancient

tombs in Etruria, which they so closely resemble not only in design but in subject, the same mythic animals and the same ornaments being introduced, that we cannot but attribute to both the

Bronze Pedestal of Figure from Polledrara.

same origin.* I have given for the sake of comparison, wood-cuts of the bronze pedestal of a figure found at Polledrara in Etruria, and now in the British Museum. The animals upon it are precisely similar to those upon the fragment of a dish brought from Nineveh, and, moreover, that peculiar Assyrian ornament, the guilloche, is introduced.

The third, 7¾ inches in diameter, and 2½ inches deep, has in the centre a star formed by the Egyptian hawk of the sun, bearing the disc, and having at its side a whip, between two rays ending in lotus flowers; on the sides are embossed figures of wild goats, lotus-shaped shrubs, and dwarf trees of peculiar form.†

Of the cups the most remarkable are : —

No. 1., diameter 5⅝ inches, and 2¼ inches deep, very elaborately ornamented with figures of animals, interlaced and grouped together in singular confusion, covering the whole inner surface; apparently representing a combat between griffins and lions; a very curious and interesting specimen, not unlike some of the Italian chasing of the cinque cento.‡

Bronze Cup, from Nimroud.

No. 2., a fragment, embossed with the figures of lions and bulls, of very fine workmanship.

Of the remaining cups many are plain but of elegant shape, one or two are ribbed, and some have simply an embossed star in the centre.

About 150 bronze vessels discovered in this chamber are now

* For the two Assyrian bowls see Plate 68. of the Monuments of Nineveh, 2nd series. These bronzes should also be compared with the vessels found at Cervetri, and engraved in Griffi's Monumenti de Ceri Antica (Roma, 1841), and with various terracottas in the British Museum.

† Monuments of Nineveh, 2nd series. C. Plate 57

‡ Id. Plate 67

in the British Museum, without including numerous fragments, which, although showing traces of ornament, are too far destroyed by decomposition to be cleaned.

I shall add, in an Appendix, some notes on the bronze and other substances discovered at Nimroud, obligingly communicated to me by Dr. Percy. It need only be observed here, that the metal of the dishes, bowls, and rings has been carefully analysed by Mr. T. T. Philips, at the Museum of Practical Geology, and has been found to contain one part of tin to ten of copper, being exactly the relative proportions of the best ancient and modern bronze. The bells, however, have fourteen per cent. of tin, showing that the Assyrians were well aware of the effect produced by changing the proportions of the metals. These two facts show the advance made by them in the metallurgic art.

The effect of age and decay has been to cover the surface of all these bronze objects, with a coating of beautiful crystals of malachite, beneath which the component substances have been converted into suboxide of copper and peroxide of tin, leaving in many instances no traces whatever of the metals.

It would appear that the Assyrians were unable to give elegant forms or a pleasing appearance to objects in iron alone, and that consequently they frequently overlaid that metal with bronze, either entirely, or partially, by way of ornament. Numerous interesting specimens of this nature are included in the collection in the British Museum. Although brass is now frequently cast over iron, the art of using bronze for this purpose had not, I believe, been introduced into modern metallurgy.* The feet of the ring-tripods previously described, furnish highly interesting specimens of this process, and prove the progress made by the Assyrians in it. The iron inclosed within the copper has not been exposed to the same decay as that detached from it, and will still take a polish.

The tin was probably obtained from Phœnicia ; and consequently that used in the bronzes in the British Museum may actually have been exported, nearly three thousand years ago, from the British Isles! We find the Assyrians and Babylonians making an extensive use of this metal, which was probably one of the chief articles of trade supplied by the cities of the Syrian coast, whose seamen sought for it on the distant shores of the Atlantic.

* Mr. Robinson of Pimlico has, I am informed, succeeded in imitating some of the Assyrian specimens.

The embossed and engraved vessels from Nimroud afford many interesting illustrations of the progress made by the ancients in metallurgy. From the Egyptian character of the designs, and especially of the drapery of the figures, in several of the specimens, it may be inferred that some of them were not Assyrian, but had been brought from a foreign people. As in the ivories, however, the workmanship, subjects, and mode of treatment are more Assyrian than Egyptian, and seem to show that the artist either copied from Egyptian models, or was a native of a country under the influence of the arts and taste of Egypt. The Sidonians, and other inhabitants of the Phœnician coast, were the most renowned workers in metal of the ancient world, and their intermediate position between the two great nations, by which they were alternately invaded and subdued, may have been the cause of the existence of a mixed art amongst them. In the Homeric poems they are frequently mentioned as the artificers who fashioned and embossed metal cups and bowls, and Solomon sought cunning men from Tyre to make the gold and brazen utensils for his temple and palaces.* It is, therefore, not impossible that the vessels discovered at Nimroud were the work of Phœnician artists†, brought expressly from Tyre, or carried away amongst the captives when their cities were taken by the Assyrians, who, we know from many passages in the Bible‡, always secured the smiths and artizans, and placed them in their own immediate dominions. They may have been used for sacrificial purposes, at royal banquets, or when the king performed certain religious ceremonies, for in the bas-reliefs he is frequently represented on such occasions with a cup or bowl in his hand; or they may have formed part of the spoil of some Syrian nation, placed in a temple at Nineveh, as the holy utensils of the Jews, after the destruction of the sanctuary, were kept in the

---

* 1 Kings, vii. 13, 14. 2 Chron. iv. The importance attached to such objects in metal, which were chiefly used for sacred purposes, is shown by its being especially recorded that Huram (or Hiram), the widow's son, was sent for to make " the pots, and the shovels, and the basons." Homer particularly mentions Sidonian goblets as used at the funeral games of Patroclus.

† It will be remembered that Phœnician characters occur on one of the plates. The discovery in Cyprus of twelve silver bowls very closely resembling those found at Nimroud, tend further to confirm the idea that many of these relics were the works of Phœnician artists ; unfortunately only two of these curious vessels have been preserved; they are now in Paris; one, the most perfect, in the collection of the Duc de Luines, the other placed by M. de Saulcy in the Louvre.

‡ 2 Kings, xxiv. 14. 16. Jeremiah, xxiv. 1.; xxix. 2.

temple of Babylon.* It is not, indeed, impossible, that some of
them may have been actually brought from the cities round Jeru-

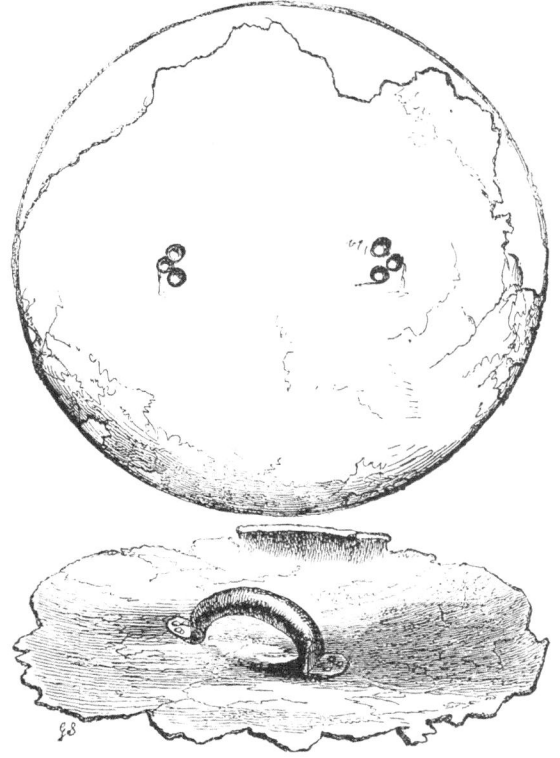

Bronze Shields, from Nimroud.

salem by Sennacherib himself, or from Samaria by Shalmaneser or
Sargon, who, we find, inhabited the palace at Nimroud, and of
whom several relics have already been discovered in the ruins.

* In ancient history, embossed or inlaid goblets are continually mentioned
amongst the offerings to celebrated shrines.   Gyges dedicated goblets, Alyattes,
a silver cup, and *an inlaid iron saucer* (the art of inlaying, having been invented, according to Herodotus, by Glaucus), and Crœsus similar vessels, in
the temple of Delphi. (Herod. i. 14. and 25. Pausanias, l. x.)   They were also
given as acceptable presents to kings and distinguished men, as we see in
2 Sam. viii. 10. and 2 Chron. ix. 23, 24.   The Lacedæmonians prepared for
Crœsus a brazen *vessel ornamented with forms of animals round the rim* (Herod.
i. 70.), like some of the bowls described in the text.   The embossings on the
Nimroud bronzes may furnish us with a very just idea of the figures and ornaments of the celebrated shield of Achilles, which were probably much the same
in treatment and execution.

Around the vessels I have described were heaped arms, remains of armour, iron instruments, glass bowls, and various objects in ivory and bronze. The arms consisted of swords, daggers, shields, and the heads of spears and arrows, which being chiefly of iron fell to pieces almost as soon as exposed to the air. A few specimens have alone been preserved, including the head of a weapon resembling a trident, and the handles of some of the swords (?), which, being partly in bronze, were less eaten away than the rest. The shields stood upright, one against the other, supported by a square piece of brick work, and were so much decayed that with great difficulty two were moved and sent to England. They are of bronze, and circular, the rim bending inwards, and forming a deep groove round the edge. The handles are of iron, and fastened by six bosses or nails, the heads of which form an ornament on the outer face of the shield.* The diameter of the largest and most perfect is 2 feet 6 inches. Although their weight must have impeded the movements of an armed warrior, the Assyrian spearmen are constantly represented in the bas-reliefs with them. Such, too, were probably the bucklers that Solomon hung on his towers.†

A number of thin iron rods, adhering together in bundles, were found amongst the arms. They may have been the shafts of arrows, which, it has been conjectured from several passages in the Old Testament, were sometimes of burnished metal. To "make bright the arrows ‡" may, however, only allude to the head fastened to a reed, or shaft of some light wood. Several such barbs, both of iron and bronze, have been found in Assyrian and Babylonian ruins, and are preserved in the British Museum.

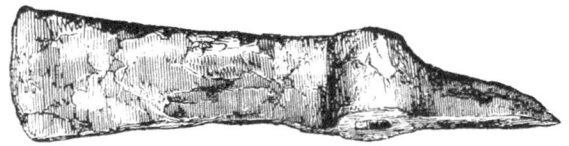

An Iron Pick, from Nimroud.

The armour consisted of parts of breast-plates (?) and of other fragments, embossed with figures and ornaments.

Amongst the iron instruments were the head of a pick, a double-

---

* Such may have been "the bosses of the bucklers" mentioned in Job, xv. 26.
† 1 Kings, x. 16, 17.; xiv. 25, 26.
‡ Jer. li. 11. Ezek. xxi. 21., and compare Isaiah, xlix. 2., where a polished *shaft* is mentioned.

handled saw (about 3 feet 6 inches in length), several objects resembling the heads of sledge-hammers, and a large blunt spear-

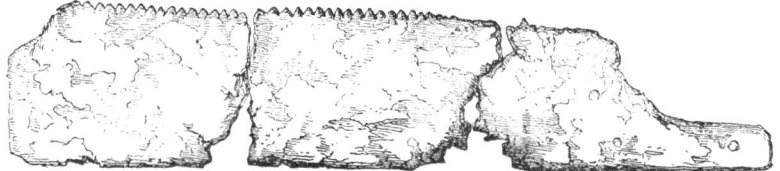

Half of a Double-handled Saw, from Nimroud.

head, such as we find from the Sculptures were used during sieges
to force stones from the walls of besieged cities.*

The most interesting of the ivory relics were, a carved staff,
perhaps a royal sceptre, part of which has been preserved, although
in the last stage of decay; and several entire elephants' tusks,
the largest being about 2 feet 5 inches long.
Amongst the smaller objects were several figures
and rosettes, and four oval bosses, with the nails
of copper still remaining, by which they were
fastened to wood or some other material.

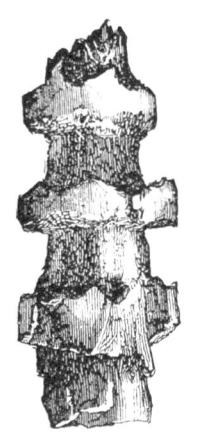

Part of Ivory Sceptre.

The ivory could with difficulty be detached
from the earth in which it was imbedded. It
fell to small fragments, and even to dust, almost
as soon as exposed to the air. Such specimens
as have been brought to this country have been
restored, and further decay checked by the same
ingenious process that was applied to the ivory
carvings first placed in the British Museum.
Parts only of the elephants' tusks have been
preserved. We find from the bas-reliefs in
the north-west palace of Nimroud †, and on
the obelisk (where captives or tribute-bearers
are seen carrying tusks), that this produce of the far East was
brought at an early period in considerable quantities to Assyria.
I have described elsewhere‡ the frequent use of ivory for the
adornment of ancient Eastern palaces and temples, as well as for

* Monuments of Nineveh, 1st series. Plate 66. All these relics are in the
British Museum.

† Id. Plate 24., where elephants' tusks are represented above the captives as
part of the spoil.

‡ Nineveh and its Remains, vol. ii. p. 420.

thrones and furniture. Ezekiel includes "horns of ivory" amongst the objects brought to Tyre from Dedan, and the Assyrians may have obtained their supplies from the same country, which some believe to have been in the Persian Gulf. *

Bronze Cubes inlaid with Gold. (Original Size.)

Amongst various small objects in bronze were two cubes, each having on one face the figure of a scarab with outstretched wings, inlaid in gold †; very interesting specimens, and probably amongst the earliest known, of an art carried in modern times to great perfection in the East.

Two entire glass bowls, with fragments of others, were also found in this chamber‡; the glass, like all that from the ruins, is covered with pearly scales, which, on being removed, leave prismatic opal-like colors of the greatest brilliancy, showing, under different lights, the most varied and beautiful tints. This is a well known effect of age, arising from the decomposition of certain component parts of the glass. These bowls are probably of the same period as the small bottle found in the ruins of the north-west palace during the previous excavations, and now in the British Museum. On this highly interesting relic is the name of Sargon, with his title of king of Assyria, in cuneiform characters, and the figure of a lion. We are, therefore, able to fix its date to the latter part of the seventh century B. C. It is, consequently, the most ancient known specimen of *transparent* glass, none from Egypt being, it is believed, earlier than the time of the Psamettici (the end of the sixth or beginning of the fifth

---

* Ezek. xxvii. 15. Ivory was amongst the objects brought to Solomon by the navy of Tharshish (1 Kings x. 22.).

† They weigh respectively 8·264 oz. and 5·299 oz., and have the appearance of weights.

‡ The larger, 5 inches in diameter, and 2¾ inches deep; the other, 4 inches in diameter, and 2½ deep.

century B. C.). Opaque colored glass was, however, manufactured at a much earlier period, and some exists of the fifteenth century, B. C. The Sargon vase was blown in one solid piece, and then shaped and hollowed out by a turning-machine, of which the marks

Glass and Alabaster Vases bearing the name of Sargon, from Nimroud.

are still plainly visible. With it were found, it will be remembered, two larger vases in white alabaster, inscribed with the name of the same king. They were all probably used for holding some ointment or cosmetic.*

With the glass bowls was discovered a rock-crystal lens, with opposite convex and plane faces. Its properties could scarcely have been unknown to the Assyrians, and we have consequently the earliest specimen of a magnifying and burning-glass.† It

---

* The height of the glass vase is $3\frac{1}{4}$ inches; of the alabaster, 7 inches. In an appendix will be found some notes by Sir D. Brewster, on the remarkable nature of the process of decomposition in the glass from Nineveh.

† I am indebted to Sir David Brewster, who examined the lens, for the following note : —" This lens is plano-convex, and of a slightly oval form, its length being $1\frac{6}{10}$ inch, and its breadth $1\frac{4}{10}$ inch. It is about $\frac{9}{10}$ths of an inch thick, and a little thicker at one side than the other. Its plane surface is pretty even, though ill polished and scratched. Its convex surface has not been ground, or polished, on a spherical concave disc, but has been fashioned on a lapidary's wheel, or by some method equally rude. The convex side is tolerably well polished, and though uneven in which it has been ground, it gives a tolerably distinct focus, at the distance of $4\frac{1}{2}$ inches from the plane side. There are about twelve cavities in the lens, that have been opened during the process of grinding it : these cavities, doubtless,

It was buried beneath a heap of fragments of beautiful blue opaque glass, apparently the enamel of some object in ivory or wood, which had perished.

In the further corner of the chamber, to the left hand, stood the royal throne. Although it was utterly impossible, from the complete state of decay of the materials, to preserve any part of it entire, I was able, by carefully removing the earth, to ascertain that it resembled in shape the chair of state of the king, as seen in the sculptures of Kouyunjik and Khorsabad, and particularly that represented in the bas-relief already described, of Sennacherib receiving the captives and spoil, after the conquest of the city of Lachish.* With the exception of the legs, which appear to have been partly of ivory, it was of wood, cased or overlaid with bronze, as the throne of Solomon was of ivory,

Fragments of Bronze Ornaments of the Throne. (Nimroud.)

overlaid with gold.† The metal was most elaborately engraved and embossed with symbolical figures and ornaments, like those

contained either naphtha, or the same fluid which is discovered in topaz, quartz, and other minerals. As the lens does not show the polarised rays at great obliquities, its plane surface must be greatly inclined to the axis of the hexagonal prism of quartz from which it must have been taken. It is obvious, from the shape and rude cutting of the lens, that it could not have been intended as an ornament; we are entitled, therefore, to consider it as intended to be used as a lens, either for magnifying, or for concentrating the rays of the sun, which it does, however, very imperfectly."

* See p. 150.

† 1 Kings, x. 18. This is a highly interesting illustration of the work in Solomon's palaces. The earliest use of metal amongst the Greeks appears also to have been as a casing to wooden objects.

embroidered on the robes of the early Nimroud king, such as winged deities struggling with griffins, mythic animals, men before the sacred tree, and the winged lion and bull. As the woodwork over which the bronze was fastened by means of small nails of the same material, had rotted away, the throne fell to pieces, but the metal casing was partly preserved. Numerous fragments of it are now in the British Museum, including the joints of the arms, and legs ; the rams' or bulls' heads, which adorned the end of the arms (some still retaining the clay and bitumen

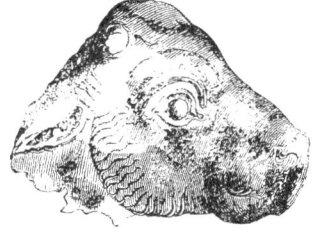
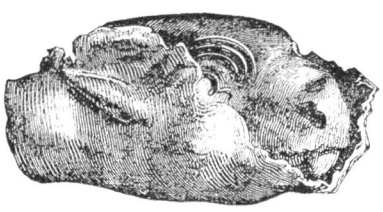

Bronze Bull's Head from Throne.   Bronze Head, part of Throne, showing bitumen inside.

with the impression of the carving, showing the substance upon which the embossing had been hammered out), and the ornamental scroll-work of the cross-bars, in the form of the Ionic volute. The legs were adorned with lion's paws resting on a pine-shaped ornament, like the thrones of the later Assyrian sculptures *, and stood on a bronze base. A rod with loose

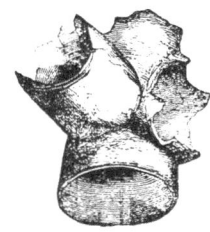

rings, to which was once hung embroidered drapery, or some rich stuff, appears to have belonged to the back of the chair, or to a frame-work raised above or behind it, though not I think, as conjectured, to a curtain concealing the monarch from those who approached him.†

In front of the throne was the foot-stool, also of wood overlaid with embossed metal, and adorned with the heads of ram or bulls. The feet ended in lion's paws and pine cones, like those of the

Bronze Binding of Joints of Throne

* I succeeded, after much trouble, in moving and packing two of these legs ; but they appear to have since fallen to pieces.

† That Eastern monarchs were, however, accustomed to conceal themselves by some such contrivances from their subjects, we know from the history of Deioces. (Herod. i. 99.) It has been even conjectured that the Hebrew word for a throne infers a veiled seat. The Assyrian kings, if we may judge from the bas-reliefs, were more accessible, and mingled more freely with their subjects.

throne. The two pieces of furniture may have been placed together in a temple as an offering to the gods, as Midas placed his throne in the temple of Delphi.* The ornaments on them were so purely Assyrian, that there can be little doubt of their having been expressly made for the Assyrian king, and not having been the spoil of some foreign nation.

Near the throne, and leaning against the mouth of the well, was a circular band of bronze, 2 feet, 4 inches in diameter, studded with nails. It appears to have been the metal casing of a wheel, or of some object of wood.

Such, with an alabaster jar †, and a few other objects in metal, were the relics found in the newly-opened room. After the examination I had made of the building during my former excavations, this accidental discovery proves that other treasures may still exist in the mound of Nimroud, and increases my regret that means were not at my command to remove the rubbish from the centre of the other chambers in the palace.

* Herod. i. 14. I need scarcely remind the reader of the frequent mention, in ancient historians, of thrones and couches ornamented with metal legs in the shape of the feet of animals.

† After my departure from Assyria, a similar alabaster jar was discovered in an adjoining chamber. Colonel Rawlinson states that the remains of preserves were found in it, and hence conjectures that the room in which the bronze objects described in this chapter were found, was a kitchen. There is nothing, however, to show that this was the case, even if the contents of the jar are such as Colonel Rawlinson supposes them to be. It is much more probable, that it was a repository for the royal arms and sacrificial vessels.

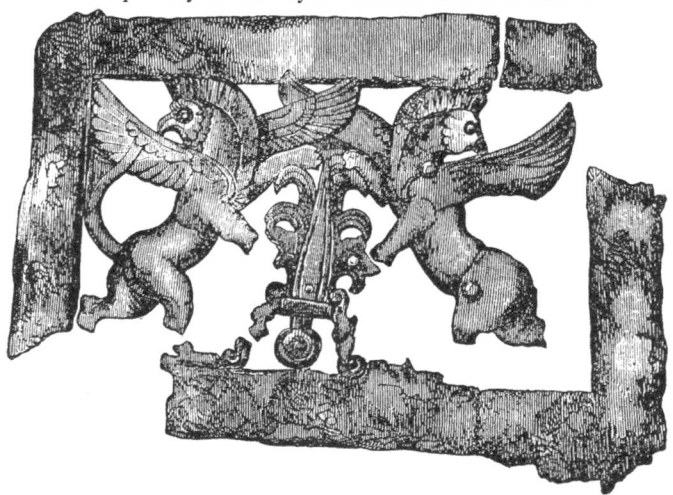

Bronze Casing from the Throne (Nimroud)

A Group of Yezidis.

CHAP. IX.

VISIT TO THE WINGED LIONS BY NIGHT. — THE BITUMEN SPRINGS. — REMOVAL
OF THE WINGED LIONS TO THE RIVER. — FLOODS AT NIMROUD — LOSS AND
RECOVERY OF LION. — YEZIDI MARRIAGE FESTIVAL. — BAAZANI. — VISIT TO
BAVIAN. — SITE OF THE BATTLE OF ARBELA. — DESCRIPTION OF ROCK-
SCULPTURES. — INSCRIPTIONS. — THE SHABBAKS.

By the 28th of January, the colossal lions forming the portal to
the great hall in the north-west palace of Nimroud were ready to
be dragged to the river-bank. The walls and their sculptured
panelling had been removed from both sides of them, and they
stood isolated in the midst of the ruins. We rode one calm cloudless night to the mound, to look on them for the last time before
they were taken from their old resting-places. The moon was at
her full, and as we drew nigh to the edge of the deep wall of
earth rising around them, her soft light was creeping over the
stern features of the human heads, and driving before it the dark
shadows which still clothed the lion forms. One by one the limbs

of the gigantic sphinxes emerged from the gloom, until the monsters were unveiled before us. I shall never forget that night, or the emotions which those venerable figures caused within me. A few hours more and they were to stand no longer where they had stood unscathed amidst the wreck of man and his works for ages. It seemed almost sacrilege to tear them from their old haunts to make them a mere wonder-stock to the busy crowd of a new world. They were better suited to the desolation around them; for they had guarded the palace in its glory, and it was for them to watch over it in its ruin. Sheikh Abd-ur-Rahman, who had ridden with us to the mound, was troubled with no such reflections. He gazed listlessly at the grim images, wondered at the folly of the Franks, thought the night cold, and turned his mare towards his tents. We scarcely heeded his going, but stood speechless in the deserted portal, until the shadows again began to creep over its hoary guardians.

Beyond the ruined palaces a scene scarcely less solemn awaited us. I had sent a party of Jebours to the bitumen springs, outside the walls to the east of the inclosure. The Arabs having lighted a small fire with brushwood awaited our coming to throw the burning sticks upon the pitchy pools. A thick heavy smoke, such as rose from the jar on the seashore when the fisherman had broken the seal of Solomon, rolled upwards in curling volumes, hiding the light of the moon, and spreading wide over the sky. Tongues of flame and jets of gas, driven from the burning pit, shot through the murky canopy. As the fire brightened, a thousand fantastic forms of light played amidst the smoke. To break the cindered crust, and to bring fresh slime to the surface, the Arabs threw large stones into the springs; a new volume of fire then burst forth, throwing a deep red glare upon the figures and upon the landscape. The Jebours danced round the burning pools, like demons in some midnight orgie, shouting their war-cry, and brandishing their glittering arms. In an hour the bitumen was exhausted for the time*, the dense smoke gradually died away, and the pale light of the moon again shone over the black slime pits.

The colossal lions were moved by still simpler and ruder means than those adopted on my first expedition. They were tilted over upon loose earth heaped behind them, their too rapid descent being checked by a hawser, which was afterwards replaced by props of wood and stone. They were then lowered, by levers and

---

* In a few hours the pits are sufficiently filled to take fire again.

jackscrews, upon the cart brought under them. A road paved with flat stones had been made to the edge of the mound, and the sculpture was, without difficulty, dragged from the trenches.

Beneath the lions, embedded in earth and bitumen, were a few bones, which, on exposure to the air, fell to dust before I could ascertain whether they were human or not. The sculptures rested simply upon the platform of sun-dried bricks without any other sub-structure, a mere layer of bitumen, about an inch thick, having been placed under the plinth.

Owing to recent heavy rains, which had left in many places deep swamps, we experienced much difficulty in dragging the cart over the plain to the river side. Three days were spent in transporting each lion. The men of Naifa and Nimroud again came to our help, and the Abou-Salman horsemen, with Sheikh Abd-ur-Rahman at their head, encouraged us by their presence. The unwieldly mass was propelled from behind by enormous levers of poplar wood; and in the costumes of those who worked, as well as in the means adopted to move the colossal sculptures, except that we used a wheeled cart instead of a sledge, the procession closely resembled that which in days of yore transported the same great figures, and which we see so graphically represented on the walls of Kouyunjik.* As they had been brought so were they taken away.

It was necessary to humor and excite the Arabs to induce them to persevere in the arduous work of dragging the cart through the deep soft soil into which it continually sank. At one time, after many vain efforts to move the buried wheels, it was unanimously declared that Mr. Cooper, the artist, brought ill luck, and no one would work until he retired. The cumbrous machine crept onwards for a few more yards, but again all exertions were fruitless. Then the Frank lady would bring good fortune if she sat on the sculpture. The wheels rolled heavily along, but were soon clogged once more in the yielding soil. An evil eye surely lurked among the workmen or the bystanders. Search was quickly made, and one having been detected upon whom this curse had alighted, he was ignominiously driven away with shouts and execrations. This impediment having been removed, the cart drew nearer to the village, but soon again came to a standstill. All the Sheikhs were now summarily degraded from their rank and honors, and a weak ragged boy having been dressed up in tawdry kerchiefs, and invested with

* See woodcut, p. 111.

a cloak, was pronounced by Hormuzd to be the only fit chief for such puny men. The cart moved forwards, until the ropes gave way, under the new excitement caused by this reflection upon the character of the Arabs. When that had subsided, and the presence of the youthful Sheikh no longer encouraged his subjects, he was as summarily deposed as he had been elected, and a greybeard of ninety was raised to the dignity in his stead. He had his turn ; then the most unpopular of the Sheikhs were compelled to lie down on the ground, that the groaning wheels might pass over them, like the car of Juggernaut over its votaries. With yells, shrieks, and wild antics the cart was drawn within a few inches of the prostrate men. As a last resource I seized a rope myself, and with shouts of defiance between the different tribes, who were divided into separate parties and pulled against each other, and amidst the deafening *tahlel* of the women, the lion was at length fairly brought to the water's edge.

The winter rains had not yet swelled the waters of the river so as to enable a raft bearing a very heavy cargo to float with safety to Baghdad. It was not until the month of April, after I had left Mosul on my journey to the Khabour, that the floods, from the melting of the snows in the higher mountains of Kurdistan, swept down the valley of the Tigris. I was consequently obliged to confide the task of embarking the sculptures to Behnan, my principal overseer, a Mosuleean stonecutter of considerable skill and experience, Mr. Vice-consul Rassam kindly undertaking to superintend the operation. Owing to extraordinary storms in the hills, the river rose suddenly and with unexampled rapidity. Mr. and Mrs. Rassam were at the time at Nimroud, and the raftmen had prepared the rafts to receive the lions. It was with difficulty that they escaped before the flood, from my house in the village to the top of the ruins. The Jaif was one vast sea, and a furious wind drove the waves against the foot of the mound. The Arabs had never seen a similar inundation, and before they could escape to the high land many persons were overwhelmed in the waters.

When the flood had subsided, the lions on the river bank, though covered with mud and silt, were found uninjured. They were speedily placed on the rafts prepared for them, but unfortunately during the operation one of them, which had previously been cracked nearly across, separated into two parts. Both sculptures were doomed to misfortune. Some person, uncovering the other during the night, broke the nose. I was unable to discover the author of this wanton mischief. He was probably a stranger, who had some feud with the Arabs working in the excavations.*

The rafts reached Baghdad in safety. After receiving the necessary repairs they floated onwards to Busrah. The waters of the Tigris throughout its course had risen far above their usual level. The embankments, long neglected by the Turkish government, had given way, and the river, bursting from its bed, spread itself over the surrounding country in vast lakes and marshes. One of the rafts was dragged into a vortex which swept through a sluice newly opened in the crumbling bank. Notwithstanding the exertions of the raftmen, aided by the crew of a boat that accompanied them, it was carried far into the interior, and left in the middle of a swamp, about a mile from the stream. The other raft fortunately escaped, and reached Busrah without accident.

For some time the stranded raft was given up for lost. Fortunately it bore the broken lion, or its recovery had probably been impossible. Captain Jones, with his usual skill and intrepidity, took his steamer over the ruined embankment, and into the unexplored morass. After great exertion, under a burning sun in the midst of summer, he succeeded in placing the two parts of the sculpture on large boats, provided for the purpose, and in conveying them to their destination.†

During my hasty visit in the autumn to Bavian, I had been unable either to examine the rock-tablets with sufficient care, or to copy the inscriptions. The lions having been moved, I seized the first leisure moment to return to those remarkable monuments.

Cawal Yusuf having invited me to the marriage of his niece at Baashiekhah, we left Nimroud early in the morning for that village, striking across the country through Tel Yakoub, Karakosh (a large village inhabited by Catholic Chaldæans, and having several churches), and Bartolli. We were met at some distance from Baashiekhah by the Cawal, followed by the principal inhabitants on horseback, and by a large concourse of people on foot, accompanied by music, and by children bringing lambs as offerings. It was already the second day of the marriage. On the previous day the parties had entered into the contract before the usual

---

* Both sculptures have, however, been completely restored in the British Museum.

† These accidents, and even still more the carelessness afterwards shown in bringing them to this country, have much injured these fine specimens of Assyrian sculpture, which now stand in a great hall of the British Museum.

witnesses, amidst rejoicings and dances. After our arrival, the
bride was led to the house of the bridegroom, surrounded by the
inhabitants, dressed in their gayest robes, and by the Cawals playing on their instruments of music. She was covered from head
to foot by a thick veil, and was kept behind a curtain in the
corner of a darkened room. Here she remained until the guests
had feasted three days, after which the bridegroom was allowed to
approach her.

The courtyard of the house was filled with dancers, and during
the day and the greater part of the night nothing was heard but
the loud signs of rejoicing of the women, and the noise of the
drum and the pipe.

On the third day the bridegroom was sought early in the morning, and led in triumph by his friends from house to house, receiving at each a trifling present. He was then placed within a
circle of dancers, and the guests and bystanders, wetting small
coins, stuck them on his forehead.* The money was collected as
it fell, in an open kerchief held by his companions under his chin.

After this ceremony a party of young men, who had attached
themselves to the bridegroom, rushed into the crowd, and carrying off the most wealthy of the guests locked them up in a
dark room until they consented to pay a ransom for their release.
This violence and restraint were cheerfully submitted to, and the
money thus collected was added to the dowry of the newly married
couple. There was feasting during the rest of the day, with raki-
drinking and music, and the usual accompaniments of an Eastern
wedding.

Leaving the revellers I rode to Baazani with Cawal Yusuf,
Sheikh Jindi (the stern leader of the religious ceremonies at Sheikh
Adi), and a few Yezidi notables, to examine the rocky valleys behind the village. I once more searched in vain for some traces of
ancient quarries from whence the Assyrians might have obtained
the slabs used in their buildings. At the entrance of one of the
deep ravines, which runs into the Gebel Makloub, a clear spring
gushes from a grotto in the hill-side. Tradition says that this is the
cave of the Seven Sleepers and their Dog, and the Yezidis have
made the spot a *ziareh,* or place of pilgrimage.†

---

* This custom of sticking coins to the forehead of a bridegroom is common
to several races of the East, amongst others to the Turcomans, who inhabit the
villages round Mosul.

† No tradition is more generally current in the East than the well known

In the sides of the same ravine are numerous excavated sepulchral chambers, with recesses or troughs in them for the reception of the dead, such as I have so frequently had occasion to describe.

Our road from Baashiekhah to Bavian lay across the rocky range of the Gebel Makloub. We found it difficult and precipitous, on the western face and scarcely practicable to laden beasts ; on the eastern, it sank gradually into a broad plain. We passed the village of Giri Mohammed Araba, built near an artificial mound of considerable size. Similar mounds are scattered here and there over the flat country, and under almost every one is a Kurdish or Arab hamlet.

A ride of seven hours brought us to the foot of the higher limestone range, and to the mouth of the ravine containing the rock-sculptures. Bavian is a mere Kurdish hamlet of five or six miserable huts on the left bank of the Ghazir. We stopped at the larger village of Khinnis ; the two being scarcely half a mile apart the place is usually called " Khinnis-Bavian." The Arab population ceases with the plains, the villages in the hills being inhabited by Kurds, and included in the district of Missouri. Adjoining Khinnis is the Yezidi district of Sheikhan.

The rock-sculptures of Bavian are the most important that have yet been discovered in Assyria.* They are carved in relief on the side of a narrow, rocky ravine, on the right bank of the Gomel, a brawling mountain torrent issuing from the Missouri hills, and one of the principal feeders of the small river Ghazir, the ancient Bumadus. The Gomel or Gomela may, perhaps, be traced in the ancient name of Gaugamela†, celebrated for that great victory which gave to the Macedonian conqueror the dominion of the Eastern world.

story of the Seven Sleepers and their Dog. There is scarcely a district without the original cave in which the youths were concealed during their miraculous slumber.

* They were first visited by the late M. Rouet, French consul at Mosul. In my Nineveh and its Remains, vol. ii. p. 142. note, will be found a short description of the sculptures by my friend Mr. Ross. These are the rock-tablets which have been recently ·described in the French papers, as a new discovery by M. Place, and as containing a series of portraits of the Assyrian kings !

† In some MSS. of Quintus Curtius, the Bumadus or Ghazir is called the " *Bumelus*," which would not be far from the modern name of the upper branch of the river. It will, of course, be remembered, that Gaugamela, according to ancient historians, signifies "a camel," as derived probably from *Gemel*, the Semitic word for that animal.

Although the battlefield was called after Arbela, a neighbouring city, we know that the river Zab intervened between them, and that the battle was fought near the village of Gaugamela, on the banks of the Bumadus or Ghazir, the Gomela of the Kurds. It is remarkable that tradition has not preserved any record of the precise scene of an event which so materially affected the destinies of the East. The history of this great battle is unknown to the present inhabitants of the country; nor does any local name, except perhaps that which I have pointed out, serve to connect it with these plains. The village, which once stood near the mound of Nimroud, was, indeed, said to have been called Dariousha, after the Persian monarch, who slept there on the night preceding the defeat that deprived him of his empire.* Some have fancied a similarity between the name of Gaugamela and that of the modern village of Karamless. The battlefield was probably in the neighbourhood of Tel Aswad, or between it and the junction of the Ghazir with the Zab, on the direct line of march to the fords of that river. We had undoubtedly crossed the very spot during our ride to Bavian. The whole of the country between the Makloub range and the Tigris is equally well suited to the operations of mighty armies, but from the scanty topographical details given by the historians of Alexander we are unable to identify the exact place of his victory. It is curious that hitherto no remains or relics have been turned up by the plough which would serve to mark the precise site of so great a battle as that of Arbela.

The principal rock-tablet at Bavian contains four figures, sculptured in relief upon the smoothed face of a limestone cliff, rising perpendicularly from the bed of the torrent. They are inclosed by a kind of frame 28 feet high by 30 feet wide, and are protected by an overhanging cornice from the water which trickles down the face of the precipice. Two deities, facing each other, are represented, as they frequently are on monuments and relics of the same period, standing on mythic animals resembling dogs. They wear the high square head-dress, with horns uniting in front, peculiar to the human-headed bulls of the later Assyrian palaces. One holds in the left hand a kind of staff surmounted by the sacred tree. To the centre of this staff is attached a ring encircling a figure,

---

* I never heard any similar tradition from the people of the country. According to the Shemutti, who inhabit the new village, the name was Darawish, i. e. the place of Dervishes. It belonged to Turcomans, who mostly died of the plague, the remainder migrating to Selamiyah.

probably that of the king. The other hand is stretched forth towards the opposite god, who carries a similar staff, and grasps in the right hand an object which is too much injured to be accurately described.* These two figures may represent but one and the same great tutelary deity of the Assyrians, as the two kings who stand in act of adoration before them are undoubtedly but one and the same king. The monarch, thus doubly portrayed, is behind the god. He raises one hand, and holds in the other the sacred mace, ending in a ball. His dress resembles that of the builder of the Kouyunjik palace, Sennacherib, with whom the inscriptions I shall presently describe, identify him. The peak projecting from the conical royal tiara is longer and more pointed than usual. The ornaments of the costumes of the four figures are rich and elaborate. The sword-scabbards end in lions, and the earrings are peculiarly elegant in design. Resting on the cornice above the sculptures, and facing the ravine, are the remains of two crouching sphinxes, probably similar in form to those at the grand entrance to the south-west palace of Nimroud.† Behind them is a narrow recess or platform in the rock.

This bas-relief has suffered greatly from the effects of the atmosphere, and in many parts the details can no longer be distinguished. But they have been still more injured by those who occupied the country after the fall of the Assyrian empire. Strangers, having no reverence for the records or sacred monuments of those who went before them, excavated in the ready-scarped rocks the sepulchral chambers of their dead.‡ In this great

* See Monuments of Nineveh, 2nd series, Plate 51. for an illustration of these rock-sculptures.

† Nineveh and its Remains, vol. i. p. 349.

‡ It is evident that these tombs are not of the Assyrian epoch, supposing even the Assyrians to have placed their dead in chambers excavated in the rocks. I have never met with rock-tombs which could be referred with any certainty to that period. In a bas-relief discovered at Khorsabad one writer (Bonomi, Nineveh and its Palaces, p. 196.) detects the representation of such excavations in a rock on which stands a castle; but I believe that houses are meant, as in a similar subject from Kouyunjik (see 2nd Series of Monuments of Nineveh, Plate 39.). It is evident that these supposed rock-tombs cannot indicate the sepulchres of the Valley of Jehoshaphat, which are of a very different period, nor, as the same writer has inferred, the city of Jerusalem. The Jews, as well as other nations of antiquity, were, however, accustomed to make such rock-chambers for their dead, as we learn from Isaiah, xxii. 16. " What hast thou here? and whom hast thou here, that thou hast hewed thee out a sepulchre here, as *he that heweth him out a sepulchre on high*, and that graveth an habitation for himself in a rock? "

P

tablet there are four such tombs. Two have been cut between the figures of the god, and have spared the sculptures. The others have destroyed the head of one king and a part of the robes of the opposite figure. The entrances to the two largest were once ornamented with columns, which have been broken away. Round the walls of these excavated chambers are the usual troughs for the bodies of the dead. I entered the tombs by means of a rope lowered from above by a party of Kurds. They were empty, their contents having, of course, been long before carried away, or destroyed.

To the left of this great bas-relief, and nearer the mouth of the

Rock Sculpture (Bavian)

ravine, is a second tablet containing a horseman at full speed, and the remains of other figures. Both horse and rider, are of colossal proportions, and remarkable for the spirit of the outline. The warrior wears the Assyrian pointed helmet, and couches a long ponderous spear, as in the act of charging the enemy. Before him is a colossal figure of the king, and behind him a deity with the horned cap; above his head a row of smaller figures of gods standing on animals of various forms, as in the rock-sculptures of Malthaiyah.

This fine bas-relief has, unfortunately, suffered even more than the other monuments from the effects of the atmosphere, and would easily escape notice without an acquaintance with its position.

Scattered over the cliff, on each side of the principal bas-reliefs, are eleven small tablets, some easily accessible, others so high up on the face of the precipice, that they are scarcely seen from below. One is on a level with the bed of the stream, and was, indeed, almost covered by the mud deposit of the floods. Each arched recess, for they are cut into the rock, contains a figure of the king, as at the Nahr-el-Kelb, near Beyrout in Syria*,

* I examined the remarkable tablets at the Nahr-el-Kelb, on my return to Europe in 1851. They were sculptured, as I stated in my first work, by Sennacherib, the king of the Bavian monuments. The only inscription partly

5 feet 6 inches high. Above his head are the sacred symbols, arranged in four distinct groups. The first group consists of three tiaras, like those worn by the gods and human-headed bulls, and of a kind of altar on which stands a staff ending in the head of a ram; the second of a crescent and of the winged disk, or globe; the third of a pedestal, on which are a trident and three staffs, one topped by a cone, another without ornament, and the last ending in two bulls' heads turned in opposite directions; and the fourth of a Maltese cross (? symbolical of the sun) and the seven stars. Some of these symbols have reference, it would seem, to the astral

Sacred Symbols or Royal Tablets (Bavian)

worship of the Assyrians; whilst others, probably, represent instruments used during sacrifices, or sacred ceremonies.

Across three of these royal tablets are inscriptions. One can be reached from the foot of the cliff, the others, being on the higher sculptures, cannot be seen from below. They are all more or less injured, but being very nearly, word for word, the same, they can to some extent be restored. I was lowered by ropes to those on the face of the precipice, which are not otherwise accessible. Standing on a ledge scarcely six inches wide, overlooking a giddy depth, and in a constrained and painful position, I had some difficulty in copying them. The stupidity and clumsiness, moreover, of the Kurds, who had never aided in such proceedings before, rendered my attempts to reach the sculptures somewhat dangerous.

preserved is unfortunately so much injured as to have hitherto defied transcription. The tablets are seven in number, and, as it is well known, are cut upon a rock near the mouth of the Nahr-el-Kelb river, adjoining three Egyptian inscriptions and bas-reliefs with the name of Rameses.

The inscriptions, the longest of which contains sixty-three lines, are in many respects of considerable importance, and have been partly translated by Dr. Hincks. They commence with an invocation to Ashur and the great deities of Assyria, the names of only eleven of whom are legible, although probably the whole thirteen are enumerated, as on the monuments from Nimroud. Then follow the name and titles of Sennacherib, Next there is an account of various great works for irrigation undertaken by this king. From eighteen districts, or villages, he declares he dug eighteen canals to the Ussur or Khusur (?), in which he collected their waters. He also dug a canal, from the borders of the town or district of Kisri to Nineveh, and brought these waters through it ; he called it the canal of Sennacherib. No traces now remain, as far as I know, of such a canal, unless the bed of the Khauser (Ussur?) was deepened by this king, and other small streams of the surrounding country led into it. Then the Ussur may mean the great ditch defending the inclosure of Kouyunjik to the east, through which the Khauser now flows. If such be the case, the canal, fed by the united streams, may have been intended for defence as well as for irrigation. Or else it may have been mainly derived from the Gomel or Ghazir, here called Ussur (?), and carried to some other part of the great city. We can then understand why the execution of this work was recorded on the rock-tablets near the source of the river. However, this part of the inscription has not yet been satisfactorily interpreted, and may hereafter be found to contain details which may help to identify the site of these artificial watercourses.

A long obscure passage precedes a very detailed account of the expedition to Babylon and Kar-Duniyas against Merodach-baladan, recorded under the first year of the annals on the Kouyunjik bulls.* After mentioning some canals which he had made in the south of Assyria, Sennacherib speaks of the army which defended the workmen being attacked by the king of Elam and the king of Babylon, with many kings of the hills and the plains who were their allies. He defeated them in the neighbourhood of Khalul (site undetermined). Many of the great people of the king of Elam and the son of the king of Kar-Duniyas were either killed or taken prisoners, while the kings themselves fled to their respective countries. Sennacherib then mentions his advance to Babylon, his conquest and plunder of it, and concludes with saying, that he brought back from that city the images of the gods which had

* See p. 140.

been taken by *Merodach-adakhe* (?), the king of Mesopotamia, from Assyria 418 years before, and put them in their places. A name imperfectly deciphered is given as that of the king of Assyria of that day. Dr. Hincks would read it Shimishti-Pal-Bithkira, but admits that the last element in particular is very doubtful. The same name is found in the inscriptions of Nimroud, as that of a predecessor of the builder of the north-west palace, as also in an inscription of the time of Tiglath Pilesar or Pul. In this place the earlier king is probably intended. Sennacherib, after his victory, appears to have transported the inhabitants of Babylon to Arakhti (? the river Araxes), but the whole passage is doubtful, owing to some important words being destroyed in the three inscriptions.

After his return from this expedition "at the *mouth* (?) of the river he had dug he set up six tablets, and beside them he put up the *full length* (?) images of the great gods."

Now, the importance of this inscription, presuming it to be correctly interpreted, will at once be perceived, for it proves almost beyond a doubt, that at that remote period the Assyrians kept an exact computation of time. We may consequently hope that sooner or later chronological tables may be discovered, which will furnish us with minute and accurate information as to the precise epoch of the occurrence of various important events in Assyrian history. It is, indeed, remarkable that Sennacherib should mark so exactly the year of the carrying away of the Assyrian gods. This very date enables us, as will hereafter be seen, to restore much of the chronology, and to place, almost with certainty, in the dynastic lists, a king whose position was before unknown.

We find also that the greater part, if not the whole, of the rock-sculptures were executed either at the end of the first, or at the beginning of the second, year of the reign of Sennacherib. As he particularly describes six tablets, it is probable that the others were added at some future period, and after some fresh victory. The mention, too, of the transportation of the inhabitants of Babylon to so remote a locality as the Araxes is highly interesting, and, if the translation of the passage may be relied on, we may perhaps trace in these colonies the origin of those Chaldæan tribes which Xenophon and Strabo describe as still, in their time, inhabiting the same region. When the whole inscription is restored we shall probably obtain many other important details which are wanting in the annals of Kouyunjik, and in the records of the same period.

Beneath the sculptured tablets, and in the bed of the Gomel, are two enormous fragments of rock, which appear to have been

Fallen Rock-Sculptures (Lavian)

torn from the overhanging cliff, and to have been hurled by some mighty convulsion of nature into the torrent below. The pent up waters eddy round them in deep and dangerous whirlpools, and when swollen by the winter rains sweep completely over them.* They still bear the remains of sculpture. One has been broken by the fall into two pieces. On them is the Assyrian Hercules strangling the lion between two winged human-headed bulls, back to back, as at the grand entrances of the palaces of Kouyunjik and Khorsabad.† Above this group is the king, worshipping between two deities, who stand on mythic animals, having the heads of eagles, the bodies and fore feet of lions, and hind legs armed with the talons of a bird of prey. The height of the whole sculpture is 24 feet, that of the winged bull 8 ft. 6 in.

* It was at this spot that Mr. Bell, the youthful artist sent out by the Trustees of the British Museum, was unfortunately drowned when bathing, in the month of July, 1851, shortly after my departure from Mosul.

† See woodcut, p. 138.

Near the entrance to the ravine the face of the cliff has been scarped for some yards to the level of the bed of the torrent. A party of Kurds were hired to excavate at this spot, as well as in other parts of the narrow valley. Remains and foundations of buildings in well-hewn stone were discovered under the thick mud deposited by the Gomel when swollen by rains. Higher up the gorge, on removing the earth, I found a series of basins cut in the

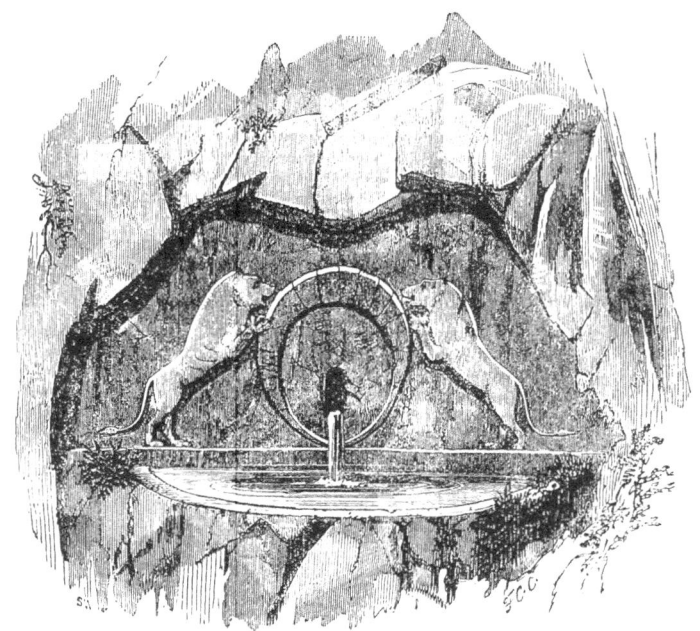

Assyrian Fountain (Bavian)

rock, and descending in steps to the stream. The water had originally been led from one to the other through small conduits, the lowest of which was ornamented at its mouth with two rampant lions in relief. These outlets were choked up, but we cleared them, and by pouring water into the upper basin restored the fountain as it had been in the time of the Assyrians.

From the nature and number of the monuments at Bavian, it would seem that this ravine was a sacred spot, devoted to religious ceremonies and to national sacrifices. When the buildings, whose remains still exist, were used for these purposes, the waters must nave been pent up between quays or embankments. They now

occasionally spread over the bottom of the valley, leaving no pathway at the foot of the lofty cliffs. The remains of a well-built
raised causeway of stone, leading to Bavian from the city of Nineveh, may still be traced across the plain to the east of the Gebel
Makloub.

The place, from its picturesque beauty and its cool refreshing
shade even in the hottest day of summer, is a grateful retreat, well
suited to devotion and to holy rites. The brawling stream almost
fills the bed of the narrow ravine with its clear and limpid waters.
The beetling cliffs rise abruptly on each side, and above them tower
the wooded declivities of the Kurdish hills. As the valley opens
into the plain, the sides of the limestone mountains are broken into
a series of distinct strata, and resemble a vast flight of steps leading up to the high lands of central Asia. The banks of the torrent are clothed with shrubs and dwarf trees, amongst which are the
green myrtle and the gay oleander, bending under the weight of
its rosy blossoms.

I remained two days at Bavian to copy the inscriptions, and to
explore the Assyrian remains. Hannah the overseer, with a party
of poor Nestorians, who, driven by want from the district of
Tkhoma, chanced to pass through the valley, was left to clear
away the earth from the lower monuments, and to excavate
amongst the ruins. No remains were discovered; and after working for a few days without results, they came to Mosul.

Wishing to visit the Yezidi chiefs, I took the road to Ain Sifni,
passing through two large Kurdish villages, Atrush and Om-es-sukr,
and leaving the entrance to the valley of Sheikh Adi to the right.
The district to the north-west of Khinnis is partly inhabited by a
tribe professing peculiar religious tenets, and known by the name
of Shabbak. Although strange and mysterious rites are, as usual,
attributed to them, I suspect that they are simply the descendants
of Kurds, who emigrated at some distant period from the Persian slopes of the mountains, and who still profess Sheeite doctrines. They may, however, be tainted with Ali-Illahism.* Their
chief, with whom I was acquainted, resides near Mosul.

* A creed professed by several tribes in Kurdistan and Louristan, and by
some of the inhabitants of the northern part of the Lebanon range in Syria. It
consists mainly in the belief, that there have been successive incarnations of the
Deity, the principal having been in the person of Ali, the celebrated son-in-law
of the prophet Mohammed. The name usually given them, Ali-Illahi, means
"believers that Ali is God." Various abominable rites have been attributed to
them, as to the Yezidis, Ansyris, and all sects whose doctrines are not known to
the surrounding Mussulman or Christian population.

We passed the night in the village of Esseeyah, where Sheikh Nasr had recently built a dwelling-house.   I occupied the same room with the Sheikh, Hussein Bey, and a large body of Yezidi Cawals, and was lulled to sleep by an interminable tale, about the prophet Mohammed and a stork, which, when we had all lain down to rest, a Yezidi priest related with the same soporific effect upon the whole party.   On the following day I hunted gazelles with Hussein Bey, and was his guest for the night at Baadri, returning next morning to Mosul.

Hussein Bey the Chief of the Yezidis, and his Brother

The Author's House at Nimroud

CHAP. X.

THE mound of Kalah Sherghat having been very imperfectly examined during my former residence in Assyria*, I had made arrangements to return to the ruins. All my preparations were complete by the 22nd of February, and I floated down the Tigris on a raft laden with provisions and tools necessary for at least a month's residence and work in the desert. I had expected to find Mohammed Seyyid, one of my Jebour Sheikhs, with a party of the Ajel, his own particular tribe, ready to accompany me. The Bedouins, however, were moving to the north, and their horsemen had already been seen in the neighbourhood of Kalah Sherghat. Nothing would consequently induce the Ajel, who were not on the best terms with the Shammar Arabs, to leave their tents, and, after much useless discussion, I was obliged to give up the journey.

Awad, with a party of Jehesn, had been for nearly six weeks

* Nineveh and its Remains, vol. ii. chap. 12.

exploring the mounds in the plan of Shomamok, the country of the Tai Arabs, and had sent to tell me that he had found remains of buildings, vases, and inscribed bricks. I determined, therefore, to make use of the stores collected for the Kalah Sherghat expedition by spending a few days in inspecting his excavations, and in carefully examining those ruins which I had only hastily visited on my previous journey. I accordingly started from Nimroud on the 2nd of March, accompanied by Hormuzd, the doctor, and Mr. Rolland. We descended the Tigris to its junction with the Zab, whose waters, swollen by the melting of the snows in the Kurdish mountains, were no longer fordable. Near the confluence of the streams, and on the southern bank of the Zab, is the lofty mound of Keshaf. This artificial platform of earth and unbaked bricks rests upon a limestone rock, projecting abruptly from the soil. Its summit is crowned by a stone wall, with an arched gateway facing the south — the remains of a deserted fort, commanding the two rivers. It was garrisoned a few years ago by an officer and a company of irregular troops from Baghdad, who were able from this stronghold to check the inroads of the Bedouins, as well as of the Tai and other tribes, who plundered the Mosul villages. Since it has been abandoned, the country has again been exposed to the incursions of these marauders, who now cross the rivers unmolested, and lay waste the cultivated districts. I could find no relics of an early date, nor did subsequent excavations lead to their discovery. The mound is, nevertheless, most probably of Assyrian origin. From the remotest period the importance of the position, at the confluence of two great rivers, must have led to the erection of a castle on this spot.

The tents of the Howar were about five miles from Keshaf. Since my last visit, he had received his cloak of investiture as Sheikh from the Pasha of Kerkouk*, and was once more the acknowledged chief of the Tai. Faras had, however, withdrawn from his rival, and, followed by his own adherents, had moved to the banks of the Lesser Zab. The Shammar Be-

---

* The great pashalic of Baghdad, formerly one of the most important and wealthy in the Turkish empire, and the first in rank, had recently been divided into several distinct governments. It once extended from Diarbekr to the Persian Gulf, and was first curtailed about fifteen years ago, when Diarbekr and Mosul were placed under independent pachas. Lately it has been reduced to the districts surrounding the city, with the Arab tribes who encamp in the neighbourhood; Kerkouk, Suleimaniyah, and Busrah being formed into separate governments. In this new division the Tai were included within the pashalic of Kerkouk.

douins, encouraged by the division in the tribe, had, only three days before our visit, crossed the Tigris and fallen suddenly upon the Kochers, or Kurdish wanderers, of the Herki clans. These nomades descend annually from the highest mountain regions to winter in the rich meadows of Shomamok. They pay a small tribute to the Tai for permission to pasture their flocks, and for protection against the desert Arabs. The Howar was consequently bound to defend them, and had sent Saleh, with his horsemen, to meet the Shammar. They had been beaten, and had lost forty of their finest mares. The Kurds appear to have little courage when attacked by the Bedouins in the plains, although they can oppose the rifle to the simple spear. A large number of them had been slain, and several thousand of their sheep and cattle had been driven across the Tigris.

We found the Howar much cast down and vexed by his recent misfortunes. The chiefs of the tribe were with him, in gloomy consultation over their losses. A Bedouin, wrapped in his ragged cloak, was seated listlessly in the tent. He had been my guest the previous evening at Nimroud, and had announced himself on a mission from the Shammar to the Tai, to learn the breed of the mares which had been taken in the late conflict. His message might appear, to those ignorant of the customs of the Arabs, one of insult and defiance. But he was on a common errand, and although there was blood between the tribes, his person was as sacred as that of an ambassador in any civilised community. Whenever a horse falls into the hands of an Arab, his first thought is how to ascertain its descent. If the owner be dismounted in battle, or if he be even about to receive his death-blow from the spear of his enemy, he will frequently exclaim, " O Fellan ! (such a one) the mare that fate has given to you, is of noble blood. She is of the breed of Sallawiyah and her dam is ridden by Awaith, a sheikh of the Fedhan" (or as the case may be). Nor will a lie come from the mouth of a Bedouin as to the race of his mare. He is proud of her noble qualities, and will testify to them as he dies. After a battle or a foray, the tribes who have taken horses from the enemy will send an envoy to ask their breed, and a person so chosen passes from tent to tent unharmed, hearing from each man, as he eats his bread, the descent and qualities of the animal he may have lost.

Amongst men who attach the highest value to the pure blood of their horses, and who have no written pedigree, for amongst the Bedouins documents of this kind do not exist, such customs are necessary. The descent of a horse is preserved by tradition, and the birth of a colt is an event known to the whole tribe. If a townsman or stranger buy a horse, and is desirous of having written evidence of its race, the seller, with his friends, will come to the nearest town to testify before a person specially qualified to take the evidence, called " the cadi of the horses," who makes out a written pedigree, accompanied by various prayers and formularies from the Koran used on such occasions, and then affixes to it his seal. It would be considered disgraceful to the character of a true Bedouin to give false testimony on such an occasion, and his word is usually received with implicit confidence.

The morning following our arrival at the tents of the Howar was ushered in by a heavy rain. I thought this a good opportunity of visiting the ruins of Mokhamour, as the Bedouins rarely leave their tents on plundering expeditions in wet weather. None of the Tai, however, would accompany me. They still dreaded the Shammar, and the Howar loudly protested against the rashness of venturing alone into the plains so recently overrun by the enemy. Awad professed to know the road, and accompanied by Hormuzd and Mr. R., I struck across the low hills under his guidance.

These ruins, of which I had so frequently received exaggerated descriptions from the Arabs, are in the deserted district between the Karachok range and the river Tigris. The plains in which they are situated are celebrated for the richness of their pastures, and are sought in spring by the Tai and the Kurdish Kochers. Even as early as the time of our visit the face of the country is usually covered with their flocks and herds. But the dread of the Shammar had now scared them from the banks of the river, and they had migrated to the inland meadows, further removed from the forays of the Bedouins. From the tents of Howar, on the low undulating hills forming the northern spur of the Karachok, to Mokhamour, a distance of some fifteen miles, we did not see a single human being.

We kept as much as possible in the broken country at the foot of the mountain to escape observation. The wooded banks of the Tigris and the white dome of the tomb of Sultan Abdallah were faintly visible in the distance, and a few artificial mounds rose in the plains. The pastures were already fit for the flocks, and luxuriant grass furnished food for our horses amidst the ruins.

The principal mound of Mokhamour is of considerable height,

and ends in a cone. It is apparently the remains of a platform built of earth and sun-dried bricks, originally divided into several distinct stages or terraces. On one side are the traces of an inclined ascent, or of a flight of steps, once leading to the summit. It stands in the centre of a quadrangle of lower mounds, about 480 paces square. I could find no remains of masonry, nor any fragments of inscribed bricks, pottery, or sculptured alabaster.

The ruins are near the southern spur of Karachok, where that mountain, after falling suddenly into low broken hills, again rises into a solitary ridge, called Bismar, stretching to the Lesser Zab, Mokhamour being between the two rivers. These detached limestone ridges, running parallel to the great range of Kurdistan, such as the Makloub, Sinjar, Karachok, and Hamrin, are a peculiar feature in the geological structure of the country lying between the ancient province of Cilicia and the Persian Gulf. Hog-backed in form, they have an even and smooth outline when viewed from a distance, but are really rocky and rugged. Their sides are broken into innumerable ravines, producing a variety of purple shadows, ever changing and contrasting with the rich golden tint of the limestone, and rendering these solitary hills, when seen from the plain, objects of great interest and beauty.* They are, for the most part, but scantily wooded with a dwarf oak, and that only on the eastern slope; their rocky sides are generally, even in spring, naked and bare of all vegetation. Few springs of fresh water being found in them, they are but thinly inhabited. In the spring months, when the rain has supplied natural reservoirs in the ravines, a few wandering Kurdish tribes pitch their tents in the most sheltered spots.

Having examined the ruins, taken bearings of the principal landmarks, and allowed our horses to refresh themselves in the high grass, I returned to the encampment of the Tai. As we rode back we spied in the desert three horses, which had been probably left by the Bedouins in their retreat, and were now quietly grazing in the pastures. After many vain efforts we succeeded in driving them before us, and on our arrival at the

---

* I take this opportunity of mentioning, with the praise it most fully deserves as a work of art, the Panorama of Nimroud, painted and exhibited by Mr. Burford, in which the Karachok and Makloub are introduced. The tints produced by the setting sun on those hills are most faithfully portrayed, and the whole scene, considering the materials from which the artist worked, is a proof of his skill as a painter, and of his feeling for Eastern scenery.

tents I presented them in due form to the Howar, who was rewarded, by this unexpected addition to his stud, for the alarm he
declared he had felt for our safety during our absence.   A ride
of three hours next morning, across the spurs of the Karachok,
brought us to the ruins of Abou-Jerdeh, near which we had
found the tents of Faras on our last visit.   The mound is of considerable size, and on its summit are traces of foundations in stone
masonry; but I could find no remains to connect it with the
Assyrian period.   The eastern base is washed by a small stream
coming from the Kordereh.

We breakfasted with our old host Wali Beg, and then continued our journey to one of the principal artificial mounds of
Shomamok, called the "Kasr," or palace.   The pastures were
covered with the flocks of the Arabs, the Kochers, and the Disdayi Kurds.   A broad and deep valley, or rather gully, worn by
a sluggish stream in the alluvial soil, crosses the plain.   The
stranger is not aware of its existence until he finds himself actually
on the brink of the lofty precipices which hem it in on both sides.
Then a long, narrow meadow of the brightest emerald green,
studded with flocks and tents, opens beneath his feet.   We crossed
this valley, called the Kordereh, and encamped for the night at the
foot of the Kasr, on the banks of a rivulet called As-surayji, which
joins the Kordereh below Abou-Jerdeh, near a village named
"Salam Aleik," or "Peace be with you."

The mound is both large and lofty, and is surrounded by the
remains of an earthen embankment.   It is divided almost into
two distinct equal parts by a ravine or watercourse, where an
ascent probably once led from the plain to the edifice on the
summit of the platform.   Above the ruins of the ancient buildings
stood a modern fort, generally garrisoned by troops belonging to
the Mutesellim of Arbil.   It was afterwards inhabited by some
families of the Jehesh tribe, who were driven away by the exactions of the chiefs of the Tai.   Awad had opened several deep
trenches and tunnels in the mound, and had discovered chambers,
some with walls of plain sundried bricks, others panelled round
the lower part with slabs of reddish limestone, about $3\frac{1}{2}$ or 4
feet high.   He had also found inscribed bricks, with inscriptions declaring that Sennacherib had here built a city, or rather palace, for
the name of which, written ⌇— ⟶⟨⟨, I cannot suggest a reading.

I observed a thin deposit, or layer, of pebbles and rubble above
the remains of the Assyrian building, and about eight feet beneath

the surface, as at Kouyunjik. It may probably have been the flooring or foundation of some edifice of a more recent date raised above the buried palaces. I could discover no traces whatever of alabaster in the ruins, although the material is common in the neighbourhood, nor could I find the smallest fragment of sculptured stone which might encourage a further search after bas-reliefs or inscriptions.

From the summit of the Kasr of Shomamok I took bearings of twenty-five considerable mounds, the remains of ancient Assyrian population *; the largest being in the direction of the Lesser Zab. Over the plain, too, were thickly scattered villages, surrounded by cultivated fields, and belonging to a tribe of Kurds called Disdayi, who move with their flocks and tents to the pastures during spring, and return to their huts in the summer to gather in the harvest and to till the soil.

Wishing to examine several ruins in the neighbourhood I left our tents early on the following morning, and rode to the mound of Abd-ul-Azeez, about eight or nine miles distant, and on the road between Baghdad and Arbil. The latter town, with its castle perched upon a lofty artificial mound, all that remains of the ancient city of Arbela, which gave its name to one of the greatest battles the world ever saw, was visible during the greater part of our day's ride. The plain abounds in villages and canals for irrigation, supplied by the As-Surayji. When the land is too high to be watered by the usual open conduits, the villagers cut subterranean passages like the Persian *Kanduks*, which are frequently at a considerable depth under ground, and are open to the air at certain regular distances by shafts sunk from above. The soil thus irrigated produces cotton, rice, tobacco, millet, melons, cucumbers, and a few vegetables. The jurisdiction of the Tai Sheikh ends at the Kasr; the villages beyond are under the immediate control of the governor of Arbil, to whom they pay their taxes. The inhabitants complained loudly of oppression, and appeared to be an active, industrious race. Upon the banks of the Lesser Zab, below Altun Kupri (or Guntera, the "Bridge," as the Arabs call the place), encamp the Arab tribe of Abou-Hamdan, renowned for the beauty of its women.

The mounds I examined, and particularly that of Abd-ul-Azeez, abound in sepulchral urns and in pottery, apparently not Assyrian.

* The names of the principal are Tel-el-Barour, Abbas, Kadreeyah, Abd-ul-Azeez, Baghurtha, Elias Tuppeb, Tarkheena, and Doghan.

The most remarkable spot in the district of Shomamok is the Gla (an Arab corruption of Kalah), or the Castle, about two miles distant from the Kasr. It is a natural elevation, left by the stream of the Kordereh, which has worn a deep channel in the soil, and dividing itself at this place into two branches forms an island, whose summit, but little increased by artificial means, is, therefore, nearly on a level with the top of the opposite precipices. The valley may be in some places about a mile wide, in others only four or five hundred yards. The Gla is consequently a natural stronghold, above one hundred feet high, furnished on all sides with outworks, resembling the artificial embankments of a modern citadel. A few isolated mounds near it have the appearance of detached forts, and nature seems to have formed a complete system of fortification. I have rarely seen a more curious place.

There are no remains of modern habitations on the summit of the Gla, which can only be ascended without difficulty from one side. Awad excavated by my directions in the mound, and discovered traces of Assyrian buildings, and several inscribed bricks, bearing the name of Sennacherib, and of a castle or palace, ⛏𝕀 ⤬, which, like that on the bricks from the Kasr, I am unable to interpret. It is highly probable that a natural stronghold, so difficult of access, almost impregnable before the use of artillery, should have been chosen at a very early period for the site of a castle. Even at this day it might become a position of some importance, especially as a check upon the Arabs and Kurds, who occasionally lay waste these rich districts. Numerous valleys, worn by the torrents, descending from the Karachok hills, open into the Kordereh. They have all the same character, deep gulleys, rarely more than half a mile in width, confined between lofty perpendicular banks, and watered during summer by small sluggish rivulets. These sheltered spots furnish the best pastures, and are frequented by the Disdayi Kurds, whose flocks were already scattered far and wide over their green meadows.

From the Gla I crossed the plain to the mound of Abou Sheetha, in which Awad had excavated for some time without making any discovery of interest. Near this ruin, perhaps at its very foot, must have taken place an event which led to one of the most celebrated episodes of ancient history. Here were treacherously seized Clearchus, Proxenus, Menon, Agias, and Socrates; and Xenophon, elected to the command of the Greek auxiliaries, commenced the ever-memorable retreat of the Ten Thousand. The

Q

camp of Tissaphernes, dappled with its many-colored tents, and glittering with golden arms and silken standards, the gorgeous display of Persian pomp, probably stood on the Kordereh, between Abou-Sheetha and the Kasr. The Greeks having taken the lower road, to the west of the Karachok range, through a plain even then as now a desert*, turned to the east, and crossed the spur of the mountain, where we had recently seen the tents of the Howar, in order to reach the fords of the Zab. I have already pointed out the probability of their having forded that river above the junction of the Ghazir †, and to this day the ford to the east of Abou-Sheetha is the best, and that usually frequented by the Arabs. Still not openly molested by the Persians, the Greeks halted for three days on the banks of the stream, and Clearchus, to put an end to the jealousies which had broken out between the two armies, sought an interview with the Persian chief. The crafty Eastern, knowing no policy but that to which the descendants of his race are still true, inveigled the Greek commanders into his power, and having seized them sent them in chains to the Persian monarch. He then put to death many of their bravest companions and soldiers, who had accompanied their chiefs. The effect which this perfidious act had on the Greek troops, surrounded by powerful enemies, wandering in the midst of an unknown and hostile country, betrayed by those they had come so far to serve, and separated from their native land by impassable rivers, waterless deserts, and inaccessible mountains, without even a guide to direct their steps, is touchingly described by the great leader and historian of their retreat: " Few ate anything that evening, few made fires, and many that night never came to their quarters, but laid themselves down, every man in the place where he happened to be, unable to sleep through sorrow and longing for their country, their parents, their wives, and children, whom they never expected to see again." But there was one in the army who was equal to the difficulties which en-

---

* Anab. b. ii. c. 4. It is remarkable that Xenophon does not mention the Lesser Zab, which he crossed near its junction with the Tigris. The Greeks must have followed the road indicated in the text, and not that to the *east* of the Karachok, now the highway between the two rivers, as Xenophon particularly mentions that the Tigris was on his left, and that he saw, at the end of the first day's journey, on its opposite bank, a considerable city named Cænæ, which must be identified with Kalah-Sherghat, as there are no other ruins to mark the site of a large place, and no open ground below it upon which one could have stood. The distance of twenty parasangs, or five days' journey, agrees very accurately with this route.

† See p. 61.

compassed them, and who had resolved to encourage his hopeless countrymen to make one great effort for their liberty and their lives. Before the break of day, Xenophon had formed his plans. Dressed in the most beautiful armour he could find, " for he thought if the gods granted him victory these ornaments would become a conqueror, and if he were to die they would decorate his fall," he harangued the desponding Greeks, and showed them how alone they could again see their homes. His eloquence and courage gave them new life. Having made their vows to the eternal gods, and singing pæans, they burnt their carriages, tents, and superfluous baggage, and prepared for the last great struggle. The sun must have risen in burning splendor over the parched and yellow plains of Shomamok, for it was early in the autumn. The world has rarely seen a more glorious sight than was witnessed on the banks of the Zab on that memorable morning. The Ten Thousand, having eaten, were permitted by the enemy, who were probably unprepared for this earnest resistance, to ford the river. Reaching the opposite bank they commenced that series of marches, directed with a skill and energy unequalled, which led them through difficulties almost insurmountable to their native shores.

Near Abou-Sheetha, too, Darius, a fugitive, urged his flying horses through the Zab, followed by the scattered remnants of an army which numbered in its ranks men of almost every race and clime of Asia. A few hours after, the Macedonian plunged into the ford in pursuit of the fallen monarch, at the head of those invincible legions which he was to lead, without almost a second check, to the banks of the Indus. The plains which stretch from the Zab below Abou-Sheetha have since been more than once the battlefield of Europe and Asia.

I gazed with deep interest upon the scene of such great events — a plain, where nothing remains to tell of the vast armies which once moved across it, of European valour, or of Eastern magnificence.

We had expected to find a raft ready for us near Abou-Sheetha. The raftmen, however, having chosen a more convenient place nearer Negoub, we had to follow the windings of the river for some miles, crossing the mouth of the Kordereh, which joins it five or six miles below Abou-Sheetha. Whilst riding through the jungle a wolf rose before me from its lair, and ran towards the plain. Following the animal, I wounded it with one barrel of my pistol, and was about to discharge the second, when my horse slipt on some wet straw left by a recent encampment, and we fell together upon the wolf. It struggled and freed itself, leaving me besmeared with its blood. The cock of the pistol fortunately broke in going off whilst the muzzle was close to my head, and I escaped without other injury than a bruised hand, the complete use of which I did not recover for some months.

On my return to Nimroud, I remained there a few days to give directions to the overseers for continuing the work during a prolonged absence which I meditated in the desert. On a level with the north-west palace, and on the south side of the high pyramidal mound, some chambers, ornamented with sculptures, had already been discovered, and it was chiefly in this part of the ruins that the excavations were now carried on; but I will defer an account of the remarkable monuments existing there until I can describe the entire building from which the earth was removed during our trip to the Khabour.

At Kouyunjik several new chambers had been opened. The western portal of the great hall, whose four sides were now completely uncovered*, led into a long narrow chamber (eighty-two feet by twenty-six), the walls of which had unfortunately been almost entirely destroyed.† On such fragments, however, as remained were traces of the usual subjects, — battles and victories. There was nothing remarkable in the dresses of the captives, or in the details, to give any clue to the conquered people, whose country was simply represented by wooded mountains and a broad river.

In the chamber beyond‡ a few slabs were still standing in their original places. In length this room was the same as that parallel to it, but in breadth it was only eighteen feet. The bas-reliefs represented the siege and sack of one of the many cities taken by the great king, and the transfer of its captives to some distant province of Assyria. The prisoners were dressed in garments falling to the calves of their legs, and the women wore a kind of turban. Although the country was mountainous, its inhabitants used the camel as a beast of burden, and in the sculptures it was represented laden with the spoil. The Assyrians, as was their custom, carried away in triumph the images of the gods of the conquered nation, which were placed on poles and borne in procession on men's shoulders. " Hath any god of the nations delivered his land out of the hand of the king of Assyria?" exclaimed the Assyrian general to the Jews. " Where are the Gods of Hamath and Arphad? where are the gods of

* No. vi. Plan 1.　　† No. ix. Same Plan.　　‡ No. x. Same Plan.

Sepharvaim?"* They had been carried away with the captives, and the very idols that were represented in this bas-relief may be amongst those to which Rabshakeh made this boasting allusion. The captured gods were three, a human figure with outstretched arms, a lion-headed man carrying a long staff in one hand, and an image inclosed by a square frame. Within a fortified camp, defended by towers and battlements, the priests were offering up the sacrifices usual upon a victory; the pontiff was distinguished by a high conical cap, and, as is always the case in the Assyrian sculptures, was beardless. By his side stood an assistant. Before the altar, on which were some sacrificial utensils, was the sacred chariot, with its elaborate yoke. On a raised band, across the centre of the castle, was inscribed the name and titles of Sennacherib.†

On the northern side of the great hall the portal formed by the winged bulls, and the two smaller doorways guarded by colossal winged figures, led into a chamber one hundred feet by twenty-four, which opened into a further room of somewhat smaller dimensions.‡ In the first, a few slabs were still standing, to show that on the walls had been represented some warlike expedition of the Assyrian king, and, as usual, the triumphant issue of the campaign. The monarch, in his chariot, and surrounded by his bodyguards, was seen receiving the captives and the spoil in a hilly country, whilst his warriors were dragging their horses up a steep mountain near a fortified town, driving their chariots along the banks of a river, and slaying with the spear the flying enemy. §

The bas-reliefs, which had once ornamented the second chamber, had been still more completely destroyed. A few fragments proved that they had recorded the wars of the Assyrians with a maritime people, whose overthrow was represented on more than one sculptured wall in the palace, and who may probably be identified with some nation on the Phœnician coast conquered by Sennacherib, and mentioned in his great inscriptions. Their galleys, rowed by double banks of oarsmen, and the high conical headdress of their women, have already been described.‖ On the best preserved slab was the interior of a fortified camp, amidst mountains. Within the walls were tents whose owners were en-

* Isaiah, xxxvi. 18, 19.
† Plate 50. 2d series of Monuments of Nineveh.
‡ Nos. vii. and viii. Plan 1.
§ Plate 29. of 2d series of Monuments of Nineveh.
‖ Nineveh and its Remains, vol. ii. p. 128.

Q 3

gaged in various domestic occupations, cooking in pots placed on stones over the fire, receiving the blood of a slaughtered sheep in a jar, and making ready the couches. Warriors were seated before a table, with their shields hung to the tent-pole above them. This bas-relief may confirm what I have elsewhere stated, that the Assyrians were accustomed to dwell in tents within the walls of their cities, as a portion of the inhabitants of many Eastern towns still do ; though it is more probable that, in this sculpture, a forti-fied camp is intended by the turretted ground-plan.*

To the south of the palace, but part of the same great building, though somewhat removed from the new excavations, and adjoining those formerly carried on, an additional chamber had been opened, in which several bas-reliefs of considerable interest had been discovered.†

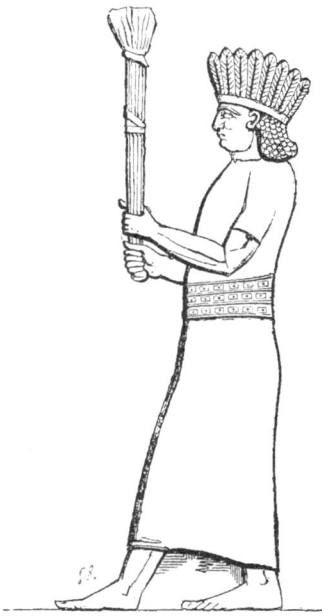

A captive (of the Tckkari ?) Kouyunjik.

Its principal entrance, facing the west, was formed by a pair of colossal human-headed lions, carved in coarse lime-stone, so much injured that even the inscriptions on the lower part of them were nearly illegible. Unfortunately the bas-reliefs were equally mutilated, four slabs only retaining any traces of sculpture. One of them represented Assyrian warriors leading captives, who differed in costume from any other conquered people hitherto found on the walls of the palaces. Their head-dress consisted of high feathers, forming a kind of tiara like that of an Indian chief, and they wore a robe confined at the waist, by an ornamented girdle. Some of them carried an object re-sembling a torch. Amongst the ene-mies of the Egyptians represented on their monuments is a tribe similarly attired. Their name has been read Tokkari, and they have been identified with an Asiatic

---

* Nineveh and its Remains, vol. ii. p. 243. It was first suggested by a recent writer on Nineveh, and, I think, for good reasons, that these ground-plans of fortifications in the bas-reliefs represent a fortified camp, and not a city. (" Assyria, her Manners and Customs, &c.," p. 327., by Mr. Goss,—a work the general accuracy of which I take the opportunity of acknowledging).

† No. xxii. Plan 1. Some of the slabs had been originally sculptured on

nation. We have seen that in the inscriptions on the bulls, the Tokkari are mentioned amongst the people conquered by Sennacherib *, and it is highly probable that the captives in the

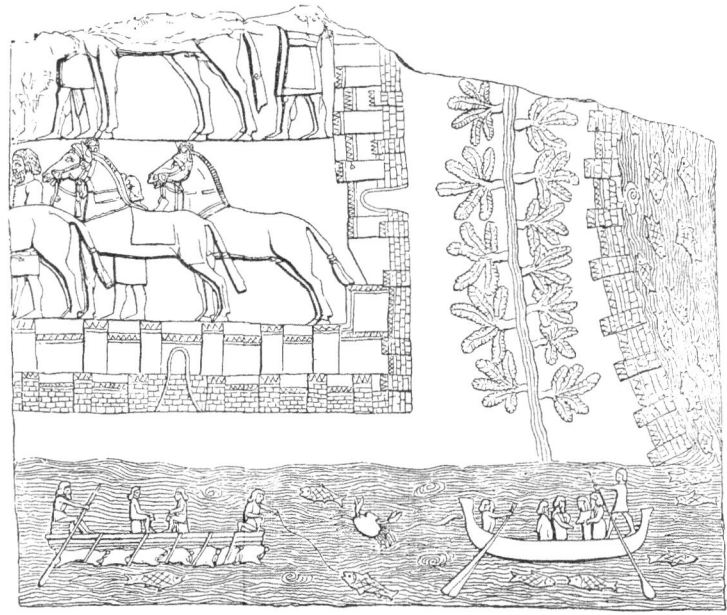

Bas-relief from Kouyunjik representing fortified City, a River with a Boat and Raft, and a Canal.

bas-reliefs I am describing belonged to them. Unfortunately no epigraph, or vestige of an inscription, remained on the sculptures themselves, to enable us to identify them.†

On a second slab, preserved in this chamber, was represented a double-walled city with arched gateways, and inclined approaches leading to them from the outer walls. Within were warriors with horses; outside the fortifications was a narrow stream or canal, planted on both sides with trees, and flowing into a broad river, on which were large boats, holding several persons, and a raft of skins,

the face now turned to the wall of sundried bricks, but they had not, I think, been brought from any other building. The style of sculpture was similar to that on the walls of Kouyunjik, and it is most probable that some error having been made in the bas-relief, it was destroyed, and the opposite face carved afresh.

* See p. 146.

† Plate 44. 2d series of Monuments of Nineveh.

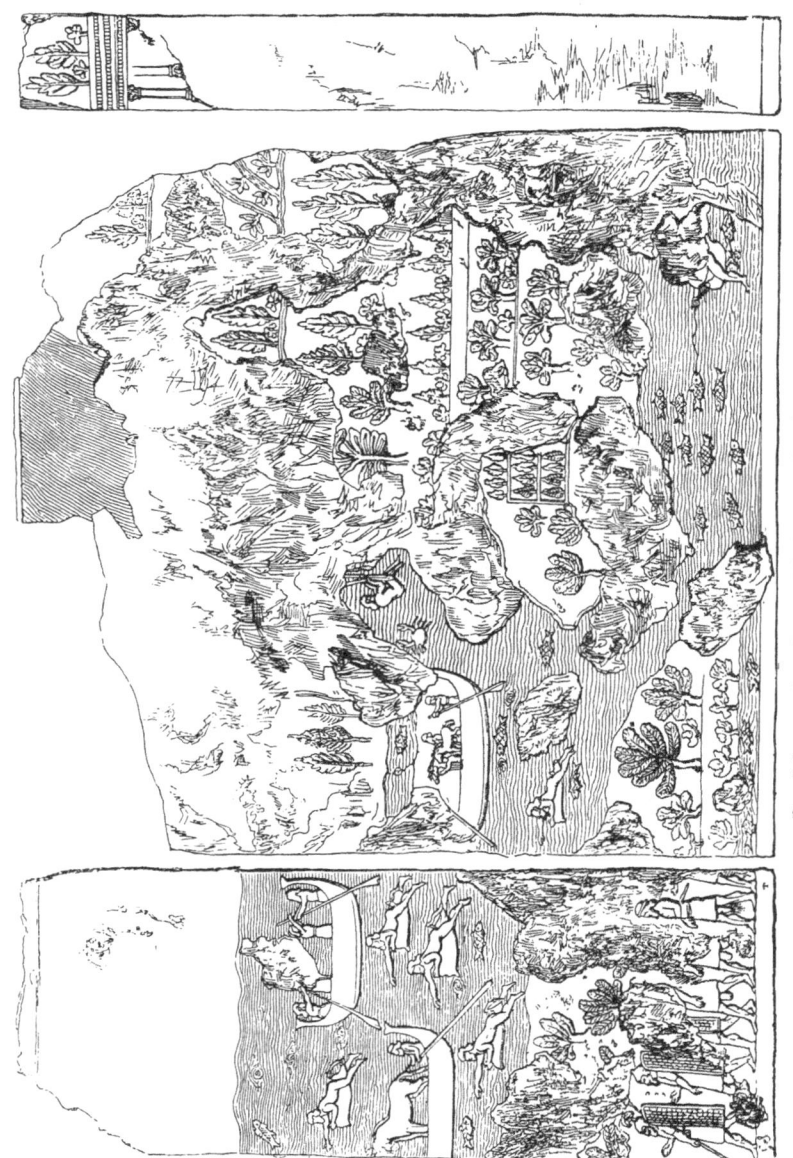

Bas-relief representing a River, and Gardens watered by Canals (Kouyunjik).

bearing a man fishing, and two others seated before a pot or caldron. Along the banks, and apparently washed by the stream, was a wall with equidistant towers and battlements. On another part of the same river were men ferrying horses across the river in boats, whilst others were swimming over on inflated skins. The water swarmed with fish and crabs. Gardens and orchards, with various kinds of trees, appeared to be watered by canals similar to those which once spread fertility over the plains of Babylonia, and of which the choked-up beds still remain. A man, suspended by a rope, was being lowered into the water. Upon the corner of a slab almost destroyed, was a hanging garden, supported upon columns, whose capitals were not unlike those of the Corinthian order. This representation of ornamental gardens was highly curious. It is much to be regretted that the bas-reliefs had sustained too much injury to be restored or removed.

Awad, Sheikh of the Jehesh

Our first Encampment in the Desert

CHAP. XI.

PREPARATIONS FOR A JOURNEY TO THE KHABOUR. — SCULPTURES DISCOVERED THERE. — SHEIKH SUTTUM. — HIS REDIFF. — DEPARTURE FROM MOSUL. — FIRST ENCAMPMENT. — ABOU KHAMEERA. — A STORM. — TEL ERMAH. — A STRANGER. — TEL JEMAL. — THE CHIEF OF TEL AFER. — A SUNSET IN THE DESERT. — A JEBOUR ENCAMPMENT. — THE BELLED SINJAR. — THE SINJAR HILL. — MIRKAN. — BUKRA. — THE DRESS OF THE YEZIDIS. — THE SHOMAL. — OSSOFA. — ALDINA. — RETURN TO THE BELLED. — A SNAKE-CHARMER. — JOURNEY CONTINUED IN THE DESERT. — RISHWAN. — ENCAMPMENT OF THE BORAIJ. — DRESS OF ARAB WOMEN. — RATHAIYAH. — HAWKING. — A DEPUTATION FROM THE YEZIDIS. — ARAB ENCAMPMENTS. — THE KHABOUR. — MOHAMMED EMIN. — ARRIVAL AT ARBAN.

I HAD long wished to visit the banks of the Khabour. This river, the Chaboras of the Greek geographers, and the Habor, or Chebar, of the Samaritan captivity *, rises in the north of Mesopotamia, and flowing to the west of the Sinjar hill, falls into the Euphrates near the site of the ancient city of Carchemish † or Circesium, still known to the Bedouins by the name of Carkeseea. As it winds through the midst of the desert, and its rich pastures are

* 2 Kings, xviii. 11.   Ezek. i. 1.   † 2 Chron. xxxv. 20.

the resort of wandering tribes of Arabs, it is always difficult of access to the traveller. It was examined, for a short distance from its mouth, by the expedition under Colonel Chesney ; but the general course of the river was imperfectly known, and several geographical questions of interest connected with it were undetermined previous to my visit.

With the Bedouins, who were occasionally my guests at Mosul or Nimroud, as well as with the Jebours, whose encamping grounds were originally on its banks, the Khabour was a constant theme of exaggerated praise. The richness of its pastures, the beauty of its flowers, its jungles teeming with game of all kinds, and the leafy thickness of its trees yielding an agreeable shade during the hottest days of summer, formed a terrestrial paradise to which the wandering Arab eagerly turned his steps when he could lead his flocks thither in safety. Ruins, too, as an additional attraction, were declared to abound on its banks and formed the principal inducement for me to undertake a long and somewhat hazardous journey. I was anxious to determine how far the influence of Assyrian art and manners extended, and whether monuments of the same period as those discovered at Nineveh existed so far to the west of the Tigris. During the winter my old friend Mohammed Emin, Sheikh of one of the principal branches of the Jebour tribe, had pitched his tents on the river. Arabs from his encampment would occasionally wander to Mosul. They generally bore an invitation from their chief, urging me to visit him when the spring rendered a march through the desert both easy and pleasant. But when a note arrived from the Sheikh, announcing that two colossal idols, similar to those of Nimroud, had suddenly appeared in a mound by the river side, I hesitated no longer, and determined to start at once for the Khabour. To avoid, however, any disappointment, I sent one of my own workmen to examine the pretended sculptures. As he confirmed, on his return, the account I had received, I lost no time in making preparations for the journey.

As the Shammar Bedouins were scattered over the desert between Mosul and the Khabour, and their horsemen continually scoured the plains in search of plunder, it was necessary that we should be protected and accompanied by an influential chief of the tribe. I accordingly sent to Suttum, a Sheikh of the Boraij, one of the principal branches of the Shammar, whose tents were at that time pitched between the river and the ruins of El Hather. Suttum was well known to me, and had already given proofs of his trustworthiness and intelligence on more than one similar occasion. He lost no time in obeying the summons. Arrangements were soon made with him. He agreed to furnish camels for our baggage, and to remain with me himself until he had seen my caravan in safety again within the gates of Mosul. He returned to the desert to fetch the camels, and to make other preparations for our journey, promising to be with me in a few days.

Punctual to his appointment, Sheikh Suttum brought his camels to Mosul on the 19th of March. He was accompanied by Khoraif, his *rediff*, as the person who sits on the dromedary * behind the principal rider is called by the Bedouins. Amongst the two great nomade tribes of the Shammar and Aneyza, the word "rediff" frequently infers a more intimate connection than a mere companionship on a camel. It is customary with them for a warrior to swear a kind of brotherhood with a person not only not related to him by blood, but frequently even of a different tribe. Two men connected by this tie are inseparable. They go together to war, they live in the same tent, and are allowed to see each other's wives. They become, indeed, more than brothers. Khoraif was of the tribe of the Aneyza, who have a deadly feud with the Shammar. Having left his own kith and kin on account of some petty quarrel, he had joined their enemies, and had become the rediff of Suttum, dwelling under his canvass, accompanying him in his expeditions, and riding with him on his deloul. Although he had deserted his tribe, Khoraif had not renounced all connection with his kindred, nor had he been cut off by them. Being thus allied to two powerful clans, he was able to render equal services to any of his old or new friends, who might fall into each other's hands. It is on this account that a warrior generally chooses his rediff from a warlike tribe with which he is at enmity, for if taken in war, he would then be *dakheel*, that is, protected, by the family, or rather particular sept, of his companion. On the other hand, should one of the rediff's friends become the prisoner of the sub-tribe into which his kinsman has been adopted, he would be under its protection, and could not be molested. Thus Khoraif would have been an important addition to our party, had we fallen

* I use the word " dromedary " for a swift riding camel, the *Deloul* of the Arabs, and *Hejin* of the Turks : it is so applied generally, although incorrectly by Europeans in the East.

in, during our journey, with Aneyza Arabs, against whom, of course, Suttum could not protect us. On warlike expeditions the rediff generally leads the mare which is to be ridden by his companion in the fight. When in face of the enemy he is left in charge of the dromedary, and takes part in the battle from its back. He rides, when travelling, on the naked back of the animal, clinging to the hinder part of the saddle, his legs crouched up almost to his chin — a very uncomfortable position for one not accustomed from childhood to a hard seat and a rough motion.

As our desert trip would probably last for more than two months, during which time we should meet with no villages, or permanent settlements, we were obliged to take with us supplies of all kinds, both for ourselves and the workmen; consequently, flour, rice, burghoul (prepared wheat, to be used as a substitute for rice), and biscuits, formed a large portion of our baggage. Two enormous boxes, each half a camel-load, were under the particular protection of Mr. Hormuzd Rassam, with whom they became a kind of hobby, notwithstanding my repeated protests against their size and inconvenience. They held various luxuries, such as sugar, coffee, tea, and spices, with robes of silk and cotton, and red and yellow boots, presents for the various chiefs whom we might meet in the desert. Baskets, tools for excavating, tents, and working utensils, formed the rest of our baggage.

I knew that I should have no difficulty in finding workmen when once in Mohammed Emin's encampment. As, however, it was my intention to explore any ruins of importance that we might see on our way, I chose about fifty of my best Arab excavators, and twelve Tiyari, or Nestorians, to accompany us. They were to follow on foot, but one or two extra camels were provided in case any were unable from fatigue to keep up with the caravan. The camels were driven into the small Mussulman burial-ground, adjoining my house in Mosul. The whole morning was spent in dividing and arranging the loads, always the most difficult part of the preparations for a journey in the East. The pack-saddles of the Bedouins, mere bags of rough canvass stuffed with straw, were ill adapted to carry anything but sacks of wheat and flour. As soon as a load was adjusted, it was sure to slip over the tail, or to turn over on one side. When this difficulty was overcome, the animals would suddenly kneel and shake off their burdens. Their owners were equally hard to please : this camel was galled, another vicious, a third weak.

Suttum and Khoraif exerted themselves to the utmost, and the inhabitants of the quarter, together with stray passers-by, joined in the proceedings, adding to the din and confusion, and of course considerably to our difficulties. At length, as the muezzin called to midday prayer, the last camel issued from the Sinjar gate. A place of general rendezvous had been appointed outside the walls, that our party might be collected together for a proper start, and that those who were good Mussulmans might go through their prayers before commencing a perilous journey.

I did not leave the town until nearly an hour and a half after the caravan, to give time for the loads to be finally adjusted, and the line of march to be formed. When we had all assembled outside the Sinjar gate, our party had swollen into a little army. The Doctor, Mr. Cooper, and Mr. Hormuzd Rassam, of course, accompanied me. Mr. and Mrs. Rolland with their servants had joined our expedition. My Yezidi fellow-traveller from Constantinople, Cawal Yusuf, with three companions, was to escort me to the Sinjar, and to accompany us in our tour through that district. Several Jebour families, whose tribe was encamped at Abou-Psera, near the mouth of the Khabour, seized this opportunity to join their friends, taking with them their tents and cattle. Thirteen or fourteen Bedouins had charge of the camels, so that, with the workmen and servants, our caravan consisted of nearly one hundred well-armed men; a force sufficient to defy almost any hostile party with which we were likely to fall in during our journey. We had about five and twenty camels, and as many horses, some of which were led. As it was spring time and the pastures were good, it was not necessary to carry much provender for our animals. Hussein Bey, the Yezidi chief, and many of our friends, as it is customary in the East, rode with us during part of our first stage; and my excellent friend, the Rev. Mr. Ford, an American missionary, then resident in Mosul, passed the first evening under our tents in the desert.

Suttum, with his rediff, rode a light fleet dromedary, which had been taken in a plundering expedition from the Aneyza. Its name was Dhwaila. Its high and picturesque saddle was profusely ornamented with brass bosses and nails; over the seat was thrown the Baghdad double bags adorned with long tassels and fringes of many-colored wools, so much coveted by the Bedouin. The Sheikh had the general direction and superintendence of our march. The Mesopotamian desert had been his home from his birth, and he knew every spring and pasture. He was of the Saadi, one of the

most illustrious families of the Shammar*, and he possessed great
personal influence in the tribe.　His intelligence was of a very high
order, and he was as well known for his skill in Bedouin intrigue,
as for his courage and daring in war.　In person he was of middle

Sheikh Suttum.

height, of spare habit, but well made, and of noble and dignified
carriage; although a musket wound in the thigh, from which the ball
had not been extracted, gave him a slight lameness in his gait.　His
features were regular and well-proportioned, and of that delicate
character so frequently found amongst the nomades of the desert.

* An Arab tribe is divided into septs, and each sept is composed of certain
families.　Thus Suttum was a Shammar, of the branch called the Boraij, and of
the family of Saadi, besides being a member of a peculiar division of the great
tribe called the Khorusseh.

A restless and sparkling eye of the deepest black spoke the inner man, and seemed to scan and penetrate everything within its ken. His dark hair was platted into many long tails ; his beard, like that of the Arabs in general, was scanty. He wore the usual Arab shirt, and over it a cloak of blue cloth, trimmed with red silk and lined with fur, a present from some Pasha as he pretended, but more probably a part of some great man's wardrobe that had been appropriated without its owner's consent. A colored kerchief, or keffieh, was thrown loosely over his head, and confined above the temples by a rope of twisted camel's hair. At his side hung a scimitar, an antique horse-pistol was held by a rope tied as a girdle round his waist, and a long spear, tufted with black ostrich feathers, and ornamented with scarlet streamers, rested on his shoulder. He was the very picture of a true Bedouin Sheikh. and his liveliness, his wit, and his singular powers of conversation, which made him the most agreeable of companions, did not belie his race.* The rest of my party, with the exception of the workmen, who were on foot, or who contrived to find places on the loads, and spare camels, were on horseback. The Bairakdar had the general management of the caravan, superintending, with untiring zeal and activity, the loading and unloading of the animals, the pitching of the tents, and the night watches, which are highly necessary in the desert.

As we wound slowly over the low rocky hills to the west of the town of Mosul, in a long straggling line, our caravan had a strange and motley appearance ; Europeans, Turks, Bedouins, town-Arabs, Tiyari, and Yezidis, were mingled in singular confusion ; each adding, by difference of costume and a profusion of bright colors, to the general picturesqueness and gaiety of the scene.

The Tigris, from its entrance into the low country at the foot of the Kurdish mountains near Jezireh, to the ruined town of Tekrit, is separated from the Mesopotamian plains by a range of low lime-

---

* Burckhardt, the English traveller best acquainted with the Bedouin character, and admirably correct in describing it, makes the following remarks : " With all their faults, the Bedouins are one of the noblest nations with which I ever had an opportunity of becoming acquainted. ... The sociable character of a Bedouin, when there is no question of profit or interest, may be described as truly amiable. His cheerfulness, wit, softness of temper, good-nature, and sagacity, which enable him to make shrewd remarks on all subjects, render him a pleasing, and often a valuable, companion. His equality of temper is never ruffled by fatigue or suffering." (Notes on the Bedouins, pp. 203. 208.) Unfortunately, since Burckhardt's time, closer intercourse with the Turks and with Europeans, has much tended to destroy many good features in the Arab character.

stone hills. We rode over this undulating ground for about an hour and a half, and then descended into the plain of Zerga, encamping for the night near the ruins of a small village, with a falling Kasr, called Sahaghi, about twelve miles from Mosul. The place had been left by its inhabitants, like all others on the desert side of the town, on account of the depredations of the Bedouins. There is now scarcely one permanent settlement on the banks of the Tigris from Jezireh to the immediate vicinity of Baghdad, with the exception of Mosul and Tekrit. One of the most fertile countries in the world, watered by a river navigable for nearly six hundred miles, has been turned into a desert and a wilderness, by continued misgovernment, oppression, and neglect.

Our tents were pitched near a pool of rain water, which, although muddy and scant, sufficed for our wants. There are no springs in this part of the plain, and the Bedouins are entirely dependent upon such temporary supplies. The remains of ancient villages show, however, that water is not concealed far beneath the surface, and that wells once yielded all that was required for irrigation and human consumption.

The loads had not yet been fairly divided amongst the camels, and the sun had risen above the horizon, before the Bedouins had arranged them to their satisfaction, and were ready to depart. The plain of Zerga was carpeted with tender grass, scarcely yet forward enough to afford pasture for our animals. Scattered here and there were tulips of a bright scarlet hue, the earliest flower of the spring.

A ride of three hours and a quarter brought us to a second line of limestone hills, the continuation of the Tel Afer and Sinjar range, dividing the small plain of Zerga from the true Mesopotamian desert. From a peak which I ascended to take bearings, the vast level country, stretching to the Euphrates, lay like a map beneath me, dotted with mounds, but otherwise unbroken by a single eminence. The nearest and most remarkable group of ruins was called Abou Khameera, and consisted of a lofty, conical mound surrounded by a square inclosure, or ridge of earth, marking, as at Kouyunjik and Nimroud, the remains of ancient walls. From the foot of the hill on which I stood there issued a small rivulet, winding amongst rushes, and losing itself in the plain. This running water had drawn together the black tents of the Jehesh, a half sedentary tribe of Arabs, who cultivate the lands around the ruined village of Abou Maria. Their flocks grazing on the plain, and the shepherds who watched them, were the only living objects in that boundless expanse. The hill and the stream are called Mohallibiyah, from the sweetness of the water, the neighbouring springs being all more or less brackish.*

As the caravan issued from the defile leading from the hills into the plain, the Arabs brought out bowls of sour milk and fresh water, inviting us to spend the night in their encampment. Eight or ten of my workmen, under a Christian superintendent, had been for some days excavating in the ruins of Abou Khameera. I therefore ordered the tents to be pitched near the reedy stream, and galloped to the mounds, which were rather more than a mile distant.

In general plan the ruins closely resemble those of Mokhamour in the Tai country.† A broad and lofty mound shows the traces of several distinct platforms or terraces rising one above the other. It is almost perpendicular on its four sides, except where, on the south-eastern, there appears to have been an inclined ascent, or a flight of steps, leading to the summit, and it stands nearly in the centre of an inclosure of earthen walls forming a regular quadrangle about 660 paces square. The workmen had opened deep trenches and tunnels in several parts of the principal ruin, and had found walls of sun-dried brick, unsculptured alabaster slabs, and some circular stone sockets for the hinges of gates, similar to those discovered at Nimroud. The baked bricks and the pieces of gypsum and pottery scattered amongst the rubbish bore no inscriptions, nor could I, after the most careful search, find the smallest fragment of sculpture. I have no hesitation, however, in assigning the ruins to the Assyrian period.

The Jehesh encamped near Abou Khameera were under Sheikh Saleh, the chief of this branch of a tribe scattered over the pashalic, and once large and powerful. They pay kowee, or black mail, to the Shammar Bedouins, and are thus able to pasture their flocks free from molestation in this part of the desert.

One of those furious and sudden storms, which frequently sweep over the plains of Mesopotamia during the spring season, burst over us in the night. Whilst incessant lightnings broke the gloom, a raging wind almost drowned the deep roll of the thunder. The

---

* There is a second spring of fresh water called Sheikh Ibrahim, beneath a high rock named Maasoud. The whole line of hills bounding the plain of Zerga to the west is called Kebritiyah, " the sulphur range," from a sulphurous spring rising at their feet. In this range are several remarkable peaks, serving as landmarks from great distances in the desert.

† See p. 225.

united strength of the Arabs could scarcely hold the flapping canvass of the tents. Rain descended in torrents, sparing us no place of shelter. Towards dawn the hurricane had passed away leaving a still and cloudless sky. When the round clear sun rose from the broad expanse of the desert, a delightful calm and freshness pervaded the air, producing mingled sensations of pleasure and repose.

The vegetation was far more forward in that part of the desert traversed during the day's journey than in the plain of Zerga. We trod on a carpet of the brightest verdure, mingled with gaudy flowers. Men and animals rejoiced equally in these luxuriant pastures, and leaving the line of march strayed over the meadows. On all sides of us rose Assyrian mounds, now covered with soft herbage. I rode with Suttum from ruin to ruin, examining each, but finding no other remains than fragments of pottery and baked bricks. The Bedouin chief had names for them all, but they were mere Arab names, derived generally from some local peculiarity; the more ancient had been long lost. From his child hood his father's tents had been pitched amongst these ruins for some weeks twice, nearly every year; when in the spring the tribe journeyed towards the banks of the Khabour, and again when in autumn they re-sought their winter camping-grounds around Babylon. These lofty mounds, seen from a great distance, and the best of landmarks in a vast plain, guide the Bedouin in his yearly wanderings.*

Tel Ermah, "the mound of the spears," had been visible from our tents, rising far above the surrounding ruins. As it was a little out of the direct line of march, Suttum mounted one of our led horses, and leaving Khoraif to protect the caravan, rode with me to the spot. The mound is precisely similar in character to Abou Khameera and Mokhamour, and, like them, stands within a quadrangle of earthen walls. On its south-eastern side also is a ravine, the remains of the ascent to the several terraces of the building. The principal ruin has assumed a conical form, like the high mound at Nimroud, and from the same cause. It was, I presume, originally square. Within the inclosure are traces of

---

* The following are the names of the principal mounds seen during this day's march: Ermah, Shibbit, Duroge, Addiyah, Abou-Kubbah, and Kharala, each name being preceded by the Arabic word Tel, *i. e.* mound. They are laid down in the map accompanying this volume, their positions having been fixed by careful bearings, and in some instances by the sextant.

ancient dwellings, but I was unable to find any inscribed fragments of stone or brick.

Whilst I was examining the ruins, Suttum, from the highest mound, had been scanning the plain with his eagle eye. At length it rested upon a distant moving object. Although with a telescope I could scarcely distinguish that to which he pointed, the Sheikh saw that it was a rider on a dromedary. He now, therefore, began to watch the stranger with that eager curiosity and suspicion always shown by a Bedouin when the solitude of the desert is broken by a human being of whose condition and business he is ignorant. Suttum soon satisfied himself as to the character of the solitary wanderer. He declared him to be a messenger from his own tribe, who had been sent to lead us to his father's tents. Mounting his horse, he galloped towards him. The Arab soon perceived the approaching horseman, and then commenced on both sides a series of manœuvres practised by those who meet in the desert, and are as yet distrustful of each other. I marked them from the ruin as they cautiously approached, now halting, now drawing nigh, and then pretending to ride away in an opposite direction. At length, recognising one another, they met, and, having first dismounted to embrace, came together towards us. As Suttum had conjectured, a messenger had been sent to him from his father's tribe. The Boraij were now moving towards the north in search of the spring pastures, and their tents would be pitched in three or four days beneath the Sinjar hill. Suttum at once understood the order of their march, and made arrangements to meet them accordingly.

Leaving the ruins of Tel Ermah, we found the caravan halting near some wells of sweet water, called Marzib. They belong to a branch of the Jebours under Sheikh Abd-ul-Azeez, and a few patches of green barley and wheat were scattered round them, but the tents of the tribe were now nearer the hills, and the cultivated plots were left unprotected.

From this spot the old castle of Tel Afer *, standing boldly on an eminence about ten miles distant, was plainly visible. Continuing our march we reached, towards evening, a group of mounds known as Tel Jemal, and pitched in the midst of them on a green lawn, enamelled with flowers, that furnished a carpet for our tents unequalled in softness of texture, or in richness of color, by the looms of Cashmere. A sluggish stream, called by the Arabs El

* Nineveh and its Remains, vol. i. p. 313

Abra, and by the Turcomans of Tel Afer, Kharala, crept through the ruins.

The tents had scarcely been raised when a party of horsemen were seen coming towards us. As they approached our encampment they played the Jerid with their long spears, galloping to and fro on their well-trained mares. They were the principal inhabitants of Tel Afer with Ozair Agha, their chief, who brought us a present of lambs, flour, and fresh vegetables. The Agha rode on a light chestnut mare of beautiful proportions and rare breed. His dress, as well as that of his followers, was singularly picturesque. His people are Turcomans, a solitary colony in the midst of the desert; and although their connection with the Bedouins has taught them the tongue and the habits of the wandering tribes, yet they still wear the turban of many folds, and the gay flowing robes of their ancestors. They allow their hair to grow long, and to fall in curls on their shoulders.

Ozair Agha was an old friend, who had more than once found refuge in my house from government oppression. He now sought my advice and protection, for he was accused of having been privy to some recent foray of the Bedouins, and was summoned to Mosul to answer the charge, of which, however, he declared himself completely innocent. I urged him to obey the summons without delay, to avoid the suspicion of rebellion against the government. I gave him, at the same time, letters to the authorities.

As the evening crept on, I watched from the highest mound the sun as it gradually sank in unclouded splendor below the sealike expanse before me. On all sides, as far as the eye could reach, rose the grass-covered heaps marking the site of ancient habitations. The great tide of civilisation had long since ebbed, leaving these scattered wrecks on the solitary shore. Are those waters to flow again, bearing back the seeds of knowledge and of wealth that they have wafted to the West? We wanderers were seeking what they had left behind, as children gather up the colored shells on the deserted sands. At my feet there was a busy scene, making more lonely the unbroken solitude which reigned in the vast plain around, where the only thing having life or motion were the shadows of the lofty mounds as they lengthened before the declining sun. Above three years before, when, watching the approach of night from the old castle of Tel Afer, I had counted nearly one hundred ruins *, now, when in the midst of them, no

---

* Nineveh and its Remains, vol. i. p. 315.

less than double that number were seen from Tel Jemal. Our tents crowning the lip of a natural amphitheatre bright with flowers, Ozair Agha and his Turcomans seated on the greensward in earnest talk with the Arab chief, the horses picketed in the long grass, the Bedouins driving home their camels for the night's rest, the servants and grooms busied with their various labors; such was the foreground to a picture of perfect calm and stillness. In the distance was the long range of the Sinjar hills, furrowed with countless ravines, each marked by a dark purple shadow, gradually melting into the evening haze.

We had a long day's march before us to the village of Sinjar. The wilderness appeared still more beautiful than it had done the day before. The recent storm had given new life to a vegetation which, concealed beneath a crust of apparently unfruitful earth, only waits for a spring shower to burst, as if by enchantment, through the thirsty soil. Here and there grew patches of a shrub-like plant with an edible root, having a sharp pungent taste like mustard, eaten raw and much relished by the Bedouins. Among them lurked game of various kinds. Troops of gazelles sprang from the low cover, and bounded over the plain. The greyhounds coursed hares; the horsemen followed a wild boar of enormous size, and nearly white from age; and the Doctor, who was the sportsman of the party, shot a bustard, with a beautiful speckled plumage, and a ruff of long feathers round its neck. This bird was larger than the common small bustard, but apparently of the same species. Other bustards, the great and the middle-sized (the Houbron and Houbara of the Arabs *), and the lesser, besides many birds of the plover kind†, rose from these tufts, which seemed to afford food and shelter to a variety of living creatures. We scanned the horizon in vain for the wild ass, which is but thinly scattered over the plains. The Arabs found many eggs of the middle bustard. They were laid in the grass without any regular nest, the bird simply making a form somewhat like that of a hare, and sitting very close, frequently not rising until it was nearly trodden under foot. One or two eggs of the great bustard were also brought to me during the day.

We still wandered amongst innumerable mounds. The largest I

---

* The Houbron is the Otis tarda, or great bustard; the Houbara, the Otis Houbara. I believe that more than one species of the lesser bustard (Otis tetrax) is found in the Mesopotamian plains.

† The most abundant was a large grey plover, called by the Bedouins "Smoug."

examined were called Hathail and Usgah. They resembled those of Abou-Khameerah and Tel Ermah, with the remains of terraces, the ascent to them being on the south-eastern side, and the inclosure of earthen walls.

We rode in a direct line to the Belled Sinjar, the residence of the governor of the district. There was no beaten track, and the camels wandered along as they listed, cropping as they went the young grass. The horsemen and footmen, too, scattered themselves over the plain in search of game. Suttum rode from group to group on his swift deloul, urging them to keep together, as the Aneyza *gazous** occasionally swept this part of the desert. But to little purpose ; the feeling of liberty and independence which these boundless meadows produced was too complete and too pleasing to be controlled by any fear of danger, or by the Sheikh's prudent counsel. All shared in the exhilarating effects of the air and scene. Hormuzd would occasionally place himself at the head of the Jebours, and chant their war songs, improvising words suited to the occasion. The men answered in chorus, dancing as they went, brandishing their weapons, and raising their bright-colored kerchiefs, as flags, on the end of their spears. The more sedate Bedouins smiled in contempt at these noisy effusions of joy, only worthy of tribes who have touched the plough ; but they indulged in no less keen, though more suppressed, emotions of delight. Even the Tiyari caught the general enthusiasm, and sung their mountain songs as they walked along.

As we drew near to the foot of the hills we found a large encampment, formed partly by Jebours belonging to Sheikh Abdul-Azeez, and partly by a Sinjar tribe called Mendka, under a chief known as the " Effendi," who enjoys considerable influence in this district. His tent is frequently a place of refuge for Bedouin chiefs and others, who have fled from successful rivals, or from the Turkish authorities. His grandfather, a Yezidi in creed, embraced Mohammedanism from political motives. The conversion was not consequently very sincere, and his descendants are still suspected of a leaning to the faith of their forefathers. This double character is one of the principal causes of the Effendi's influence. His tribe, which inhabits the Belled and adjoining villages on the south side of the mountain, consists almost entirely of Yezidis The chief himself resides during the winter and

---

* A plundering party, the *chappou* of the Persian tribes.

spring in tents, and the rest of the year in a village named Soulak. The Yezidis of the Sinjar are divided into ten distinct tribes, the Heska, Mendka, Houbaba, Merkhan, Bukra, Beit-Khaled, Amera, Al Dakhi, Semoki, and Kerani.

I dismounted at a short distance from the encampment, to avoid a breach of good manners, as to refuse to eat bread, or to spend the night, after alighting near a tent, would be thought a grave slight upon its owner. The caravan continued its journey towards the village. I was soon surrounded by the principal people of the camp; amongst them was one of my old workmen, Khuther, who now cultivated a small plot of ground in the desert.

It was with difficulty that I resisted the entreaties of the Effendi to partake of his hospitality. We did not reach the Belled until after the sun had gone down, the caravan having been ten hours in unceasing march. The tents were pitched on a small plot of ground, watered by numerous rills, and in the centre of the ruins. Although almost a swamp, it was the only spot free from stones and rubbish. In front of the tent door rose a leaning minaret, part of a mosque, and other ruins of Arab edifices. To the right was an old wall with a falling archway, from beneath which gushed a most abundant stream of clear sweet water, still retained for a moment in the stone basins once the fountains and reservoirs of the city.

I had scarcely entered my tent when the governor of the district, who resides in a small modern castle built on the hill-side, came to see me. He was a Turkish officer belonging to the household of Kiamil Pasha, and complained bitterly of his solitude, of the difficulties of collecting the taxes, and of dealing with the Bedouins who haunted the plains. The villages on the northern side of the mountain were not only in open rebellion to his authority, but fighting one with the other; all, however, being quite of one mind in refusing to contribute to the public revenues. He was almost shut up within the walls of his wretched fort, in company with a garrison of a score of half-starved Albanians. This state of things was chiefly owing to the misconduct of his predecessor, who, when the inhabitants of the Sinjar were quiet and obedient, had treacherously seized two of their principal chiefs, Mahmoud and Murad, and had carried them in chains to Mosul, where they had been thrown into prison. A deputation having been sent to obtain their release, I had been able to intercede with Kiamil Pasha in their behalf, and now bore to their followers the welcome news of their speedy return to their homes.

The tent was soon filled with the people of the Belled, and they remained in animated discussion until the night was far spent.

Early on the following morning, I returned the visit of the governor, and, from the tower of the small castle, took bearings of the principal objects in the plain. The three remarkable peaks rising in the low range of Kebriteeyah, behind Abou Khameera, were still visible in the extreme distance, and enabled me to fix with some accuracy the position of many ruins. They would be useful landmarks in a survey of this part of the desert. About four or five miles distant from the Belled, which, like the fort, is built on the hill side, is another large group of mounds, resembling that of Abou Khameera, called by the Bedouins simply the "Hosh," the courtyard or inclosure.

The ruins of the ancient town, known to the Arabs as " El Belled," or *the* city, are divided into two distinct parts by a range of rocky hills, which, however, are cleft in the centre by the bed of a torrent, forming a narrow ravine between them. This ravine is crossed by a strong well-built wall, defended by a dry ditch cut into the solid rock. An archway admits the torrent into the southern part of the city, which appears to have contained the principal edifices. The northern half is within the valley, and is surrounded by ruined fortifications. I could find no traces of remains of any period earlier than the Mohammedan, unless the dry ditch excavated in the rock be more ancient; nor could I obtain any relics, or coins, from the inhabitants of the modern village. The ruins are, undoubtedly, those of the town of Sinjar, the capital of an Arab principality in the time of the Caliphs. Its princes frequently asserted their independence, coined money, and ruled from the Khabour and Euphrates to the neighbourhood of Mosul. The province was included within the dominions of the celebrated Saleh-ed-din (the Saladin of the Crusades), and was more than once visited by him.

The ruins of Sinjar are also believed to represent the Singara of the Romans. On coins struck under the Emperor Gordian, and bearing his effigy with that of the empress Tranquillina, this city is represented by a female wearing a mural crown surmounted by a centaur, seated on a hill *with a river at her feet* (?).* Ac-

---

* There were also coins of Alexander Severus, struck in Singara. It is to be remarked that, in consequence of considerable discrepancies in the accounts of ancient geographers, several authors have been inclined to believe that there were two cities of the same name ; one, according to Ptolemy, on the Tigris, the other under the mountain. It was long a place of contention between the Romans and Parthians.

cording to the Arab geographers, the Sinjar was celebrated for its palms. This tree is no longer found there, nor does it bear fruit, I believe, anywhere to the north of Tekrit in Mesopotamia.

Roman Coin of Gordian and Tranquillina, struck at Singara (British Museum

Wishing to visit the villages of the *Shomal*, or northern side of the mountain, and at the same time to put an end, if possible, to the bloodshed between their inhabitants, and to induce them to submit to the governor, I quitted the Belled in the afternoon, accompanied by Cawal Yusuf and his Yezidi companions, Mr. and Mrs. R., the Doctor, and Mr. Cooper. The tents, baggage, and workmen were left under the charge of the Bairakdar. Suttum went to his tribe to make further arrangements for our journey to the Khabour.

We followed a precipitous pathway along the hill side to Mirkan, the village destroyed by Tahyar Pasha on my first visit to the Sinjar.* This part of the mountain is coated with thin strata of a white fossiliferous limestone, which detach themselves in enormous flakes, and fall into the valleys and ravines, leaving an endless variety of singular forms in the rocks above. In some places the declivities are broken into stupendous flights of steps, in others they have the columnar appearance of basalt. This limestone produces scarcely a blade of vegetation, and its milk-white color, throwing back the intense glare of the sun's rays, is both painful and hurtful to the sight.

Mirkan was in open rebellion, and had refused both to pay taxes and to receive the officer of the Pasha of Mosul. I was, at first, somewhat doubtful of our reception. Esau, the chief, came out, however, to meet me, and led us to his house. We were soon surrounded by the principal men of the village. They were also

* Nineveh and its Remains, vol. i. p. 317

at war with the tribes of the " Shomal." A few days before they had fought with the loss of several men on both sides. Seconded by Cawal Yusuf, I endeavored to make them feel that peace and union amongst themselves was not only essential to their own welfare, but to that of the Yezidis of Kurdistan and Armenia, who had, at length, received a promise of protection from the Turkish government, and who would suffer for their misdeeds. After a lengthened discussion the chief consented to accompany me to the neighbouring village of Bukra, with whose inhabitants his people had been for some time at war.

Mirkan had been partly rebuilt since its destruction three years before; but the ruins and charred timbers of houses still occupied much of its former site. We crossed the entrance to the ravine filled with caverns into which the Yezidis had taken refuge, when they made the successful defence I have elsewhere described.

There are two pathways from Mirkan to the " Shomal," one winding through narrow valleys, the other crossing the shoulder of the mountain. I chose the latter, as it enabled me to obtain an extensive view of the surrounding country, and to take bearings of many points of interest. The slopes around the villages are most industriously and carefully cultivated. Earth, collected with great labor, is spread over terraces, supported by walls of loose stones, as on the declivities of Mount Lebanon. These stages, rising one above the other, are planted with fig-trees, between which is occasionally raised a scanty crop of wheat or barley. The neatness of these terraced plots conveys a very favorable impression of the industry of the Yezidis.

Near the crest of the hill we passed a white conical building, shaded by a grove of trees. It was the tomb of the father of Murad, one of Yusuf's companions, a Cawal of note, who had died near the spot of the plague some years before. The walls were hung with the horns of sheep, slain in sacrifice, by occasional pilgrims.

I had little anticipated the beauty and extent of the view which opened round us on the top of the pass. The Sinjar hill is a solitary ridge rising abruptly in the midst of the desert; from its summit, therefore, the eye ranges on one side over the vast level wilderness stretching to the Euphrates, and on the other over the plain bounded by the Tigris and the lofty mountains of Kurdistan. Nisibin and Mardin were both visible in the distance. I could distinguish the hills of Baadri and Sheikh Adi, and many well-known peaks of the Kurdish Alps. Behind the lower ranges,

each distinctly marked by its sharp, serrated outline, were the
snow-covered heights of Tiyari and Bohtan. Whilst to the south
of the Sinjar artificial mounds appeared to abound, to the north
I could distinguish but few such remains. We dismounted to
gaze upon this truly magnificent scene lighted up by the setting
sun. I have rarely seen any prospect more impressive than these
boundless plains viewed from a considerable elevation. Besides
the idea of vastness they convey, the light and shade of passing

Interior of a Yezidi House at Bukra, in the Sinjar

clouds flitting over the face of the land, and the shadows as they
lengthen towards the close of day, produce constantly changing
effects of singular variety and beauty.*

It was night before we reached Bukra, where we were welcomed with great hospitality. The best house in the village had

* The traveller who has looked down from Mardin, for the first time, upon the
plains of Mesopotamia, can never forget the impression which that singular scene
must have made upon him. The view from the Sinjar hill is far more beautiful
and varied.

been made ready for us, and was scrupulously neat and clean, as the houses of the Yezidis usually are. It was curiously built, being divided into three principal rooms, opening one into the other. They were separated by a wall about six feet high, upon which were placed wooden pillars supporting the ceiling. The roof rested on trunks of trees, raised on rude stone pedestals at regular intervals in the centre chamber, which was open on one side to the air, like a Persian Iwan. The sides of the rooms were honeycombed with small recesses like pigeon-holes, tastefully arranged. The whole was plastered with the whitest plaster, fancy designs in bright red being introduced here and there, and giving the interior of the house a very original appearance.

The elders of Bukra came to me after we had dined, and seated themselves respectfully and decorously round the room. They were not averse to the reconciliation I proposed, received the hostile chief without hesitation, and promised to accompany me on the morrow to the adjoining village of Ossofa, with which they were also at war. Amongst those who had followed us was an active and intelligent youth, one of the defenders of the caverns when the Turkish troops under Tahyar Pasha attacked Mirkan. He related with great spirit and zest the particulars of the affair, and assured me that he had killed several men with his own gun. He was then but a boy, and it was the first time he had seen war. His father, he said, placed a rifle in his hand, and pointing to a soldier who was scaling the rocks exclaimed, " Now, show me whether thou art a man, and worthy of me. Shoot that enemy of our faith, or I will shoot you!" He fired, and the assailant rolled back into the ravine.

In the morning we visited several houses in the village. They were all built on the same plan, and were equally neat and clean. The women received us without concealing their faces, which are, however, far from pleasing, their features being irregular, and their complexion sallow. Those who are married dress entirely in white, with a white kerchief under their chins, and another over their heads held by the *agal*, or woollen cord, of the Bedouins. The girls wear white shirts and drawers, but over them colored *zabouns,* or long silk dresses, open in front, and confined at the waist by a girdle ornamented with pieces of silver. They twist gay kerchiefs round their heads, and adorn themselves with coins, and glass and amber beads, when their parents are able to procure them. But the Yezidis of the Sinjar are now very poor, and nearly all the trinkets of the women have long since fallen into the hands of the Turkish soldiery, or have been sold to pay taxes and arbitrary fines. The men have a dark complexion, black and piercing eyes, and frequently a fierce and forbidding countenance. They are of small stature, but have well proportioned limbs strongly knit together, and are muscular, active, and capable of bearing great fatigue. Their dress consists of a shirt, loose trowsers and cloak, all white, and a black turban, from beneath which their hair falls in ringlets. Their long rifles are rarely out of their hands, and they carry pistols in their girdle, a sword at their side, and a row of cartouche cases, generally made of cut reeds, on their breast. These additions to their costume, and their swarthy features, give them a peculiar look of ferocity, which, according to some, is not belied by their characters.

The Yezidis are, by one of their religious laws, forbidden to wear the common Eastern shirt open in front, and this article of their dress is always closed up to the neck. This is a distinctive mark of the sect by which its members may be recognised at a glance. The language of the people of Sinjar is Kurdish, and few speak Arabic. According to their traditions they are the descendants of a colony from the north of Syria, which settled in Mesopotamia at a comparatively recent period, but I could obtain no positive information on the subject. It is probable, however, that they did not migrate to their present seats before the fall of the Arab principality, and the invasion of Timourleng, towards the end of the fourteenth century.

The north side of the mountain is thickly inhabited, and well-cultivated as far as the scanty soil will permit. Scarcely three quarters of a mile to the west of Bukra is the village of Naksi, the interval between the two being occupied by terraces planted with fig-trees. We did not stop, although the inhabitants came out to meet us, but rode on to Ossofa, or Usifa, only separated from Naksi by a rocky valley. The people of this village were at war with their neighbours, and as this was one of the principal seats of rebellion and discontent, I was anxious to have an interview with its chief.

The position of Ossofa is very picturesque. It stands on the edge of a deep ravine; behind it are lofty crags and narrow gorges, whose sides are filled with natural caverns. On over-hanging rocks, towering above the village, are two *ziarehs*, or holy places, of the Yezidis, distinguished from afar by their white fluted spires.

Pulo, the chief, met us at the head of the principal inhabitants

and led me to his house where a large assembly was soon collected to discuss the principal object of my visit. The chiefs
of Mirkan and Bukra were induced to make offers of peace,
which were accepted, and after much discussion the terms of an
amicable arrangement were agreed to and ratified by general consent. Sheep were slain to celebrate the event. The meat, after
the Yezidi fashion, was boiled in onions, and a kind of parched
pea, and afterwards served up, like porridge, in large wooden
bowls. The mess is not unsavoury, and is the principal dish
of the Sinjar. Dried figs, strung in rows and made up into grotesque figures, were brought to us as presents. After the
political questions had been settled, the young men adjourned
to an open spot outside the village to practise with their rifles.
They proved excellent shots, seldom missing the very centre of
the mark.

The villages of Bouran (now deserted), Gundi-Gayli, Kushna,
and Aldina, follow to the west of Ossofa, scarcely half a mile in-
tervening between each. They are grouped together on the
mountain side, which, above and below them, is divided into terraces and planted with fig-trees. The loose stones are most carefully removed from every plot of earth, however small, and built
up into walls ; on the higher slopes are a few vineyards.

We passed the night at Aldina, in the house of Murad, one of
the imprisoned chiefs, whose release I had obtained before leaving
Mosul. I was able to announce the good tidings of his approaching return to his wife, to whom he had been lately married, and
who had given birth to a child during his absence.

Below Aldina stands a remarkable *ziareh*, inclosed by a wall of
cyclopean dimensions. In the plain beneath, in the midst of a
grove of trees, is the tomb of Cawal Hussein, the father of Cawal
Yusuf, who died in the Sinjar during one of his periodical visitations. He was a priest of sanctity and influence, and his
grave is still visited as a place of pilgrimage. Sacrifices of sheep
are made there, but they are merely in remembrance of the
deceased, and have no particular religious meaning attached to
them. The flesh is distributed amongst the poor, and a sum of
money is frequently added. Approving the ceremony as one
tending to promote charity and kindly feeling, I gave a sheep to
be sacrificed at the tomb of the Cawal, and one of my fellow
travellers added a second, the carcases being afterwards divided
among the needy.

All the villages we had passed during our short day's journey stand high on the mountain side, where they have been built for security against the Bedouins. They command extensive views of the plain, the white barracks of Nisibin, although certainly between twenty and thirty miles distant, being visible from them, and the snowy range of Kurdistan forming a magnificent back ground to the picture. The springs, rising in the hill, are either entirely absorbed in irrigation, or are soon lost in the thirsty plain beneath. Parallel to the Sinjar range is a long narrow valley, scarcely half a mile in width, formed by a bold ridge of white limestone rocks, so friable that the plain for some distance is covered with their fragments.

A messenger brought me word during the night that Suttum had returned from his tribe, and was waiting with a party of horsemen to escort us to his tents. I determined, therefore, to cross at once to the Belled by a direct though difficult pass. The Doctor and Mr. R., leaving the pathway, scaled the rocks in search of the ibex, or wild goat, which abounds in the highest ridges of the mountain.

We visited Nogray and Ameera, before entering the gorge leading to the pass. Only two other villages of any importance, Semoka and Jafri, were left unseen. The ascent of the mountain was extremely precipitous, and we were nearly two hours in reaching the summit. We then found ourselves on a broad green platform thickly wooded with dwarf oak. I was surprised to see snow still lying in the sheltered nooks. On both sides of us stretched the great Mesopotamian plains. To the south, glittering in the sun, was a small salt lake about fifteen miles distant from the Sinjar, called by the Arabs, Munaif. From it the Bedouins, when in their northern pastures, obtain their supplies of salt.

We descended to the Belled through a narrow valley thick with oak and various shrubs. Game appeared to abound. A Yezidi, who had accompanied us from Aldina, shot three wild boars, and we put up several coveys of the large red partridge. The Doctor and Mr. R., who joined us soon after we had reached our tents, had seen several wild goats, and had found a carcase half devoured by the wolves.

In the valley behind the Belled we passed the ruins of a large deserted village, whose inhabitants, according to Cawal Yusuf, had been entirely destroyed by the plague. We were nearly five hours in crossing the mountain.

Suttum and his Bedouin companions were waiting for us, but

were not anxious to start before the following morning. A Yezidi snake-charmer, with his son, a boy of seven or eight years old, came to my tents in the afternoon, and exhibited his tricks in the midst of a circle of astonished beholders. He first pulled from a bag a number of snakes knotted together, which the bystanders declared to be of the most venomous kind. The child took the reptiles fearlessly from his father, and placing them in his bosom allowed them to twine themselves round his neck and arms. The Bedouins gazed in mute wonder at these proceedings, but when the Sheikh, feigning rage against one of the snakes which had drawn blood from his son, seized it, and biting off its head with his teeth threw the writhing body amongst them, they could no longer restrain their horror and indignation. They uttered loud curses on the infidel snake-charmer and his kindred to the remotest generations. Suttum did not regain his composure during the whole evening, frequently relapsing into profound thought, then suddenly breaking out in a fresh curse upon the Sheikh, who, he declared, had a very close and unholy connection with the evil one. Many days passed before he had completely got over the horror the poor Yezidi's feats had caused him.

The poisonous teeth of the snakes which the Sheikh carried with him had probably been drawn, although he offered to practise upon any specimens we might procure for him. I did not, however, deem it prudent to put him to the test. The ruins of the Sinjar abound with these reptiles, and I had seen many amongst them. That most commonly found is of a dark brown color, nearly approaching to black, and, I believe, harmless. I have met with them above six feet in length. Others, however, are of a more dangerous character, and the Bedouins are in great dread of them.

Suttum had changed his deloul for a white mare of great beauty, named Athaiba. She was of the race of Kohaila, of exquisite symmetry, in temper docile as a lamb, yet with an eye of fire, and of a proud and noble carriage when excited in war or in the chase. His saddle was the simple stuffed pad generally used by the Bedouins, without stirrups. A halter alone served to guide the gentle animal. Suttum had brought with him several of the principal members of his family, all of whom were mounted on high-bred mares. One youth rode a bay filly, for which, I was assured, one hundred camels had been offered.

We followed a pathway over the broken ground at the foot of the Sinjar, crossing deep watercourses worn by the small streams, which lose themselves in the desert. The villages, as on the opposite slope, or " Shomal," are high up on the hill-side. The first we passed was Gabara, inhabited by Yezidis and Mussulmans. Its chief, Ruffo, with a party of horsemen, came to us, and intreated me to show him how to open a spring called *Soulak*, which, he said, had suddenly been choked up, leaving the village almost without water. Unfortunately, being ignorant of the arts for which he gave me credit, I was unable to afford him any help. Beyond Gabara, and nearer to the plain, we saw some modern ruins named Werdiyat, and encamped, after a short ride, upon a pleasant stream beneath the village of Jedaila.

We remained here a whole day in order to visit Suttum's tribe, which was now migrating towards the Sinjar. Early in the morning a vast crowd of moving objects could be faintly perceived on the horizon. These were the camels and sheep of the Boraij, followed by the usual crowd of men, women, children, and beasts of burden. We watched them as they scattered themselves over the plain, and gradually settled in different pastures. By midday the encampment had been formed and all the stragglers collected. We could scarcely distinguish the black tents, and their site was only marked by curling wreaths of white smoke.

In the afternoon Suttum's father, Rishwan, came to us, accompanied by several Sheikhs of the Boraij. He rode on a white deloul celebrated for her beauty and swiftness. His saddle and the neck of the animal were profusely adorned with woollen tassels of many colors, glass beads, and small shells, after the manner of the Arabs of Nejd. The well-trained dromedary having knelt at the door of my tent, the old man alighted, and throwing his arms round my neck kissed me on both shoulders. He was tall, and of noble carriage. His beard was white with age, but his form was still erect and his footsteps firm. Rishwan was one of the bravest warriors of the Shammar. He had come, when a child, with his father from the original seat of the tribe in northern Arabia. As the leader of a large branch of the Boraij he had taken a prominent part in the wars of the tribe, and the young men still sought him to head their distant forays. But he had long renounced the toils of the *gazou*, and left his three sons, of whom Suttum was the second, to maintain the honor of the Saadi. He was a noble specimen of the true Bedouin, both in character and appearance. With the skill and daring of the Arab warrior he united the hospitality, generosity, and good faith of a hero of Arab romance. He spoke in the rich dialect of the desert tongue, with the eloquence peculiar to his race. He sat with me during the greater part of the afternoon, and having eaten bread returned to his tent.

The Yezidi chiefs of Kerraniyah or Sekkiniyah (the village is known by both names) came to our encampment soon after Rishwan's arrival. As they had a feud with the Bedouins, I took advantage of their visit to effect a reconciliation, both parties swearing on my hospitality to abstain from plundering one another hereafter. The inhabitants of this village and of Semokiyah give tithes of produce (and also of property taken in forays) to Hussein Bey alone; whilst others pay tithes to Sheikh Nasr as well as to the chief.

Being anxious to reach the end of our journey I declined Suttum's invitation to sleep in his tent, but sending the caravan to the place appointed for our night's encampment, I made a detour to visit his father, accompanied by Mr. and Mrs. R., the Doctor, Mr. C., and Hormuzd. Although the Boraij were above six miles from the small rivulet of Jedaila, they were obliged to send to it for water.* As we rode towards their tents we passed their camels and sheep slowly wandering towards the stream. The camels, spreading far and wide over the plain, were divided according to their colors; some herds being entirely white, some yellow, and others brown or black. Each animal bore the well-known mark of the tribe branded on his side. The Arabs, who drove them, were mounted on dromedaries carrying the capacious *rouwis*, or buckets made of bullock skins, in which water is brought to the encampment for domestic purposes.

A Bedouin warrior, armed with his long tufted spear, and urging his fleet deloul, occasionally passed rapidly by us leading his high-bred mare to water, followed by her colt gambolling unrestrained over the greensward. In the throng we met Sahiman, the elder brother of Suttum. He was riding on a bay horse, whose fame had spread far and wide amongst the tribes, and whose exploits were a constant theme of praise and wonder with the Shammar. He was of the race of Obeyan Sherakh, a breed now almost

---

* In the spring months, when the pastures are good, the sheep and camels of the Bedouins require but little water, and the tents are seldom pitched near a well or stream; frequently as much as half a day's journey distant. Suttum assured me that at this time of the year the camels need not be watered for two months, such is the richness of the grass of the Desert.

extinct, and perhaps more highly prized than any other of the Desert. He had established his fame when but two years old. Ferhan, with the principal warriors of the Khurusseh*, had crossed the Euphrates to plunder the Aneyza. They were met by a superior force, and were completely defeated. The best mares of the tribe fell into the hands of the enemy, and the bay colt alone, although followed by the fleetest horses of the Aneyza, distanced his pursuers.† Such noble qualities, united with the purest blood, rendered him worthy to be looked upon as the public property of the Shammar, and no sum of money would induce his owner to part with him. With a celebrated bay horse belonging to the Hamoud, a branch of the same tribe, he was set apart to propagate the race of the finest horses in Mesopotamia. In size he was small, but large in bone and of excellent proportions. On all sides I heard extraordinary instances of his powers of endurance and speed.

Near the encampment of the Boraij was a group of mounds resembling in every respect those I have already described. The Bedouins call them Abou-Khaima. Are these singular ruins those of towns or of temples? Their similarity of form, — a centre mound divided into a series of terraces, ascended by an inclined way or steps, and surrounded by equilateral walls,—would lead to the conjecture that they were fire temples, or vast altars, destined for Astral worship. It will be seen hereafter that the well-known ruin of the Birs Nimroud, on or near the site of ancient Babylon, is very nearly the same in shape. When I come to describe those remarkable remains, I will add some further observations upon their original form.

---

* Five sects or subdivisions of the great tribe of Shammar, renowned for their bravery and virtues, and supposed to be descended from the same stock, are so called. Their hereditary chief is Ferhan. To belong to the Khurusseh is an honorable distinction amongst the Shammar. The five septs are the Boraij, the Fedagha, the Alayian, the Ghishm, and the Hathba; of this last, and of the family of Ahl-Mohammed, was the celebrated Bedouin chief Sofuk. The other clans forming the tribe of Shammar are the Abde, Assaiyah (divided into As-Subhi and Al-Aslam), Thabet, Hamoud, Theghavgheh, Ghatha, Dhirayrie, Ghufayla, and Azumail. All these tribes are again divided into numerous septs. The Assaiyah have nearly all crossed the Euphrates, owing to a blood feud with the rest of the Shammar, and have united with the Aneyza. The Raffidi, however, a large section of the Aneyza, have left their kindred, and are now incorporated with the Shammar.

† It is an error to suppose that the Bedouins never ride horses; for several reasons, however, they seldom do so.

The Bedouins who accompanied us galloped to and fro, engaging in mimic war with their long quivering spears, until we reached the encampment of the Boraij.   The tents were scattered far and wide over the plain ; for so they are pitched during this season of the year when the pastures are abundant, and no immediate danger is apprehended from hostile tribes.   At other times they are ranged in parallel lines close together, the Sheikh always occupying the foremost place, facing the side from which the guest, as well as the enemy, is expected, that he may be the first to exercise hospitality, and the first to meet the foe.   This position, however, varies in winter, when the tent must be closed completely on one side, according to the prevailing wind, so that when the wind changes, the whole camp suddenly, as it were, turns round, the last tent becoming the foremost.   It is thought unmannerly to approach by the back, to step over the tent-ropes, or to ride towards the woman's compartment, which is almost always on the right.   During warm weather the whole canvass is raised on poles to allow the air to circulate freely, a curtain being used in the morning and evening to ward off the rays of the sun.   The Bedouin can tell at once, when drawing near to an encampment, the tent of the Sheikh.   It is generally distinguished by its size, and frequently by the spears standing in front of it.   If the stranger be not coming directly towards it, and wishes to be the guest of the chief, he goes out of his way, that on approaching he may ride at once to it without passing any other, as it is considered uncourteous and almost an insult to go by a man's tent without stopping and eating his bread.   The owner of a tent has even the right to claim any one as his guest who passes in front of it on entering an encampment.

Rishwan, Suttum, Mijwell his younger brother, and the elders of the tribe, were standing before the tent ready to receive us.   All the old carpets and coverlets of the family, and ragged enough they were, had been spread out for their guests.   As we seated ourselves two sheep were slain before us for the feast ; a ceremony it would not have been considered sufficiently hospitable to perform previous to our arrival, as it might have been doubtful whether the animals had been slain wholly for us.   The chief men of the encampment collected round us, crouching in a wide circle on the grass.   We talked of Arab politics and Arab war, *ghazous*, and Aneyza mares stolen or carried off in battle by the Shammar.   Huge wooden platters, heavy with the steaming messes of rice and boiled meat, were soon brought in and placed on the ground before us.   Immense lumps of fresh butter were then heaped upon them, and allowed to melt, the chief occasionally mixing and kneading the whole up together with his hands.   When the dishes had cooled * the venerable Rishwan stood up in the centre of the tent, and called in a loud voice upon each person by name and in his turn to come to the feast.   We fared first with a few of the principal Sheikhs.   The most influential men were next summoned, each however resisting the honor, and allowing himself to be dragged by Suttum and Mij- well to his place.   The children, as is usual, were admitted last, and wound up the entertainment by a general scramble for the frag-

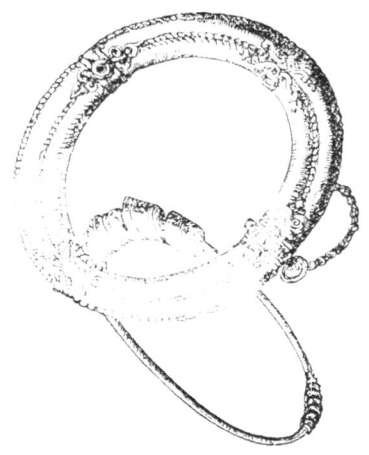

ments and the bones.   Neither Rishwan nor his sons would eat of the repast they had prepared, the laws of hospitality requiring that it should be left entirely to their guests.

After we had eaten, I accom- panied Mrs. R. to the harem, where we found assembled the wives and daughters of Rishwan, of his sons, and of the elders of the tribe, who had met together to see the Frank lady.   Amongst them were several of considerable beauty.   The wife of Sahiman, the eldest of the three brothers, was most distinguished for her

Arab Nose Ring and Bracelet of Silver

good looks.   They were all dressed in the usual long blue shirt, and striped, or black, abba, with a black headkerchief, or keffieh, confined by a band of spun camel's wool.   Massive rings of silver, adorned with gems and coral, hung from their noses †, and bracelets

---

* It is considered exceedingly inhospitable amongst the Shammar to place a hot dish before guests, as they are obliged to eat quickly out of consider- ation for others, who are awaiting their turn, which they cannot do, unless the mess be cool, without burning their mouths, or wasting half their time pick- ing out the colder bits.   On one occasion, Ferhan, the great chief of the Sham- mar, and a large number of horsemen having alighted at my tent, I prepared a dinner for them.   The Sheikh was afterwards heard to say that the Bey's feast was sumptuous, but that he had not treated his guests with proper hospitality, as the dishes were so hot nobody could eat his fill.

† These are "the rings and nose jewels," which Isaiah (iii. 21.) describes as worn by the Jewish women.   It is curious that no representation of them has

in the same metal, and also set with precious stones, encircled their wrists and ankles. Some wore necklaces of coins, coarse amber, agate, cornelian beads and cylinders, mostly Assyrian relics picked up amongst ruins after rain. These ornaments were confined to the unmarried girls, and to the youngest and prettiest wives, who on waxing old are obliged to transfer them to a more favored successor.

When Bedouin ladies leave their tents, or are on a march, they sometimes wear a black kerchief over the lower part of the face, showing only their sparkling eyes. Like the men they also use the keffieh, or head-kerchief, to cover their features. Their complexion is of a dark rich olive. Their eyes are large, almond-shaped, expressive, and of extraordinary brilliancy and fire. They suffer their black, and luxuriant hair to fall in clusters of curls. Their carriage in youth is erect and graceful. They are able to bear much fatigue, and show great courage and spirit in moments of difficulty and danger. But their beauty is only the companion of extreme youth. With few exceptions, soon after twenty, and the birth of one or two children, they rapidly change into the most hideous of old hags, the lightning-like brightness of the eye alone surviving the general wreck. When young, the daughters and wives of the chiefs are well cared for; they move with the tribe in the covered camel-saddle, shaded by carpets from the rays of the sun. Daughters are looked upon in the Desert* as a source of strength and advantage, from the alliances they enable the father to make with powerful and influential chiefs, being frequently the means of healing feuds which have existed for many years.

The children of Rishwan's family were naked, and, of course, dirty. One who, singularly enough for a Bedouin, had light flaxen hair and blue eyes, was on this account supposed to bear a striking likeness to Mr. C., and had, consequently, been nicknamed the *Musauer*, the artist, a name by which he will probably be known for the rest of his days.

Before we left the encampment Suttum led before me as a pre-

hitherto been found in the Assyrian sculptures. I take this opportunity of mentioning, that I saw a finger-ring sculptured on a fragment at Khorsabad.

* Amongst the inhabitants of towns, a daughter is considered a kind of flaw in the family, and the death of a girl, too frequently purposely brought about, is rarely a cause of grief.

sent a handsome grey colt, which was as usual returned with a request to take care of it until it was required, the polite way to decline a gift of this nature.*

Suttum having saddled his deloul was ready to accompany us on our journey. As he was to be for some time absent from his tents, he asked to take his wife with him, and I willingly consented. Rathaiyah was the sister of Suttâm el Meekh, chief of the powerful tribe of the Abde, one of the principal divisions of the Shammar. Although no longer young she still retained much of her early beauty. There was more than the usual Bedouin fire in her large black eyes, and her hair fell in many ringlets on her shoulders. Her temper was haughty and imperious, and she evidently held more sway over Suttum than he liked to acknowledge, or was quite consistent with his character as a warrior. He had married her from motives of policy, as cementing an useful alliance with a powerful tribe. She appears to have soon carried matters with a high hand, for poor Suttum had been compelled, almost immediately after his marriage, to send back a young and beautiful wife to her father's tent. This prior claimant upon his affections was now on the Khabour with her tribe, and it was probably on this account that Rathaiyah, knowing the direction he was about to take, was so anxious to accompany her husband. She rode on the dromedary behind her lord, a comfortable seat having been made for her with a rug and a coverlet. The Sheikh carried his hawk, Hattab, on his wrist, guiding the deloul by a short hooked stick held in the right

---

* As this was known to be a mere matter of form with me, as I made it a rule never to accept presents of this kind, Suttum might have offered me his bay colt, the most valuable horse amongst the Shammar, to increase the display of hospitality. The reason he did not was this, that although he knew I would have returned the horse, I might have expressed a wish to buy it, and have offered a price. An offer of this kind would have at once injured the value of the animal in the eyes of the Bedouins, and its owner might have been ultimately compelled to sell it. On one occasion, when I was amongst the Shammar, at Al Hather, an Arab rode into my encampment on a beautiful grey colt. I was so much struck with the animal, that I at once expressed a wish to its rider to purchase it. He merely intimated that the sum I named was beneath the value. I increased it, but he only shook his head, and rode off. Nevertheless, the report spread amongst the tribes that he had bargained for the sale of his horse. Although of the best blood, the animal was looked upon with suspicion by the Bedouins, and the owner was, some months after, obliged to sell him at a lower price than I had bid, to a horse-dealer of Mosul! A knowledge of such little prejudices and customs is very necessary in dealing with the Arabs of the Desert, who are extremely sensitive, and easily offended.

hand. Khoraif, his rediff, rode on this occasion a second dromedary named Sheaila, with a Shammar Bedouin.

The true Sinjar mountain ends about nine miles from Jedaila,
the high ridge suddenly subsiding into low broken hills. From all
parts of the plain it is a very beautiful object. Its limestone rocks,
wooded here and there with dwarf oak, are of a rich golden color;
and the numberless ravines, which furrow its sides, form ribs of
deep purple shadow. The western part of the Sinjar is inhabited
by the Yezidi tribe of Kherraniyah. We rode over the plain in a
parallel line to the mountain, and about seven or eight miles from
it. Towards nightfall we skirted a ridge of very low hills rising to
our left. They are called Alouvi and Yusuf Beg.

The Desert abounded in the houbara, or middle-sized bustard,
the bird usually hawked by the Arabs, and esteemed by them a
great delicacy. Hattab had been principally trained to this game,
and sat on the raised wrist of Suttum, scanning the plain with his
piercing eye. He saw the crouching quarry long before we could
distinguish it, and spreading his wings struggled to release himself
from the tresses. Once free he made one straight, steady swoop
towards the bustard, which rose to meet the coming foe, but was
soon borne down in his sharp talons. A combat ensued, which was
ended by a horseman riding up, substituting the lure for the game,
and hooding the hawk, which was again placed on its master's wrist.

Thus we rode joyously over the plain, night setting in before we
could see the tents. No sound except the mournful note of the
small desert owl, which has often misled the weary wanderer*,
broke the deep silence, nor could we distinguish the distant fires
usually marking the site of an encampment. Suttum, however,
well knew where the Bedouins would halt, and about an hour after
dark we heard the well-known voice of Dervish, and others of my
workmen, who, anxious at our delay, had come out to seek us.
The tents stood near a muddy pool of salt water, thick with loath-
some living things and camels' dung. The Arabs call the place
Om-el-Dhiban, "the mother of flies," from the insects which
swarm around it, and madden by their sting the camels and horses
that drink at the stagnant water.

Our encampment was full of Yezidis of the Kherraniyah tribe,
who had ridden from the tents to see me, bringing presents of
sheep, flour, and figs. They were at war, both with the Bedouins

---

* Its note resembles the cry of the camel-driver, when leading the herds
home at night, for which it is frequently mistaken.

and the inhabitants of the northern side of the mountain. My large tent was soon crowded with guests. They squatted down on the ground in double ranks. For the last time I spoke on the advantage of peace and union amongst themselves, and I exacted from them a solemn promise that they would meet the assembled tribes at the next great festival in the valley of Sheikh Adi, referring their differences in future to the decision of Hussein Bey, Sheikh Nasr, and the Cawals, instead of appealing to arms. I also reconciled them with the Bedouins, Suttum entering into an engagement for his tribe, and both parties agreeing to abstain from lifting each other's flocks when they should again meet in the pastures at the foot of the hills. The inhabitants of the Sinjar are too powerful and independent to pay *kowee**, or black-mail, to the Shammar, who, indeed, stand in much awe of their Yezidi enemies. They frequently raise their annual revenues, and enrich themselves almost entirely, at the expense of the Arabs. They watch their opportunity, when the tribes are migrating in the spring and autumn, and falling by night on their encamp-ments, plunder their tents, and drive off their cattle. Return-ing to the hills, they can defy in their fastnesses the revenge of the Bedouins.

The Yezidis returned to their encampment late at night, but about a hundred of their horsemen were again with me before the tents were struck in the morning. They promised to fulfil the engagements entered into on the previous evening, and accom-panied me for some miles on our day's journey. Cawal Yusuf returned with them on his way back to Mosul. It was agreed that he should buy, at the annual auction, the Mokhatta, or revenues of the Sinjar †, and save the inhabitants from the tyranny and exac-

---

* Literally, "strength-money:" the small tribes, who wander in the Desert, and who inhabit the villages upon its edge, are obliged to place themselves under the protection of some powerful tribe to avoid being utterly destroyed. Each great division of the Shammar receives a present of money, sheep, camels, corn, or barley, from some tribe or another for this protection, which is always respected by the other branches of the tribe. Thus the Jehesh paid *kowee* to the Boraij, the Jebours of the Khabour to Ferhan (the hereditary chief of all the Shammar), the people of Tel Afer to the Assaiyah. Should another branch of the Shammar plunder, or injure, tribes thus paying kowee, their protectors are bound to make good, or revenge, their losses.

† The revenues, *i. e.* the different taxes, tithes, &c. of some pashalics are sold by auction in the spring to the highest bidders, who pay the purchase-money, or give sufficient security, and collect the revenues themselves. This is a sys-tem which has contributed greatly to the ruin of some of the finest provinces in the empire

tions of the Turkish tax-gatherer. I wrote letters for him to the
authorities of Mosul, recommending such an arrangement, as
equally beneficial to the tranquillity of the mountain and the treasury of the Pasha.*

After leaving Om-el-Dhiban we entered an undulating country
crossed by deep ravines, worn by the winter torrents. Veins of
Mosul marble, the alabaster of the Assyrian sculptures, occasionally
appeared above the soil, interrupting the carpet of flowers spread
over the face of the country. We drew near to the low hills into
which the Sinjar subsides to the west. They are called Jeraiba, are
well wooded with the ilex and dwarf oak, and abound in springs,
near which the Shammar Bedouins encamp during the summer.
Skirting them we found a beaten path, the first we had seen since
entering the Desert, leading to the Jebour encampments on the
Khabour, and we followed it for the rest of the day. It seemed
irksome after wandering, as we had listed, over the boundless un-
trodden plain, to be again confined to the narrow track of the footsteps of man. However, the Bedouins declared that this pathway
led to the best water, and we had committed ourselves to their
guidance. Four hours' ride brought us to a scanty spring; half
an hour beyond we passed a second; and in five and a half hours
pitched the tents, for the rest of the day, near a small stream. All
these springs are called Maalaga, and rising in the gypsum or Mosul
marble, have a brackish and disagreeable taste. The Bedouins
declare that, although unpalatable, they are exceedingly wholesome, and that even their mares fatten on the waters of Jeraiba.

Near our tents were the ruins of an ancient village surrounded
by a wall. The spring once issued from the midst of them, but its
source had been choked by rubbish, which, as some hours of day-
light still remained, Hormuzd employed the Jebours and Tiyari in
removing. Before sunset the supply and quality of the water had
much improved. Suttum, who could not remain idle, wandered
over the plain on his deloul with his hawk in search of game, and
returned in the evening with a bag of bustards. He came to me
before nightfall, somewhat downcast in look, as if a heavy weight
were on his mind. At length, after various circumlocutions, he said
that his wife would not sleep under the white tent which I had

---

* Cawal Yusuf actually became the farmer of the revenues for a sum scarcely
exceeding 350*l*. The inhabitants of the Sinjar were greatly pleased by this con-
cession to one of their own faith, and were encouraged to cultivate the soil, and
to abstain from mutual aggressions.

lent her, such luxuries being, she declared, only worthy of city ladies, and altogether unbecoming the wife and daughter of a Bedouin. "So determined is she," said Suttum, "in the matter, that, Billah! she deserted my bed last night and slept on the grass in the open air; and now she swears she will leave me and return on foot to her kindred, unless I save her from the indignity of sleeping under a white tent." It was inconvenient to humour the fancies of the Arab lady, but as she was inexorable, I gave her a black Arab tent, used by the servants for a kitchen. Under this sheet of goat-hair canvass, open on all sides to the air, she said she could breathe freely, and feel again that she was a Bedouin.

As the sun went down we could distinguish, in the extreme distance, a black line marking the wooded banks of the Khabour, beyond which rose the dark hills of Abd-ul-Azeez. Columns of thin curling smoke showed that there were encampments of Bedouins between us and the river, but we could neither see their tents nor their cattle. The plains to the south of our encampment was bounded by a range of low hills, called Rhoua and Haweeza.

We crossed, during the following evening, a beautiful plain covered with sweet smelling flowers and aromatic herbs, and abounding in gazelles, hares, and bustards. We reached in about two hours the encampments, whose smoke we had seen during the preceding evening. They belonged to Bedouins of the Hamoud branch of the Shammar. The tents were pitched closely together in groups, as if the owners feared danger. We alighted at some distance from them to avoid entering them as guests. The chiefs soon came out to us, bringing camels' milk and bread. From them we learnt that they had lately plundered, on the high road between Mosul and Mardin, a caravan conveying, amongst other valuable loads, a large amount of government treasure. The Turkish authorities had called upon Ferhan, as responsible chief of the Shammar, to restore the money, threatening, in case of refusal, an expedition against the whole tribe. The Hamoud, unwilling to part with their booty, and fearing lest the rest of the Shammar might compel them to do so in order to avoid a war, were now retreating towards the north, and, being strong in horsemen, had openly defied Ferhan. They had been joined by many families from the Assaiyah, who had crossed the Euphrates, and united with the Aneyza on account of a blood feud with the Nejm. The

Hamoud are notorious for treachery and cruelty, and certainly the
looks of those who gathered round us, many of them grotesquely
attired in the plundered garments of the slaughtered Turkish soldiery, did not belie their reputation. They fingered every article
of dress we had on, to learn its texture and value.

Leaving their encampments, we rode through vast herds of
camels and flocks of sheep belonging to the tribe, and at length
came in sight of the river.

The Khabour flows through the richest pastures and meadows.
Its banks were now covered with flowers of every hue, and its
windings through the green plain were like the coils of a mighty
serpent. I never beheld a more lovely scene. An uncontrollable
emotion of joy seized all our party when they saw the end of
their journey before them. The horsemen urged their horses to
full speed; the Jebours dancing in a circle, raised their colored
kerchiefs on their spears, and shouted their war cry, Hormuzd
leading the chorus; the Tiyari sang their mountain songs and fired
their muskets into the air.

Trees in full leaf lined the water's edge. From amongst them
issued a body of mounted Arabs. As they drew nigh we recognised at their head Mohammed Emin, the Jebour Sheikh, and his
sons, who had come out from their tents to welcome us. We dismounted to embrace, and to exchange the usual salutations, and
then rode onwards, through a mass of flowers, reaching high above
the horses' knees, and such as I had never before seen, even in the
most fertile parts of the Mesopotamian wilderness.

The tents of the chief were pitched under the ruins of Arban,
and on the right or northern bank of the river, which was not at
this time fordable. As we drew near to them, after a ride of
nearly two hours, Mohammed Emin pointed in triumph to the
sculptures, which were the principal objects of my visit. They
stood a little above the water's edge, at the base of a mound of
considerable size. We had passed several *tels* and the double banks
of ancient canals, showing that we were still amidst the remains
of ancient civilisation. Flocks of sheep and herds of camels
were spread over the meadows on both sides of the river. They
belonged to the Jebours, and to a part of the Boraij tribe under
Moghamis, a distinguished Arab warrior, and the uncle of Suttum.
Buffaloes and cattle tended by the Sherabbeen and Buggara, small
clans pasturing under the protection of Mohammed Emin, stood

lazily in the long grass, or sought refuge in the stream from the flies and noonday heat.

At length we stopped opposite to the encampment of the Jebour Sheikh, but it was too late to cross the river, some time being required to make ready the rafts. We raised our tents, therefore, for the night on the southern bank. They were soon filled by a motley group of Boraij, Hamoud, Assaiyah, and Jebour Arabs. Moghamis himself came shortly after our arrival, bringing me as a present a well-trained hawk and some bustards, the fruits of his morning's sport. The falcon was duly placed on his stand in the centre of the spacious tent, and remained during the rest of my sojourn in the East a member of my establishment. His name was Fawaz, and he was a native of the hills of Makhhoul, near Tekrit, celebrated for their breed of hawks. He was of the species called "chark," and had been given by Sadoun-el-Mustafa, the chief of the great tribe of Obeid, to Ferhan, the sheikh of the Shammar, who had bestowed him in token of friendship on Moghamis.

A Sheikh of the Hamoud also brought us a wild ass-colt, scarcely two months old, which had been caught whilst following its dam, and had been since fed upon camel's milk.* Indeed, nearly all

---

* I am indebted to Mr. Grey for the following remarks on the skin of a young wild ass brought by me to this country : — "It is, I have no doubt, the wild ass, or onager of the ancients. It is evidently the same as the ass without a stripe, which has been described by several authors as the *Equus Hemionus*, found in Cutch, and quite distinct from the *Equus Hemionus* described by Pallas as found in the snowy mountains of Asia, and called by Mr. Hodgson *Equus Kiang* and *E. polyodon*. The wild ass, or onager, was one of the desiderata of zoologists, as it was only described from some specimens seen at a distance, and not from the examination of specimens, and is characterised by being said to have larger and more acute ears than the *Hemione* of Pallas. I do not find this to be the case in the young specimen you have sent to the Museum. The great difference between the wild ass of the plains of Mesopotamia and the Hemione of Tibet is, that the former is a yellowish white, and the latter a bright bay in summer, both being greyish white in winter. There is also some difference in the forms of the skull, and in the disposal of the hole for the transmission of the bloodvessels and nerves of the face." The Arabs of Mesopotamia frequently capture this beautiful animal when young, and generally kill it at once for food. It is almost impossible to take it when full grown. The colt mentioned in the text died before we returned to Mosul. A second, after living eight or nine months, also died ; and a third met with the same fate. I was desirous of sending a live specimen to England, but thus failed in all my attempts to rear one. They became very playful and docile. That which I had at Mosul followed like a dog.

those who came to my tent had some offering, either sheep, milk, curds, or butter; even the Arab boys had caught for us the elegant jerboa, which burrows in vast numbers on the banks of the river.   Suitable presents were made in return.   Dinner was cooked for all our guests, and we celebrated our first night on the Khabour by general festivities.

Suttum, with his Wife, on his Dromedary.

Sheikh Mohammed Emin

CHAP. XII.

ARBAN. — OUR ENCAMPMENT. — SUTTUM AND MOHAMMED EMIN. — WINGED
BULLS DISCOVERED. — EXCAVATIONS COMMENCED. — THEIR RESULTS. — DIS-
COVERY OF SMALL OBJECTS — OF SECOND PAIR OF WINGED BULLS — OF LION —
OF CHINESE BOTTLE — OF VASE — OF EGYPTIAN SCARABS — OF TOMBS. —
THE SCENE OF THE CAPTIVITY.

ON the morning after our arrival in front of the encampment of
Sheikh Mohammed Emin we crossed the Khabour on a small raft,
and pitched our tents on its right, or northern, bank. I found the
ruins to consist of a large artificial mound of irregular shape,

F. Cooper

John Murray, Albemarle Street, 1852.

N. Chevalier, lith.

Mound of Arban on the Khabour.

washed, and indeed partly carried away by the river which was gradually undermining the perpendicular cliff left by the falling earth. The Jebours were encamped to the west of it. I chose for our tents a recess, like an amphitheatre, facing the stream. We were thus surrounded and protected on all sides. Be-hind us and to the east rose the mound, and to the west were the family and dependents of Mohammed Emin. In the Desert, beyond the ruins, were scattered far and wide the tents of the Jebours, and of several Arab tribes who had placed themselves under their protection; the Sherabeen, wandering keepers of herds of buffaloes; the Buggara, driven by the incursions of the Aneyza from their pasture grounds at Ras-al-Ain (the source of the Khabour); and some families of the Jays, a large clan residing in the district of Orfa, whose sheikh having quarrelled with his brother chiefs had now joined Mohammed Emin. From the top of the mound the eye ranged over a level country bright with flowers, and spotted with black tents, and innumerable flocks of sheep and camels. During our stay at Arban the color of these great plains was undergoing a continual change. After being for some days of a golden yellow, a new family of flowers would spring up, and it would turn almost in a night to a bright scarlet, which would again as suddenly give way to the deepest blue. Then the meadows would be mottled with various hues, or would put on the emerald green of the most luxuriant of pastures. The glowing de-scriptions I had so frequently received from the Bedouins of the beauty and fertility of the banks of the Khabour were more than realised. The Arabs boast that its meadows bear three distinct crops of grass during the year, and the wandering tribes look upon its wooded banks and constant greensward as a paradise during the summer months, where man can enjoy a cool shade, and beast can find fresh and tender herbs, whilst all around is yellow, parched, and sapless.

In the extreme distance, to the east of us, rose a solitary conical elevation, called by the Arabs, Koukab. In front, to the south, was the beautiful hill of the Sinjar, ever varying in color and in outline as the declining sun left fresh shadows on its furrowed sides. Behind us, and not far distant, was the low, wooded range of Abd-ul-Azeez. Artificial mounds, smaller in size than Arban, rose here and there above the thin belt of trees and shrubs skirting the river bank.

I had brought with me a tent large enough to hold full two hundred persons, and intended as a "museef," or place of reception.

T

always open to the wayfarer and the Arab visitor; for the first duty of a traveller wishing to mix with true Bedouins, and to gain an influence over them, is the exercise of hospitality. This great pavilion was pitched in the centre of my encampment, with its entrance facing the river. To the right were the tents of the Cawass and servants; one fitted up expressly for the Doctor to receive patients, of whom there was no lack at all times, and the black Arab tent of Rathaiyah, who would not mix with the Jebours. To the left were those of my fellow travellers, and about 200 yards beyond, near the excavations, my own private tent, to which I retired during the day, when wishing to be undisturbed, and to which the Arabs were not admitted. In it, also, we usually breakfasted and dined, except when there were any Arab guests of distinction with whom it was necessary to eat bread. In front of our encampment, and between it and the river, was a small lawn, on which were picketed our horses. Suttum and Mohammed Emin usually eat with us, and soon became perfectly reconciled to knives and forks, and the other restraints of civilised life. Suttum's tact and intelligence were indeed remarkable. Nothing escaped his hawk-like eye. A few hours had enabled him to form a correct estimate of the character of each one of the party, and he had detected peculiarities which might have escaped the notice of the most observant European. The most polished Turk would have been far less at home in the society of ladies, and during the whole of our journey he never committed a breach of manners, only acquired after a few hours' residence with us. As a companion he was delightful, — full of anecdote, of unclouded spirits, acquainted with the history of every Bedouin tribe, their politics and their wars, and intimate with every part of the Desert, its productions and its inhabitants. Many happy hours I spent with him, seated, after the sun went down, on a mound overlooking the great plain and the winding river, listening to the rich flow of his graceful Bedouin dialect, to his eloquent stories of Arab life, and to his animated descriptions of forays, wars, and single combats.

Mohammed Emin, the Sheikh of the Jebours, was a good-natured portly Arab, in intelligence greatly inferior to Suttum, and wanting many of the qualities of the pure Bedouin. During our intercourse I had every reason to be satisfied with his hospitality and the cordial aid he afforded me. His chief fault was a habit of begging for every thing. Always willing to give he was equally ready to receive. In this respect, however, all Arabs are alike, and when the habit is understood it is no longer a source of inconvenience, as on a refusal no offence is taken. The Jebour chief was a complete patriarch in his tribe, having no less than sixteen children, of whom six sons were horsemen and the owners of mares. The youngest, a boy of four years old named Sultan, was his favorite. His usual costume consisted simply of a red Turkish skull cap, or fez, on his head. He scarcely ever left his father, who always brought the child with him when he came to our tent. He was as handsome and dirty as the best of Arab children. His mother, who had recently died, was the beautiful sister of Abd-rubbou. I chanced to be her brother's guest when the news of her death was brought to him. An Arab of the tribe, weary and wayworn, entered the tent and seated himself without giving the usual salutation ; all present knew that he had come from the Khabour and from distant friends. His silence argued evil tidings. By an indirect remark, immediately understood, he told his errand to one who sat next him, and who in turn whispered it to Sheikh Ibrahim, the chief's uncle. The old man said aloud, with a sigh, "It is the will and mercy of God; she is not dead but released!" Abd-rubbou at once understood of whom he spake. He arose and went forth, and the wailing of the mother and of the women soon issued from the inner recesses of the tent.

We were for a day or two objects of curiosity to the Arabs who assembled in crowds around our tents. Having never before seen an European, it was natural that they should hasten to examine the strangers. They soon, however, became used to us, and things went on as usual. It is a circumstance well worthy of mention, and most strongly in favor of the natural integrity of the Arab when his guests are concerned, that during the whole of our journey and our residence on the Khabour, although we lived in open tents, and property of all kinds was scattered about, we had not to complain of a single loss from theft.

My first care, after crossing to Arban, was to examine the sculptures described by the Arabs. The river having gradually worn away the mound had, during the recent floods, left uncovered a pair of winged human-headed bulls, some six feet above the water's edge, and full fifty beneath the level of the ruin. Only the forepart of these figures had been exposed to view, and Mohammed Emin would not allow any of the soil to be removed before my arrival. The earth was soon cleared away, and I found them to be of a coarse limestone, not exceeding 5½ feet in height by 4½ in length. Between them was a pavement slab of the same material. They resembled in general form the well-known winged bulls of Nineveh, but in the style of art they differed considerably from them. The outline and treatment was bold and angular, with an archaic feeling conveying the impression of great antiquity. They bore the same relation to the more delicately finished and highly ornamented sculptures of Nimroud, as the earliest remains of Greek art do to the exquisite monuments of Phidias and Praxiteles. The human features were unfortunately much injured, but such parts as remained were sufficient to show that the countenance had a peculiar character, differing from the Assyrian type. The sockets of the eyes were deeply sunk,

probably to receive the white and the ball of the eye in ivory or glass. The nose was flat and large, and the lips thick and overhanging like those of a negro. Human ears were attached to the head, and bull's ears to the horned cap, which was low and square at the top, not high and ornamented like those of Khorsabad and Kouyunjik, nor rounded like those of Nimroud. The hair was elaborately curled, as in the pure Assyrian sculptures, though more rudely carved. The wings were small in proportion to the size of the body, and had not the majestic spread of those of the bulls that adorned the palaces of Nineveh. Above the figure were the following characters*, which are purely Assyrian.

Front View of Winged Bull at Arban.

It would appear from them that the sculptures belonged to the palace of a king whose name has been found on no other monument. No titles are attached to it, not even that of "king;" nor is the country over which he reigned mentioned; so that some doubt may exist as to whether it really be a royal name.

The great accumulation of earth above these sculptures proves that, since the destruction of the edifice in which they stood, other

---

* The last letter is in one instance omitted. For a drawing of the bull see woodcut at the end of the chapter.

habitations have been raised upon its ruins. Arban, indeed, is mentioned by the Arab geographers as a flourishing city, in a singularly fertile district of the Khabour. Part of a minaret, whose walls were cased with colored tiles, and ornamented with cufic inscriptions in relief, like that of the Sinjar, and the foundations of buildings, are still seen on the mound; and at its foot, on the western side, are the remains of a bridge which once spanned the stream. But the river has changed its course. The piers, adorned with elegantly shaped arabesque characters, are now on the dry land.

I will describe, at once, the results of the excavations carried on during the three weeks our tents were pitched at Arban. To please the Jebour Sheikh, and to keep around our encampment, for greater security, a body of armed men, when the tribe changed their pastures, I hired about fifty of Mohammed Emin's Arabs, and placed them in parties with the workmen who had accompanied me from Mosul. Tunnels were opened behind the bulls already uncovered, and in various parts of the ruins on the same level. Trenches were also dug into the surface of the mound.

Behind the bulls were found various Assyrian relics; amongst them a copper bell, like those from Nimroud, and fragments of bricks with arrow-headed characters painted yellow with white outlines, upon a pale green ground. In other parts of the mound were discovered glass and pottery, some Assyrian, others of a more doubtful character. Several fragments of earthenware, ornamented with flowers and scrollwork, and highly glazed, had assumed the brilliant and varied iridescence of ancient glass.*

It was natural to conclude, from the usual architectural arrangement of Assyrian edifices, that the two bulls described stood at an entrance to a hall, or chamber. We searched in vain for the remains of walls, although digging for three days to the right and left of the sculptures, a work of considerable difficulty in consequence of the immense heap of superincumbent earth. I then directed a tunnel to be carried towards the centre of the mound, hoping to find a corresponding doorway opposite. I was not disappointed. On the fifth day a similar pair of winged bulls were discovered. They were of the same size, and inscribed with the same characters. A part of one having been originally broken off, either in carving the sculpture or in moving it, a fresh piece of stone had

* These relics are now in the British Museum.

been carefully fitted into its place. I also dug to the right and
left of these sculptures for remains of walls, but without success,
and then resumed the tunnelling towards the centre of the mound.
In a few days a lion, with extended jaws, sculptured in the same
coarse limestone, and in the same bold archaic style as the bulls,
was discovered. It had five legs, and the tail had the claw at

Lion discovered at Arban.

the end, as in the Nineveh bas-reliefs. In height it was nearly
the same as the bulls. I searched in vain for the one which must
have formed the opposite side of the doorway.

With the exception of these sculptures no remains of building
were found in this part of the mound. In another tunnel, opened
at some distance from the bulls, half of a human figure in relief
was discovered.* The face was in full. One hand grasped a sword
or dagger; the other held some object to the breast. The hair and
beard were long and flowing, and ornamented with a profusion of

* The height of this fragment was 5 ft. 8 in.

curls as in the Assyrian bas-reliefs. The head-dress appeared to consist of a kind of circular helmet, ending in a sharp point. The

treatment and style marked the sculpture to be of the same period as the bull and lion.

Such were the sculptures discovered in the mound of Arban. Amongst smaller objects of different periods were some of considerable interest, jars, vases, funeral urns, highly-glazed pottery, and fragments of glass. In a trench, on the south side of the ruin, was found a small green and white bottle, inscribed with Chinese characters. A similar relic was brought to me subsequently by an Arab from a barrow in the neighbourhood. Such bottles have been discovered in Egyptian tombs, and considerable doubt exists as to their antiquity, and as to the date and manner of their importa-

Bas-relief discovered at Arban.

tion into Egypt.* The best opinion now is that they are comparatively modern, and that they were probably brought by the Arabs, in the eighth or ninth century, from the kingdoms of the far East,

with which they had at that period extensive commercial intercourse. Bottles precisely similar are still offered for sale in the bazars at Cairo, and are used to hold the kohl, or powder for staining the eyes of ladies.

A jar, about four feet high, in coarse half-baked clay, was dug out of the centre of the mound. The handles were formed by rudely-designed human figures, and the sides covered with grotesque representations of men and animals, and arabesque ornaments in relief.

Chinese Bottle discovered at Arban.

Vases of the same material, ornamented with figures, are frequently discovered in digging the foundations of houses in the

* Wilkinson, in his "Ancient Egyptians," vol. iii. p. 107., gives a drawing of a bottle precisely similar to that described in the text, and mentions one which, according to Rosellini, had been discovered in a *previously* unopened tomb, believed to be of the 18th dynasty ; but there appears to be considerable doubt on the subject.

modern town of Mosul. They appear to belong to a comparatively recent period, later probably than the Christian era, but previous to the Arab occupation. As they have upon them human figures, dressed in a peculiar costume, consisting of a high cap and embroi-

dered robes, I should attribute them to the Persians. A vase, similar in size and shape to that of Arban, and also covered with grotesque representations of monstrous animals, the finest specimen I have seen of this class of antiquities, was found beneath the foundations of the very ancient Chaldæan church of Meskinta at Mosul, when that edifice was pulled down and rebuilt two years ago.* It was given to me by the Catholic Chaldæan Patriarch to whom it belonged as chief of the community, but was unfortunately destroyed, with other interesting relics, by the Arabs, who plundered a raft laden with antiqui-

Figure in Pottery from Mosul.

ties, on its way to Baghdad, after my return to Europe.

Amongst other relics discovered at Arban were, a large copper ring, apparently Assyrian; an ornament in earthenware, resembling the pine-cone of the Assyrian sculptures; a bull's head in terracotta; fragments of painted bricks, probably of the same period; and several Egyptian scarabæi. It is singular that engraved stones and scarabs bearing Egyptian devices, and in some

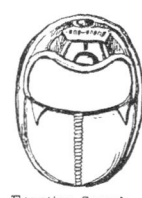

instances even royal cartouches, should have been found on the banks of the Khabour. Similar objects were subsequently dug up at Nimroud, and brought to me by the Arabs from various ruins in Assyria. I will take this opportunity of adding the following remarks by Mr. Birch on those deposited in the British Museum.

Egyptian Scarab, from Arban.

* In laying the foundations of the new church, the tombs of two of the early Chaldæan patriarchs were discovered amongst other objects of interest. The bodies, being still preserved, were, of course, canonised at once, and turned into a source of profit by the bishop, the faithful paying a small sum for permission to touch the sacred relics. One had been head of the Oriental church before the Arab invasion. By his side was his crozier ending in a silver crook, on which was an inscription in Chaldee letters. The second was of a rather later period. His crozier was of ebony, surmounted by a ball of glass, and inscribed with the earliest cufic characters. I examined these interesting relics immediately after their discovery.

1. A scarabæus, having on the base *Ra-men-chepr*, the prenomen of Thothmes III. Beneath is a scarab between two feathers, placed on the basket *sub*.

2. A scarabæus in dark steaschist, with the figure of the sphinx (the sun), and an emblem between the forepaws of the monster. The sphinx constantly appears on the scarabæi of Thothmes III., and it is probably to this monarch that the one here described belongs.* After the sphinx on this scarab, are the titles of the king, " the sun placer of creation," of Thothmes III.

3. Small scarabæus of white steaschist, with a brownish hue ; reads *Neter nefer nebta Ra-neb-ma*, " The good God, the Lord of the earth, the sun, the Lord of truth, rising in all lands." This is of Amenophis III., one of the last kings of the eighteenth dynasty, who flourished about the fifteenth century B. C., and who records amongst his conquests As-su-ru (Assyria), Naharaina (Mesopotamia), the Saenkar (Shinar or Sinjar), and Pattana (Padan Aram). The expression, " who rises in all lands," refers to the solar character of the king, and to his universal dominion.

4. Scarabæus in white steaschist, with an abridged form of the prenomen of Thothmes III., *Ra men cheper at en Amen*, " The sun-placer of creation, the type of Ammon." This monarch was the greatest monarch of the eighteenth dynasty, and conquered Naharaina and the Saenkar, besides receiving tribute from Babel or Babylon and Assyria.

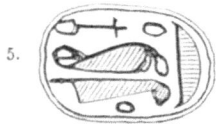

5. Scarabæus in pale white steaschist, with three emblems that cannot well be explained. They are the sun's disk, the ostrich feather, the uræus, and the guitar nabluim. They may mean " Truth the good goddess," or " lady," or *ma nefer*, " good and true."

6. Scarabæus in the same substance, with a motto of doubtful meaning.

7. Scarabæus, with a hawk, and God holding the emblem of life, and the words *ma nefer*, " good and true." The meaning very doubtful.

---

* On many scarabæi in the British Museum, and on those figured by Klaproth from the Palin Collection, in Leeman's Monuments, and in the " Description de l'Égypte," Thothmes is represented as a sphinx treading foreign prisoners under him

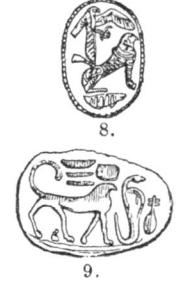

8.

10.

11.

8. A scarabæus, with a hawk-headed gryphon, emblem of *Menta-Ra*, or Mars. Behind the monster is the goddess Sati, or Nuben. The hawk-headed lion is one of the shapes into which the sun turns himself in the hours of the day. It is a common emblem in the Aramæan religion.

9. Scarabæus, with hawk-headed gryphon, having before it the uræus and the " *nabla* " or guitar, hieroglyphic of good. Above it are the hieroglyphs " Lord of the earth."

10. Small scarabæus in dark steaschist, with a man in adoration to a king or deity, wearing the crown of the upper country, and holding in the left hand a lotus flower. Between them is the emblem of life.

11. Scarabæus, with the hawk-headed scarabæus, emblem of *Ra-cheper*, " the Creator Sun," flying with expanded wings, four in number, which do not appear in Egyptian mythology till after the time of the Persians, when the gods assume a more Pantheistic form. Such a representation of the sun, for instance, is found on the Torso Borghese.

It will be observed that most of the Egyptian relics discovered in the Assyrian ruins are of the time of the 18th Egyptian dynasty, or of the 15th century before Christ; a period when, as we learn from Egyptian monuments, there was a close connection between Assyria and Egypt.

Several tombs were also found in the ruins, consisting principally of boxes, or sarcophagi, of earthenware, like those existing above the Assyrian palaces near Mosul. Some, however, were formed by two large earthen jars, like the common Eastern vessel for holding oil, laid horizontally, and joined mouth to mouth. These terracotta coffins appear to be of the same period as those found in all the great ruins on the banks of the rivers of Mesopotamia, and are not Assyrian.* They contained human remains turned to dust, with the exception of the skull and a few of the larger bones, and generally three or four urns of highly-glazed blue pottery.

Fewer remains and objects of antiquity were discovered in the mounds on the Khabour than I had anticipated. They were sufficient, however, to prove that the ruins are, on the whole, of the same character as those on the banks of the Tigris. That the Assyrian empire at one time embraced the whole of Mesopotamia, including the country watered by the Khabour, there can be no

* Most of the small objects described in the text are now in the British Museum.

doubt, as indeed is shown by the inscriptions on the monuments of Nineveh. Whether the sculptures at Arban belong to the period of Assyrian domination, or to a distinct nation afterwards conquered, or whether they may be looked upon as cotemporary with, or more ancient than, the bas-reliefs of Nimroud, are questions not so easily answered. The archaic character of the treatment and design, the peculiar form of the features, the rude though forcible delineation of the muscles, and the simplicity of the details, certainly convey the impression of greater antiquity than any monuments hitherto discovered in Assyria Proper.*

A deep interest, at the same time, attaches to these remains from the site they occupy. To the Chebar were transported by the Assyrian king, after the destruction of Samaria, the captive children of Israel, and on its banks " the heavens were opened " to Ezekiel, and " he saw visions of God," and spake his prophecies to his brother exiles.† Around Arban may have been pitched the tents of the sorrowing Jews, as those of the Arabs were during my visit. To the same pastures they led their sheep, and they drank of the same waters. Then the banks of the river were covered with towns and villages, and a palace-temple still stood on the mound, reflected in the transparent stream. We have, however, but one name connected with the Khabour recorded in Scripture, that of Tel-Abib, "the mound of Abib, or, of the heaps of ears of corn," but whether it applies to a town, or to a simple artificial elevation, such as still abound, and are still called " tels," is a matter of doubt. I sought in vain for some trace of the word amongst the names now given by the wandering Arab to the various ruins on the Khabour and its confluents.‡

---

* A lion very similar to that discovered at Arban, though more colossal in its dimensions, exists near Seroug. (Chesney's Expedition, vol. i. p. 114.)

† 2 Kings, xvii. 6. Ezek. i. 1. In the Hebrew text the name of this river is spelt in two different ways. In Kings we have חָבוֹר, Khabour, answering exactly to the Chaboras of the Greeks and Romans, and the Khabour of the Arabs. In Ezekiel it is written כְּבָר, Kebar. There is no reason, however, to doubt that the same river is meant.

‡ The name occurs in Ezekiel, iii. 15. " Then I came to them of the captivity at *Tel-Abib*, that dwelt by the river of Chebar." In the Theodosian tables we find *Thallaba* on the Khabour, with which it may possibly be identified. (Illustrated Commentary on the Old and New Testaments, published by Charles Knight, a very useful and well-digested summary, in note to word.) It is possible that Arbonad, a name apparently given to the Khabour in Judith, ii. 24., may be connected with Arban : however, it is not quite clear what river is really meant, as there appears to be some confusion in the geographical details. The cities on the Khabour, mentioned by the Arab geographers, are Karkisia (Circesium, at the junction of the river with the Euphrates), Makeseen (of which I could find no trace), Arban, and Khabour. I have not been able to discover the site of

We know that Jews still lingered in the cities of the Khabour until long after the Arab invasion; and we may perhaps recognise in the Jewish communities of Ras-al-Ain, at the sources of the river, and of Karkisia, or Carchemish, at its confluence with the Euphrates, visited and described by Benjamin of Tudela, in the latter end of the twelfth century of the Christian æra, the descendants of the captive Israelites.

But the hand of time has long since swept even this remnant away, with the busy crowds which thronged the banks of the river. From its mouth to its source, from Carchemish to Ras-al-Ain, there is now no single permanent human habitation on the Khabour. Its rich meadows and its deserted ruins are alike become the encamping places of the wandering Arab.

any ruin of the same name as the river. Karkisia, when visited in the twelfth century by Benjamin of Tudela, contained about 500 Jewish inhabitants, under two Rabbis. According to Ibn Haukal, it was surrounded by gardens and cultivated lands. The spot is now inhabited by a tribe of Arabs.

Winged Bull discovered at Arban.

Arab Women grinding Corn with a Handmill rolling out the Dough, and baking the Bread.

CHAP. XIII.

In the preceding chapter I have given an account of the discoveries made in the ruins of Arban, I will now add a few notes of our residence on the Khabour. A sketch of Arab life, and a description of a country not previously visited by European travellers, may be new and not uninteresting to my readers.

During the time we dwelt at Arban, we were the guests and under the protection of Mohammed Emin, the Sheikh of the Jebours. On the day we crossed the river, he celebrated our arrival by a feast after the Arab fashion, to which the notables of the tribe were invited. Sheep, as usual, were boiled and served up piecemeal in large wooden bowls, with a mass of butter and bread soaked in the gravy. The chief's tent was spacious, though poorly furnished. It was the general resort of those who chanced to wander, either on business or for pleasure, to the Khabour, and was, consequently, never without a goodly array of guests ; from a company of Shammar horsemen out on a foray to the solitary Bedouin who was seeking to become a warrior in his tribe, by first stealing a mare from some hostile encampment.

Amongst the strangers partaking, at the time of our visit, of the Sheikh's hospitality, were Serhan, a chief of the Agaydat, and Dervish Agha, the hereditary Lord of Nisibin, the ancient Nisibis. The tents of the former were at the junction of the Khabour and Euphrates, near Karkisia (the ancient Carchemish), or, as it is more generally called by the Arabs, Abou-Psera.* The fertile meadows near the confluence of the two rivers formerly belonged to the Jebours, who occupied the banks of the Khabour throughout nearly the whole of its course. An old feud kept them at con-tinual war with the great tribe of the Aneyza. They long suc-cessfully struggled with their enemies, but having at length been overpowered by superior numbers, they lost their horses, their flocks, their personal property, and even their tents. Thus left naked and houseless, they sought refuge in the neighbourhood of Mosul, and learnt to cultivate the soil and to become subjects of the Turks. The Agaydat, who before dwelt principally on the western banks of the Euphrates, crossed the river and seized the deserted pastures. The Jebours who had returned to the Khabour, claimed their former encamping grounds, and threat-ened to reoccupy them by force of arms. It was to settle these differences that Serhan had visited Mohammed Emin. After remaining two or three days, he went back to his tents without, however, having succeeded in his mission. I learnt from him that there were many artificial mounds near the confluence of the rivers, but he had never heard, nor had Mohammed Emin, of any sculptures, or other monuments of antiquity, having been found in them.

Dervish Agha, of Kurdish descent, was the representative of an ancient family, whose members were formerly the semi-inde-pendent chiefs of Nisibin and the surrounding districts. He was still the recognised Mutesellim, or governor of that place, and had been sent to Mohammed Emin by the commander of the Turkish

---

* Col. Chesney states that the real name is " Abou Serai," " the father (or chief) of palaces;" such may be the case.

troops, one Suleiman Agha, who was at this time encamped in the plain beneath Mardin. His business was to prevail upon the Jebour Sheikh to assist Ferhan in recovering the plundered treasure from the Hamoud, and to visit afterwards the encampment of the Agha, with both which requests his host had good reasons not to comply.

My own large tent was no less a place of resort than that of Mohammed Emin, and as we were objects of curiosity, Bedouins from all parts flocked to see us. With some of them I was already acquainted, having either received them as my guests at Mosul, or met them during excursions in the Desert. They generally passed one night with us, and then returned to their own tents. A sheep was always slain for them, and boiled with rice, or prepared wheat, in the Arab way : if there were not strangers enough to consume the whole, the rest was given to the workmen or to the needy, as it is considered derogatory to the character of a truly hospitable and generous man to keep meat until the following day, or to serve it up a second time when cold. Even the poorest Bedouin who kills a sheep, invites all his friends and neighbours to the repast, and if there be still any remnants, distributes them amongst the poor and the hungry, although he should himself want on the morrow.

We brought provision of flour with us, and the Jebours had a little wheat raised on the banks of the river. The wandering Arabs have no other means of grinding their corn than by handmills, which they carry with them wherever they go. They are always worked by the women, for it is considered unworthy of a man to engage in any domestic occupation. These handmills are simply two circular flat stones, generally about eighteen inches in diameter, the upper turning loosely upon a wooden pivot, and moved quickly round by a wooden handle. The grain is poured through the hole of the pivot, and the flour is collected in a cloth spread under the mill. It is then mixed with water, kneaded in a wooden bowl, and pressed by the hand into round balls ready for baking. During these processes, the women are usually seated on the bare ground : hence, in Isaiah *, is the daughter of Babylon told to sit in the dust and on the ground, and " to take the millstones to grind meal."

The tribes who are always moving from place to place bake their bread on a slightly convex iron plate, called a *sadj*, mode-

* xlvii. 1, 2.

rately heated over a low fire of brushwood or camels' dung. The lumps of dough are rolled, on a wooden platter, into thin cakes, a foot or more in diameter, and laid by means of the roller upon the iron. They are baked in a very short time, and should be eaten hot.* The Kurds, whose flour is far whiter and more carefully prepared than that of the Arabs, roll the dough into large cakes, scarcely thicker than a sheet of paper. When carefully baked by the same process, it becomes crisp and exceedingly agreeable to the taste. The Arab tribes, that remain for many days in one place, make rude ovens by digging a hole about three feet deep, shaping it like a reversed funnel, and plastering it with mud. They heat it by burning brushwood within, and then stick the lumps of dough, pressed into small cakes about half an inch thick, to the sides with the hand. The bread is ready in two or three minutes. When horsemen go on an expedition, they either carry with them the thin bread first described, or a bag of flour, which, when they come to water, they moisten and knead on their cloaks, and then bake by covering the balls of dough with hot ashes. All Arab bread is unleavened.

If a Bedouin tribe be moving in great haste before an enemy, and should be unable to stop for many hours, or be making a forced march to avoid pursuit over a desert where the wells are very distant from each other, the women sometimes prepare bread whilst riding on camels. The fire is then lighted in an earthen vessel. One woman kneads the flour, a second rolls out the dough, and a third bakes, boys or women on foot passing the materials, as required, from one to the other. But it is very rare that the Bedouins are obliged to have recourse to this process, and I have only once witnessed it.

The fuel used by the Arabs consists chiefly of the dwarf shrubs, growing in most parts of the Desert, of dry grass and of camels' dung. They frequently carry bags of the latter with them when in summer they march over very arid tracts. On the banks of the great rivers of Mesopotamia, the tamarisk and other trees furnish them with abundant firewood. They are entirely dependent

---

* See woodcut at the head of this chapter. Such was probably the process of making bread mentioned in 2 Sam. xiii. 8, 9. "So Tamar went to her brother Amnon's house; and he was laid down. And she took flour and kneaded it, and made cakes *in his sight*, and did bake the cakes. And she took a pan and poured them out before him." It will be observed that the bread was made at once, without leaven; such also was probably the bread that Abraham commanded Sarah to make for the three angels. (Gen. xviii. 6.)

for their supplies of wheat upon the villages on the borders of the Desert, or on the sedentary Arabs, who, whilst living in tents, cultivate the soil.   Sometimes a tribe is fortunate enough to plunder a caravan laden with corn, or to sack the granaries of a village; they have then enough to satisfy their wants for some months. But the Bedouins usually draw near to the towns and cultivated districts soon after the harvest, to lay in their stock of grain.   A party of men and women, chosen by their companions, then take with them money, or objects for sale or exchange, and drive the camels to the villages, where they load them and return to their tents.   Latterly a new and very extensive trade has been opened with the Bedouins for the wool of their sheep, much prized for its superior quality in European markets.   As the time for shearing is soon after the harvest, the Arabs have ready means of obtaining their supplies, as well as of making a little money, and buying finery and arms.

Nearly the whole revenue of an Arab Sheikh, whatever it may be, is laid out in corn, rice, and other provisions.   The quantity of food consumed in the tents of some of the great chiefs of the Bedouins is very considerable.   Almost every traveller who passes the encampment eats bread with the Sheikh, and there are generally many guests dwelling under his canvas.   In times of difficulty or scarcity, moreover, the whole tribe frequently expects to be fed by him, and he considers himself bound, even under such circumstances, by the duties of hospitality, to give all that he has to the needy.   The extraordinary generosity displayed on such occasions by their chiefs forms some of the most favourite stories of the Arabs.

The common Bedouin can rarely get meat.   His food consists almost exclusively of wheaten bread with truffles, which are found in great abundance during the spring, a few wild herbs, such as asparagus, onions, and garlic, fresh butter, curds, and sour milk. But, at certain seasons, even these luxuries cannot be obtained; for months together he often eats bread alone.   The Sheikhs usually slay a sheep every day, of which their guests, a few of their relatives, and their immediate adherents partake.   The women prepare the food, and always eat after the men, who rarely leave them much wherewith to satisfy their hunger.

The dish usually seen in a Bedouin tent is a mess of boiled meat, sometimes mixed with onions, upon which a lump of fresh butter is placed and allowed to melt.   The broad tail of the Mesopotamian

U

sheep is used for grease when there is no butter. Sometimes cakes of bread are laid under the meat, and the entertainer tearing up the thin loaves into small pieces, soaks them in the gravy with his hands. The Aneyza make very savory dishes of chopped meat and bread mixed with sour curds, over which, when the huge platter is placed before the guest, is poured a flood of melted butter. Roasted meat is very rarely seen in a Bedouin tent. Rice is only eaten by the Sheikhs, except amongst the tribes who en-camp in the marshes of Southern Mesopotamia, where rice of an inferior quality is very largely cultivated. There it is boiled with meat and made into pilaws.

The Bedouins do not make cheese. The milk of their sheep and goats is shaken into butter or turned into curds : it is rarely or never drank fresh, new milk being thought very unwholesome, as by experience I soon found it to be, in the Desert. I have frequently had occasion to describe the process of making butter by shaking the milk in skins. This is also an employment confined to the women, and one of a very laborious nature. The curds are formed by boiling the milk, and then putting some of the curds made on the previous day into it and allowing it to stand. When the sheep no longer give milk, some curds are dried, to be used as leaven on a future occasion. This preparation, called leben, is thick and acid, but very agreeable and grateful to the taste in a hot climate. The sour milk, or sheneena, an universal beverage amongst the Arabs, is either butter-milk pure and diluted, or curds mixed with water. Camel's milk is drank fresh. It is pleasant to the taste, rich, and exceedingly nourishing. It is given in large quantities to the horses. The Shammar and Aneyza Bedouins have no cows or oxen, those animals being looked upon as the peculiar property of tribes who have forgotten their independence, and degraded themselves by the cultivation of land. The sheep are milked at dawn, or even before daybreak, and again in the evening on their return from the pastures. The milk is immediately turned into leben, or boiled to be shaken into butter. Amongst the Bedouins and Jebours it is considered derogatory to the character of a man to milk a cow or a sheep, but not to milk a camel.

The Sheikhs occasionally obtain dates from the cities. They are either eaten dry with bread and leben, or fried in butter, a very favorite dish of the Bedouin.*

---

* In speaking of the Bedouins I mean the Aneyza, Shammar, Al Dhefyr, and

To this spare and simple dish the Bedouins owe their freedom from sickness, and their extraordinary power of bearing fatigue. Diseases are rare amongst them; and the epidemics, which rage in the cities, seldom reach their tents. The cholera, which has of late visited Mosul and Baghdad with fearful severity, has not yet struck the Bedouins, and they have frequently escaped the plague, when the settlements on the borders of the Desert have been nearly depopulated by it. The small pox, however, occasionally makes great havoc amongst them, vaccination being still unknown to the Shammar, and intermittent fever prevails in the autumn, particularly when the tribes encamp near the marshes in Southern Mesopotamia. Rheumatism is not uncommon, and is treated, like most local complaints, with the actual cautery, a red hot iron being applied very freely to the part affected. Another cure for rheumatism consists in killing a sheep and placing the patient in the hot reeking skin.

Ophthalmia is common in the desert as well as in all other parts of the East, and may be attributed as much to dirt and neglect as to any other cause.

The Bedouins are acquainted with few medicines. The Desert yields some valuable simples, which are, however, rarely used. Dr. Sandwith hearing from Suttum that the Arabs had no opiates, asked what they did with one who could not sleep. " Do ! " answered the Sheikh, " why, we make use of him, and set him to watch the camels." If a Bedouin be ill, or have received a wound, he sometimes comes to the nearest town to consult the barbers, who are frequently not unskilful surgeons. Hadjir, one of the great chiefs of the Shammar, having been struck by a musket ball which lodged beneath the shoulder-blade, visited the Pasha of Mosul to obtain the aid of the European surgeons attached to the Turkish troops. They declared an operation to be impossible, and refused to undertake it. The Sheikh applied to a barber, who in his shop, in the open bazar, quietly cut down to the ball, and taking it out brought it to the Pasha in a plate, to claim a reward for his skill. It is true that the European surgeons in the service of the Porte are not very eminent in their profession. The Bedouins set broken limbs by means of rude splints.

The women suffer little in labor, which often takes place during

other great tribes inhabiting Mesopotamia and the Desert to the north of the Gebel Shammar. With the Arabs of the Hedjaz and Central Arabia I am unacquainted.

a march, or when they are far from the encampment watering the flocks or collecting fuel. They allow their children to remain at the breast until they are nearly two and even three years old, and, consequently, have rarely many offspring.

Soon after our arrival at the Khabour I bought a deloul, or dromedary, as more convenient than a horse for making excursions in the Desert. Her name was Sahaima, and she belonged to Moghamis, the uncle of Suttum, having been taken by him from the Aneyza; she was well trained, and swift and easy in her paces. The best delouls come from Nedjd and the Gebel Shammar. They are small and lightly made, the difference between them and a common camel being as great as that between a high-bred Arab mare and an English cart-horse. Their powers of endurance are very great. Suttum mentioned the following as well authenticated instances. With a companion, each being on his own dromedary, he once rode from Ana to Rowah in one day, one of the animals, however, dying soon after they reached their journey's end. An Arab of the Hamoud, leaving an encampment about five miles inland from Dair, on the west bank of the Euphrates, reached Koukab within twenty-four hours. Suttum rode from Mosul to Khatouniyah in two days.[*]

The deloul is much prized, and the race is carefully preserved. The Arabs breed from them once in two years, and are very particular in the choice of the male. An ordinary animal can work for twenty years. Suttum assured me that they could travel in the spring as many as six days without water. Their color is generally light brown and white, darker colors and black are more uncommon. Their pace is a light trot kept up for many hours together without fatigue; they can increase it to an unweildly gallop, a speed they cannot long maintain. A good deloul is worth at the most 10l., the common price is about 5l.

After the day's work at Arban I generally rode with Suttum into the Desert on our delouls, with the hawks and greyhounds.

---

* Burckhardt (Notes on the Bedouins, &c. p. 262.) mentions as the best authenticated instance of the wonderful speed and endurance of a deloul which had come to his knowledge, a journey for a wager, of 115 miles in eleven hours, including twenty minutes in crossing the Nile twice in a ferry-boat. As that traveller, however, justly remarks, it is by the ease with which they can carry their rider during an uninterrupted journey of several days and nights at a kind of easy amble of five, or five and a half miles, an hour, that they are unequalled by any other animal.

During these rides over the flowered greensward, the Arab Sheikh would entertain me with stories of his tribe, of their wars and intrigues, their successful plundering expeditions, and their occasional defeats. In the evening Mohammed Emin would join our party in the tent, remaining until the night was far spent. Both the Arab chiefs were much troubled by the report of an expedition against the tribes, to which the approach of Suleiman Agha, with a considerable body of troops, to the upper part of the Khabour, had given rise. However, the season was too far advanced for the march of an army through the waterless plains of Mesopotamia. A general campaign against the Bedouins must be undertaken in the winter, or very early in the spring, and even then, if organised by the Turks, would probably fail. The Shammar would at once leave Mesopotamia, and take refuge in the deserts of Nedjd, where no troops could follow them. They would, of course, abandon their flocks and the greater part of their camels, but they would be ready to return as soon as the enemy retreated from the open country, and to revenge themselves amply for their losses upon the unprotected population of the cultivated districts. To bring the Bedouins under subjection, a regular system, steadily pursued, and well selected military posts, are essentially necessary.

The grass around Arban having been eaten by the flocks, the Jebours struck their tents at dawn on the 4th of April, and wandered down the Khabour in search of fresh pastures. The Boraij, too, moved further inland from the river. During the whole morning the Desert around the ruins was a busy scene; sheep, cattle, beasts of burden, men, women, and children being scattered far and wide over the plain. By midday the crowd had disappeared, and the meadows, which a few hours before had been teeming with living things, were now again left lonely and bare. I know no feeling more melancholy than that caused by the sudden breaking up of a large tribe, and by the sight of the spent fires and rubbish-heaps of a recent encampment; the silence and solitude which have suddenly succeeded to the busy scene of an Arab community. Mohammed Emin alone, with a few Sherabeen Arabs, remained to protect us.

Soon after our arrival at the Khabour, Adla, Suttum's first wife, came to us with her child. After the Sheikh's marriage with Rathaiyah, she had been driven from her husband's tent by the imperious temper of his new bride, and had returned to Moghamis her father. Her eldest sister was the wife of Suttum's eldest brother Sahiman, and her youngest, Maizi, was betrothed to Sut-tum's youngest brother Midjwell. The three were remarkable for their beauty; their dark eyes had the true Bedouin fire, and their long black hair fell in clusters on their shoulders. Their cousins, the three brothers, had claimed them as their brides according to Bedouin law.* Adla now sought to be reconciled through me to her husband. Rathaiyah, the new wife, whose beauty was already on the wane, dreaded her young rival's share in the affections of her lord, over whom she had established more influence than a lady might be supposed to exercise over her spouse amongst independent Arabs. The Sheikh was afraid to meet Adla, until, after much negotiation, Hormuzd acting as ambassador, the proud Rathaiyah consented to receive her in her tent. Then the injured lady refused to accept these terms, and the matter was only finished by Hormuzd taking her by the arm and dragging her by force over the grass to her rival. There all the outward forms of perfect reconciliation were satisfactorily gone through, although Suttum evidently saw that there was a different reception in store for himself when there were no European eye-witnesses. Such are the trials of married life in the Desert !

I may here mention that polygamy is very common amongst the Bedouins. It is considered disgraceful for a man to accept money for his daughter, according to the custom in towns and amongst the cultivating tribes; and a girl cannot be forced against her will to marry a man unless he be her cousin, and legally entitled to demand her hand.

On the sixth of April we witnessed a remarkable electrical pheno-menon. During the day heavy clouds had been hanging on the horizon, foreboding one of those furious storms which at this time of the year occasionally visit the Desert. Late in the afternoon these clouds had gathered into one vast circle, which moved slowly round like an enormous wheel, presenting one of the most extra-ordinary and awful appearances I ever saw. From its sides leaped, without ceasing, forked flames of lightning. Clouds springing up from all sides of the heavens, were dragged hurriedly into the vortex, which advanced gradually towards us, and threatened soon to break over our encampment. Fortunately, however, we only felt the very edge of the storm,—a deluge of rain and of hail of the size of pigeons' eggs. The great rolling cloud, attracted by the

---

* Amongst the Bedouins a man has a right to demand his cousin in marriage, and she cannot refuse him.

Sinjar hill, soon passed away, leaving in undiminished splendor the setting sun.

Monday, 8th of April. The Mogdessi, one of my servants, caught a turtle in the river measuring three feet in length. The Arabs have many stories of the voracity of these animals, which attain, I am assured, to even a larger size, and Suttum declared that a man had been pulled under water and devoured by one, probably an Arab exaggeration.

A Bedouin, who had been attacked by a lion whilst resting, about five hours lower down on the banks of the river, came to our encampment. He had escaped with the loss of his mare. The lion is not uncommon in the jungles of the Khabour, and the Bedouins and Jebours frequently find their cubs in the spring season.

In the afternoon, Mohammed Emin learned that the Sherabeen buffalo keepers, who lived under his protection paying a small annual tribute, were about to leave him for the Tai of Nisibin, with whom the Jebours had a blood feud. The Sheikh asked the help of my workmen to bring back the refractory tribe, who were encamped about three hours up the river, and the party marched in the evening singing their war songs.

April 9th. Messengers arrived during the night for further assistance, and Suttum mounting his mare joined the combatants. Early in the morning the Jebours returned in triumph, driving the flocks and buffaloes of the Sherabeen before them. They were soon followed by the tribe, who were compelled to pitch their tents near our encampment.

A Bedouin youth, thin and sickly, though of a daring and resolute countenance, sat in my guest tent. His singular appearance at once drew my attention. His only clothing was a kerchief, very dirty and torn, falling over his head, and a ragged cloak, which he drew tightly round him, allowing the end of a knotted club to appear above its folds. His story, which he was at length induced to tell, was characteristic of Bedouin education. He was of the Boraij tribe, and related to Suttum. His father was too poor to equip him with mare and spear, and he was ashamed to be seen by the Arabs on foot and unarmed. He had now become a man, for he was about fourteen years old, and he resolved to trust to his own skill for his outfit as a warrior. Leaving in his father's tent all his clothes, except his dirty keffieh and his tattered aba, and, without communicating his plans to his friends, he bent his way to the Euphrates. For three months his family hearing nothing of him, believed him to be dead. During that time, however, he had lived in the river jungle, feeding on roots and herbs, hiding himself during the day in the thickets, and prowling at night round the tents of the Aneyza in search of a mare that might have strayed, or might be less carefully guarded than usual. At length the object of his ambition was found, and such a mare had never been seen before; but, alas! her legs were bound with iron shackles, and he had brought no file with him. He succeeded in leading her to some distance from the encampment, where, as morning dawned, to avoid detection, he was obliged to leave his prize and return to his hiding-place. He was now on his way back to his tents, intending to set forth again, after recruiting his strength, on new adventures in search of a mare and spear, promising to be wiser in future and to carry a file under his cloak. Suttum seemed very proud of his relative, and introduced him to me as a promising, if not distinguished, character.* It is thought no disgrace thus to steal a mare as long as the thief has not eaten bread in the tent of her owner.

April 11*th.* The waters of this river had been rising rapidly since the recent storm, and had now spread over the meadows. We moved our tents, and the Arabs took refuge on the mound, which stood like an island in the midst of the flood. The Jebours killed four beavers, and brought three of their young to us alive. They had been driven from their holes by the swollen stream. Mohammed Emin eagerly accepted the musk bags, which are much valued as *majouns* by the Turks, and, consequently, fetch a large price in the towns. The Arabs eat the flesh, and it was cooked for us, but proved coarse and tough. The young we kept for some days on milk, but they eventually died. Their cry resembled that of a newborn infant. The Khabour beavers appeared to me to differ in several respects from the American. The tail, instead of being large and broad, was short and pointed. They do not build huts, but burrow in the banks, taking care to make the entrance to their holes below the surface of the stream to avoid detection, and the chambers above, out of reach of the ordinary floods.

Beavers were formerly found in large numbers on the Khabour, but in consequence of the value attached to the musk bag, they have

---

* The title of haraymi (thief), so far from being one of disgrace, is considered evidence of great prowess and capacity in a young man. Like the Spartans of old he only suffers if caught in the act. There was a man of the Assaiyah tribe who had established an immense renown by stealing no less than ninety horses, amongst which was the celebrated mare given by Sofuk to Beder Khan Bey.

been hunted almost to extermination by the Arabs. Mohammed Emin assured me that for several years not more than one or two had been seen. Sofuk, the great Shammar Sheikh, used to consider the musk bag of a beaver the most acceptable present he could send to a Turkish Pasha, whose friendship he wished to secure.

Two Sheikhs of the Buggara Arabs, who inhabit the banks of the Euphrates opposite Dair, visited our encampment. They described some large mounds near their tents, called Sen, to which they offered to take me; but I was unable to leave my party. The tribe is nominally under the Pasha of Aleppo, but only pay him taxes when he can send a sufficient force to collect them.

Our encampment was further increased by several families of Jays, who had fled from the north on account of some quarrel with the rest of the tribe. They inhabit the country round the ancient Harran and Orfa, the Ur of the Chaldees, and still called Urrha by the Bedouins.

April 12th. We rode this morning with Mohammed Emin, Suttum, and the Sheikhs of the Buggara, Jays, and Sherabeen, to the tents of the Jebours, which had now been moved some miles down the river. Rathaiyah remained behind. The large tents and the workmen were left under the care of the Bairakdar. The chiefs were mounted on well-bred mares, except one of the Jays Sheikhs, who rode a handsome and high-mettled horse. He was gaily dressed in a scarlet cloak lined with fur, a many-colored keffieh, and new yellow boots. His steed, too, was profusely adorned with silken tassels, and small bells, chains, and other ornaments of silver, reminding me forcibly of the horses of the Assyrian sculptures. He had been in the service of the Turks, whose language he had learned, and from whom he had acquired his taste for finery. He was a graceful rider, and managed his horse with great dexterity.

About three miles from Arban we passed a small artificial mound called Tel Hamer (the red); and similar ruins abound on the banks of the river. Near it we met four Shammar Bedouins, who had turned back empty-handed from a thieving expedition to the Aneyza, on account of the floods of the Euphrates, which they described as spreading over the surrounding country like a sea.

Three hours from Arban we reached a remarkable artificial mound called Shedadi, washed by the Khabour. It consists of a lofty platform, nearly square, from the centre of which springs a cone. On the top are the tombs of several Jebour chiefs, marked by the raised earth, and by small trees now dry, fixed upright in the graves. I found fragments of pottery and bricks, but no trace of inscriptions.

Between Shedadi and Arban we saw several ruined bridges, probably of the time of the Caliphs. The mounds are evidently the remains of a much earlier civilisation, when the Assyrian empire extended far beyond the Khabour, and when, as we learn from the inscriptions, the whole face of the country was covered with cities, and with a thriving and wealthy population.

We did not reach the encampment of Mohammed Emin, spreading three or four miles along the Khabour, until after sunset. The chief's tents were pitched near a mound called Ledjmiyat, on a bend of the river, and opposite to a very thick *zor* or jungle, known to the Arabs as El Bostan " the garden," a kind of stronghold of the tribe, which the Sheikh declared could resist the attack of any number of *nizam* (regular troops), if only defended by Jebours. Suttum looked upon the grove rather as a delicious retreat from the rays of the summer's sun, to which the Boraij occasionally resorted, than as a place for war.

During the evening, the different Sheikhs assembled in my tent to plan a *ghazou*, or plundering expedition, for the following day, against the Agaydat, encamped at Abou Psera (Carchemish). Suttum was much cast down at not being able to join them, and mourned over his life of inactivity. I urged him to go, but he vowed that, as long as we were under his protection, he would not leave us. I should have taken this opportunity to visit the Khabour to its mouth, but did not wish to appear to mix myself up with the broils of the tribes.*

On the following morning, Mohammed Emin, with two of his sons, the horsemen of the tribe, and the Sheikhs who were his guests, started on their *ghazou.* They were all mounted on mares, except the Jays chief and one of Mohammed Emin's sons, who rode a beautiful white horse of the Khalawi race. I accompanied them as far as a large ruin called Shemshani. Suttum came with us carrying his hawk, Hattab, on his wrist.

The plain, like all the country watered by the Khabour, was one vast meadow teeming with flowers. Game abounded, and the falcon soon flew towards a bustard, which his piercing eye had seen lurking in the long grass. The sun was high in the heavens,

---

* The confluence of the Euphrates and Khabour is, according to Arab reckoning, one day's journey from Ledjmiyat, and two short from Arban. Arban is two long days from Nisibin, three from Orfa, and four from Severek.

already soaring in the sky, was the enemy of the trained hawk, the " agab" a kind of kite or eagle, whose name, signifying " butcher," denotes his bloody propensities.*   Although far beyond our ken, he soon saw Hattab, and darted upon him in one swoop. The affrighted falcon immediately turned from his quarry, and with shrill cries of distress flew towards us.   After circling round, unable from fear to alight, he turned towards the Desert, still followed by his relentless enemy.   In vain his master, following as long as his mare could carry him, waved the lure, and called the hawk by his name ; he saw him no more.   Whether the noble bird escaped, or fell a victim to the " butcher," we never knew.

Suttum was inconsolable at his loss.   He wept when he returned without his falcon on his wrist, and for days he would suddenly exclaim, " O Bej ! Billah ! Hattab was not a bird, he was my brother."   He was one of the best trained hawks I ever saw amongst the Bedouins, and was of some substantial value to his owner, as he would daily catch six or seven bustards, except during the hottest part of summer, when the falcon is unable to hunt.

About a mile and a half below Ledjmiyat, but on the opposite bank of the river, was another large mound called Fedghami.   We reached Shemshani in an hour and three quarters.   It is a considerable ruin on the Khabour, and consists of one lofty mound, surrounded on the Desert side by smaller mounds and heaps of rubbish. It abounds in fragments of glazed and plain pottery, bricks, and black basaltic stone, but I could find no traces of sculpture or inscription.   The remains of walls protrude in many places from the soil.   Above the ancient ruins once stood a castle, the foundations of which may still be seen.

The Arabs have many traditions attaching to these ruins. Among others, that they are the remains of the capital of an infidel king, whose daughter, at the time of the first Mussulman invasion, eloped with a true believer.   The lovers were pursued by the father, overtaken, and killed (the lady having, of course, first embraced Islamism), in a narrow valley of the neighbouring hills.   A flickering flame, still distinctly seen to rise from the earth on Friday nights, marks the spot of their martyrdom.   The city soon fell into the hands of the Mussulmans, who took a signal revenge upon its idolatrous inhabitants.

The Jebours some years ago cultivated the lands around Shem-

---

* Easterns never hawk, if they can avoid it, when the sun is high, as the bird of prey described in the text then appears in search of food.

shani, and there are still many traces of watercourses, and of the square plots set apart for rice.*

Leaving Mohammed Emin to continue his journey we returned to our tents. On our road we met Moghamis, and a large party of Bedouins on their way to join the Jebour horsemen, for they also had been invited to take part in the attack on the Agaydat, and to share in the spoil. They rode their swift dromedaries, two men on each, the *rediff* leading the mare of his companion ; that of the Sheikh was of the Obeyan race, and far famed in the Desert. She was without saddle or clothes, and we could admire the exquisite symmetry and beauty of her form.

We dismounted, embraced, and exchanged a few words. The Bedouins then continued their rapid course over the Desert. We passed other riders on delouls and mares, hastening to join the main body, or to meet their friends at the rendezvous for the night near Abou Psera. The attack on the tents was to be made at dawn on the following morning, the true Bedouin never taking an unfair advantage of his enemy in the dark.

April 14*th.* We were awoke long before dawn by the Jebours striking their tents. By sunrise the whole encampment had disappeared, and we were left almost alone. They were returning towards Arban, fearing lest the Agaydat, assisted by the Aneyza, might seek a speedy revenge after the attack upon them. We breakfasted, and then soon overtook the line of march. For two hours we amused ourselves by riding through the dense and busy throng. I have already described the singular spectacle of a great Arab tribe changing its pastures, — its mingled crowd of women and girls, some with burdens, others without, of warriors on highbred mares and on fleet camels, of shepherds with their knotted clubs, of sheep, goats, camels, beasts of burden, children, lambs, and all the various appendages of Arab life. A more stirring and joyous scene can scarcely be imagined.

The family of the chief, as is usual, moved in front of the tribe. We left them pitching their tents near the mound of Shedadi, and rode to our own encampment at Arban.

On the 16th of April, Mohammed Emin and his sons returned from their expedition, driving before them their spoil of cows, oxen, and mares. The Agaydat were taken by surprise, and made but a

---

* Between Shemshani and the mouth of the Khabour, according to Mohammed Emin, are the following mounds—El Murgadeh (about five miles distant), El Hussain, Sheikh Ahmed, Suor, and El Efdaya.

feeble defence; there was, consequently, little bloodshed, as is usually the case when Arabs go on these forays. The fine horse of the Jays chief had received a bad gunshot wound, and this was the only casualty amongst my friends. Mohammed Emin brought me one or two of the captured mares as an offering. They were, of course, returned, but they involved the present of silk dresses to the Sheikh and his sons.

April 18th. To-day we visited the tents of Moghamis and his tribe; they were pitched about five miles from the river. The face of the Desert was as burnished gold. Its last change was to flowers of the brightest yellow hue *, and the whole plain was dressed with them. Suttum rioted in the luxuriant herbage and scented air. I never saw him so exhilarated. "What Kef (delight)," he continually exclaimed, as his mare waded through the flowers, "has God given us equal to this? It is the only thing worth living for. Ya Bej! what do the dwellers in cities know of true happiness, they have never seen grass or flowers? May God have pity on them!"

The tents were scattered far and wide over the plain. The mares recently returned from the foray wandered loose in the midst of them, cropping the rich grass. We were most hospitably received by Moghamis. Such luxuries, in the way of a ragged carpet and an old coverlet, as his tent could afford, had been spread for Mrs. R., whose reputation had extended far and wide amongst the Arabs, and who was looked upon as a wonder, but always treated with the greatest consideration and respect. The wild Bedouin would bring a present of camel's milk or truffles, and the boys caught jerboas and other small animals for the Frank lady. During the whole of our journey she was never exposed to annoyance, although wearing, with the exception of the Tarboush, or an Arab cloak, the European dres

Moghamis clad himself in a coat of chain mail, of ordinary materials and rude workmanship, but still strong enough to resist the coarse iron spear-heads of the Arab lance, though certainly no protection against a well-tempered blade. The Arabs wear their armour beneath the shirt, because an enemy would otherwise strike at the mare and not at her rider.†

---

* I have already mentioned the changes in the colors of the Desert. Almost in as many days white had succeeded to pale straw color, red to white, blue to red, lilac to blue, and now the face of the country was as described in the text.

† One of the principal objects of Bedouins in battle being to carry off their adversaries' mares, they never wound them if they can avoid it, but endeavour to kill or unhorse the riders.

After we had enjoyed all the luxuries of an Arab feast, visited
the women's compartments, where most of the ladies of the tribe
had assembled to greet us, examined the "chetab," or camel saddle,
used by the wives of the chiefs*, and enquired into various details
of the harem, we returned as we came, through the flowers and long
grass to our tents at Arban.

* See woodcut, p. 63., of the abridged edition of my "Nineveh and its
Remains," for a sketch of this extraordinary contrivance.

Saddling a Deloul, or Dromedary

Kurdish Women.

CHAP. XIV.

THE hot weather was rapidly drawing near. Enough had not been discovered in the mound of Arban, nor were there ruins of sufficient importance near the river, to induce me to remain much longer on the Khabour. I wished, however, to explore the stream, as far as I was able, towards its principal source, and to visit Suleiman Agha, the Turkish commander, who was now encamped on its banks. In answer to a letter, he urged me to come to his tents, and to bring the Sheikh of the Jebours with me, pledging himself to place no restraint whatever on the perfect liberty of the Arab chief. With such a guarantee, I ventured to invite Mohammed Emin to accompany me. After much hesitation, arising from a very natural fear of treachery, he consented to do so.

On the 19th of April we crossed the Khabour, and encamped for the night on its southern bank. On the following morning we turned from the ruins of Arban, and commenced our journey to the eastward. The Jebours were now dwelling higher up the stream, and Mohammed Emin, with his two sons, and Abdullah his nephew, met us on our way. He was still in doubt as to whether he should go with me or not; but at last, after more than once turning back, he took a desperate resolution, and pushed his mare boldly forward. His children commended him, with tears, to my protection, and then left our caravan for their tents.

We rode from bend to bend of the river, without following its tortuous course. Its banks are belted with poplars, tamarisks, and brushwood, the retreat of wild boars, francolins, and other game, and studded with artificial mounds, the remains of ancient settlements. This deserted though rich and fertile district must, at one time, have been the seat of a dense population. It is only under such a government as that of Turkey that it could remain a wilderness. The first large ruin above Arban, and some miles from it on the left bank of the river, is called Mishnak. According to a tradition preserved by the Jebours, the Persians were defeated near it, with great slaughter, in the early days of Islam, by the celebrated Arab tribe of the Zobeide. About one mile and a half beyond is another ruin called Abou Shalah, and three miles further up the stream a third, called Taaban, upon which are the remains of a modern fort. Near Taaban, Mohammed Emin had recently built a small enclosure of rude stone walls, a place of refuge in case of an attack from the Aneyza Bedouins. Around it the Jebours sow corn and barley, re-opening the ancient watercourses to bring water to their fields. The wheat was almost ready for the sickle even at this early season of the year.

After a short day's journey of four hours and a half we raised our tents for the night amongst luxuriant herbage, which afforded abundant pasture for our horses and camels. The spot was called Nahab. The river, divided into two branches by a string of small wooded islands, is fordable except during the freshes. Near our encampment was a large mound named Mehlaibiyah, and in the stream I observed fragments of stone masonry, probably the remains of ancient dams for irrigation.

Next morning Suttum returned to his tents with Rathaiyah, leaving us under the care of his younger brother Mijwell. After I had visited the Turkish commander, whom he did not appear over anxious to meet, he was to join us in the Desert, and accompany me to Mosul. Mijwell was even of a more amiable disposition than his brother; was less given to diplomacy, and troubled himself little with the politics of the tribes. A pleasant smile lighted up his features, and a fund of quaint and original humor made him at all times an agreeable companion. Although he could neither read nor write, he was one of the cadis or judges of the Shammar, an office hereditary in the family of the Saadi, at the head of which is Rishwan. The old man had delegated the dignity to his younger son, who, by the consent of his brothers, will enjoy it after their father's death. Disputes of all kinds are referred to these recognised judges. Their decrees are obeyed with readiness, and the other members of the tribe are rarely called upon to enforce them. They administer rude justice; and, although pretending to follow the words of the Prophet, are rather guided by ancient custom than by the law of the Koran, which binds the rest of the Mohammedan world. The most common source of litigation is, of course, stolen property. They receive for their decrees, payment in money or in kind; and he who gains the suit has to pay the fee. Amongst the Shammar, if the dispute relates to a deloul, the cadi gets two gazees, about eight shillings; if to a mare, a deloul; if to a man, a mare.* Various ordeals, such as licking a red-hot iron, are in use, to prove a man's innocence. If the accused's tongue is burnt, no doubt exists as to his guilt.

One of the most remarkable laws in force amongst the wandering Arabs, and one probably of the highest antiquity, is the law of blood, called the Thar, prescribing the degrees of consanguinity within which it is lawful to revenge a homicide. Although a law, rendering a man responsible for blood shed by any one related to him within the fifth degree, may appear to members of a civilised community one of extraordinary rigour, and involving almost manifest injustice, it must nevertheless be admitted, that no power vested in any one individual, and no punishment however severe, could tend more to the maintenance of order and the prevention of bloodshed amongst the wild tribes of the Desert. As Burckhardt

---

* Burckhardt gives a somewhat different table of fees as existing amongst the Bedouin tribes with which he was acquainted. His whole account of Arab law is singularly interesting and correct; there is, indeed, very little to be added to it. (See his Notes on the Bedouins, p. 66.)

has justly remarked, " this salutary institution has contributed in
a greater degree than any other circumstance, to prevent the warlike tribes of Arabia from exterminating one another."

If a man commit a homicide, the cadi endeavours to prevail
upon the family of the victim to accept a compensation for the blood
in money or in kind, the amount being regulated according to
custom in different tribes. Should the offer of " blood-money "
be refused, the " Thar " comes into operation, and any person within the " khomse," or the fifth degree of blood of the homicide, may
be legally killed by any one within the same degree of consanguinity to the victim.*

This law is enforced between tribes remote from one another,
as well as between families, and to the blood revenge may be attributed many of the bitter feuds which exist amongst the Arab
clans. It affects, in many respects, their social condition, and has
a marked influence upon their habits, and even upon their manners.
Thus an Arab will never tell his name, especially if it be an uncommon one, to a stranger, nor mention that of his father or of his
tribe, if his own name be ascertained, lest there should be Thar
between them. Even children are taught to observe this custom,
that they may not fall victims to the blood revenge. Hence the
extreme suspicion with which a Bedouin regards a stranger in the
open country, or in a tent, and his caution in disclosing anything
relating to the movements, or dwelling-place, of his friends. In
most encampments are found refugees, sometimes whole families,
who have left their tribe on account of a homicide for which they
are amenable. In case, after a murder, persons within the " Thar "
take to flight, three days and four hours are by immemorial custom
allowed to the fugitives before they can be pursued. Frequently
they never return to their friends, but remain with those who give
them protection, and become incorporated into the tribe by which
they are adopted. Thus there are families of the Harb, Aneyza,

* Burckhardt has thus defined the terms of this law : " The Thar rests with
the khomse, or fifth generation, those only having a right to revenge a slain
parent, whose fourth lineal ascendant is, at the same time, the fourth lineal ascendant of the person slain ; and, on the other side, only those male kindred of
the homicide are liable to pay with their own for the blood shed, whose fourth
lineal ascendant is at the same time the fourth lineal ascendant of the homicide.
The present generation is thus comprised within the number of the khomse.
The lineal descendants of all those who are entitled to revenge at the moment of
the manslaughter inherit the right from their parents. The right to blood-
revenge is never lost ; it descends on both sides to the latest generation." (Notes
on Arabs, p. 85.)

Dhofyr, and other great clans, who for this cause have joined the Shammar, and are now considered part of them. Frequently the homicide himself will wander from tent to tent over the Desert, or even rove through the towns and villages on its borders, with a chain round his neck and in rags, begging contributions from the charitable to enable him to pay the apportioned blood-money. I have frequently met such unfortunate persons who have spent years in collecting a small sum. I will not weary the reader with an account of the various rules observed in carrying out this law, where persons are killed in private dissensions, or slain in the act of stealing, in war, or in the ghazou. In each case the cadi determines, according to the ancient custom of the tribe, the proper compensation.

Mijwell now took Suttum's place in the caravan, and directed the order of our march. Four miles from Nahab we passed a large mound called Thenenir, at the foot of which is a spring much venerated by the Arabs. Around it the Jebours had sown a little wheat. Near this ruin an ancient stone dam divides the Khabour into several branches: it is called the "*Saba Sekour*," or the seven rocks.

Leaving the caravan to pursue the direct road, I struck across the country to the hill of Koukab, accompanied by Mohammed Emin and Mijwell. This remarkable cone, rising in the midst of the plain, had been visible from our furthest point on the Khabour. Some of the Arabs declared it to be an artificial mound; others said, that it was a mountain of stones. Mohammed Emin would tell me of a subterranean lake beneath it, in a cavern large enough to afford refuge to any number of men. As we drew nearer, the plain was covered with angular fragments of black basalt, and crossed by veins, or dykes, of the same volcanic rock. Mohammed Emin led us first to the mouth of a cave in a rocky ravine not far from the foot of the hill. It was so choked with stones that we could scarcely squeeze ourselves through the opening, but it became wider, and led to a descending passage, the bottom of which was lost in the gloom. We advanced cautiously, but not without setting in motion an avalanche of loose stones, which, increasing as it rolled onwards, by its loud noise disturbed swarms of bats that hung to the sides and ceiling of the cavern. Flying towards the light, these noisome beasts almost compelled us to retreat. They clung to our clothes, and our hands could scarcely prevent them settling on our faces. The rustling of their wings was like the noise of a great wind, and an

abominable stench arose from the recesses of the cave. At length
they settled again to their daily sleep, and we were able to go
forward.

After descending some fifty feet, we found ourselves on the
margin of a lake of fresh water. The pitchy darkness prevented
our ascertaining its size, which could not have been very great,
although the Arabs declared that no one could reach the opposite
side. The cave is frequently a place of refuge for the wandering
Arabs, and the Bedouins encamp near it in summer to drink the
cool water of this natural reservoir. Mohammed Emin told me
that last year he had found a lion in it, who, on being disturbed,
merely rushed out and fled across the plain.

Leaving the cavern and issuing from the ravine, we came to
the edge of a wide crater, in the centre of which rose the remarkable cone of Koukab. To the left of us was a second crater,
whose lips were formed by the jaggy edges of basaltic rocks, and
in the plain around were several others smaller in size. They
were all evidently the remains of an extinct volcano, which had
been active within a comparatively recent geological period, even
perhaps within the time of history, or tradition, as the name of
the mound amongst the Arabs denotes a jet of fire or flame, as well
as a constellation.

I ascended the cone, which is about 300 feet high, and composed
entirely of loose lava, scoria, and ashes, thus resembling precisely
the cone rising in the craters of Vesuvius and Ætna. It is steep
and difficult of ascent, except on one side, where the summit is
easily reached even by horses. Within, for it is hollow, it resembles an enormous funnel, broken away at one edge, as if a molten
stream had burst through it. Anemonies and poppies, of the
brightest scarlet hue, covered its side; although the dry lava and
loose ashes scarcely seemed to have collected sufficient soil to
nourish their roots. It would be difficult to describe the richness
and brilliancy of this mass of flowers, the cone from a distance
having the appearance of a huge inverted cup of burnished copper,
over which poured streams of blood.

From the summit of Koukab I gazed upon a scene as varied as
extensive. Beneath me the two principal branches of the Khabour
united their waters. I could track them for many miles by the
dark line of their wooded banks, as they wound through the golden
plains. To the left, or the west, was the true Khabour, the Chaboras of the ancients; a name it bears from its source at Ras-al-ain

(*i. e.* the head of the spring).* The second stream, that to the
east, is called by the Arabs the Jerujer (a name, as uttered by the
Bedouins, equally difficult to pronounce and to write), and is the
ancient Mygdonius, flowing through Nisibin.† Khatouniyah and
its lake were just visible, backed by the solitary hill of the Sinjar.
The Kurdish mountains bounded the view to the east. In the
plain, and on the banks of the rivers, rose many artificial mounds;
whilst, in the extreme distance to the north could be distinguished
the flocks and black tents of a large wandering tribe. They were
those of the Chichi and Milli Kurds, encamped with the Turkish
commander Suleiman Agha.

On some fragments of basaltic rock projecting from the summit
of the cone, were numerous rudely-cut signs, which might have
been taken for ancient and unknown characters. They were the
devices of the Shammar, carved there on the visit of different
Sheikhs. Each tribe, and, indeed, each subdivision and family,
has its peculiar mark, to be placed upon their property and burnt
upon their camels. Mijwell identified the signs; that of his own
family, the Saadi, being amongst them. In little recesses, carefully sheltered by heaped-up stones, were hung miniature cradles,
like those commonly suspended to the poles of a Bedouin tent.
They had been placed there as exvotos by Shammar women who
wished to be mothers.

After I had examined the second large crater, — a deep hollow,
surrounded by basaltic rocks, but without a projecting cone of
lava, — we rode towards the Jerujer, on whose banks the caravan
was to await us. The plain was still covered with innumerable
fragments of basalt embedded in scarlet poppies. We found our
companions near the junction of the rivers, where a raft had been
constructed to enable us to cross the smaller stream. I had sent
the Bairakdar two days before to apprise Suleiman Agha of my
intended visit, and to learn how far I could with safety take Mohammed Emin with me to the Turkish camp. He had returned,

---

* One of the sources of this branch of the Khabour is, I am told, in the
Kharej Dagh, to the west of Mardin. This small stream, called Ajjurgub, falls
into the river near Ras-al-Ain.

† The name of Hawali, by which this branch of the Khabour appears to have
been called by the Arab geographers, and which is retained in our maps, appears to be derived from the " Hol," which will be described hereafter. The
course of the stream is also erroneously laid down in all the maps; and, what is
more curious, is as wrongly described by the Arab writers, some of whom
place a branch of it to the south-east of the Sinjar, confounding it apparently
with the Thathar.

and was waiting for me.   The Agha had given a satisfactory
guarantee for the Sheikh's safety, and had sent an officer, with a
party of irregular troops, to receive me.

We had scarcely crossed the river before a large body of horsemen were seen approaching us.   As they drew nigh I recognised in
the Turkish commander an old friend, "the Topal," or lame, Suleiman Agha, as he was generally called in the country.   He had
been Kiayah or lieutenant-governor, to the celebrated Injeh Bairakdar Mohammed Pasha, and, like his former master, possessed
considerable intelligence, energy, and activity.   From his long
connection with the tribes of the Desert, his knowledge of their
manners, and his skill in detecting and devising treacheries and
stratagems, he was generally chosen to lead expeditions against the
Arabs.   He was now, as I have stated, endeavoring to recover
the government treasure plundered by the Hamoud Bedouins.

He was surrounded by Hyta-Bashis, or commanders of irregular
cavalry, glittering with gold and silver-mounted arms, and rich in
embroidered jackets, and silken robes, by Aghas of the Chichi and
Milli Kurds, and by several Arab chiefs.   About five hundred
horsemen, preceded by their small kettle-drums, crowded behind
him.   His tents were about six miles distant; and, after exchang-
ing the usual salutations, we turned towards them.   Many fair
speeches could scarcely calm the fears of the timid Jebour Sheikh.
Mijwell, on the other hand, rode boldly along, casting contemp-
tuous glances at the irregular cavalry, as they galloped to and fro
in mimic combat.

The delta, formed by the two streams, was covered with tents.
We wended our way through crowds of sheep, horses, cattle, and
camels.   The Chichi and Milli Kurds, who encamp during the
spring at the foot of the mountains of Mardin, had now sought,
under the protection of the Turkish soldiery, the rich pastures of
the Khabour, and many families of the Sherabbeen, Buggara, and
Harb Arabs had joined the encampment.*

Suleiman Agha lived under the spacious canvas of the Chichi
chief.   The tents of the Kurdish tribes, who wander in the low
country at the foot of the mountains in winter and spring, and seek
the hill pastures in the summer, and especially those of the principal men, are remarkable for their size, and the richness of their

---

* The Harb is a branch of the great tribe of the same name inhabiting the
northern part of the Hedjaz, which, in consequence of some blood-feud, migrated many years ago to Mesopotamia.

F Cooper.

John Murray, Albemarle Street, 1852.

N. Chevalier, lith.

Encampment on the Khabour.

carpets and furniture. They are often divided into as many as four or five distinct compartments, by screens of light cane or reeds, bound together with many-coloured woollen threads, disposed in elegant patterns and devices. Carpets hung above these screens complete the divisions. In that set aside for the women a smaller partition incloses a kind of private room for the head of the family and his wives. The rest of the harem is filled with piles of carpets, cushions, domestic furniture, cooking utensils, skins for making butter, and all the necessaries of a wandering life. Here the handmaidens prepare the dinner for their master and his guests. In the tents of the great chiefs there is a separate compartment for the servants, and one for the mares and colts.

I sat a short time with Suleiman Agha, drank coffee, smoked, and listened patiently to a long discourse on the benefits of *tanzimat*, which had put an end to bribes, treachery, and irregular taxation, especially intended for Mohammed Emin, who was however by no means reassured by it. I then adjourned to my own tents, which had been pitched upon the banks of the river opposite a well-wooded island, and near a ledge of rocks forming one of those beautiful falls of water so frequent in this part of the Khabour. Around us were the pavilions of the Hytas, those of the chiefs marked by their scarlet standards. At a short distance from the stream the tents of the Kurds were pitched in parallel lines forming regular streets, and not scattered, like those of the Bedouins, without order over the plain. Between us and them were picketed the horses of the cavalry, and as far as the eye could reach beyond, grazed the innumerable flocks and herds of the assembled tribes.

We were encamped near the foot of a large artificial Tel called Umjerjeh; and on the opposite side of the Khabour were other mounds of the same name. My Jebour workmen began to excavate in these ruins the day after our arrival. I remained in my tent to receive the visits of the Kurdish chiefs and of the commanders of the irregular cavalry. From these freebooters I have derived much curious and interesting information relating to the various provinces of the Turkish empire and their inhabitants, mingled with pleasant anecdotes and vivid descriptions of men and manners. They are generally very intelligent, frank, and hospitable. Although too often unscrupulous and cruel, they unite many of the good qualities of the old Turkish soldier with most of his vices. They love hard-drinking and gambling, staking their horses, arms, and even clothes, on the most childish game of chance.

Their pay, at the same time, is miserably small, rarely exceeding a few shillings a month, and they are obliged to plunder the peaceable inhabitants to supply their actual wants. The race is now fast disappearing before the Nizam, or regular troops.

On the second day, accompanied by Mijwell, I visited a large mound called Mijdel, on the right bank of the river about five miles above Umjerjeh. We rode through the golden meadows, crossing the remains of ancient canals and watercourses, and passing the ruins of former habitations. A Sheikh of the Buggara was with us, an intelligent Arab, whose tribe in times of quiet encamp at Ras-al-Ain near the sources of the Khabour. The Aneyza were out on this side of the Euphrates, and were prowling over the Desert in search of plunder. As Suleiman Agha declared that, without an escort of at least one hundred horsemen, I could not go to Ras-al-Ain, I was unable to visit the extensive ruins which are said to exist there.

Ras-al-Ain was once a place of considerable importance. It was known to the ancients under the name of Rasina. Benjamin of Tudela found two hundred Jews dwelling there in the 12th century.* The Arabs assured me that columns and sculptures still mark the site of the ancient city. Their accounts are, however, probably exaggerated.

Mijdel is a lofty platform, surrounded by groups of smaller mounds, amongst which may still be traced the lines of streets and canals. It is about four or five miles from the ridge of Abd-ul-Azeez. These low hills, scantily wooded with dwarf oak, are broken into innumerable valleys and ravines, which abound, it is said, with wild goats, boars, leopards, and other animals. According to my Bedouin informants, the ruins of ancient towns and villages still exist, but they could only give me the name of one, Zakkarah. The hills are crossed in the centre by a road called Maghliyah, from an abundant spring. On the opposite side of the Khabour, and running parallel with the Abd-ul-Azeez range, is another line of small hills, called Hamma, in which there are many wells †

---

* The name is by some error omitted in the Hebrew text, but it is evident, from the distance to Harran, that Ras-al-Ain is meant. Asher (Benjamin of Tudela's Itinerary), note to passage, vol. ii. p. 128.) points out that it should be the *sources* of the Khabour, not the *mouth*, as usually translated.

† The Buggara chief gave me the following names for mounds, in the order in which they occur, between Mijdel and Ras-al-Ain. The Gla (Kalah) or Tel Romana, a large mound visible from Mijdel; El Mogas, near a ford and a place called El Auja; El Tumr, about four hours from Umjerjeh, at the junc-

The Shammar Bedouins encamp on the banks of this part of the Khabour during the hot months. The mound of Mijdel is a favorite resort of the Boraij in the "eye of the summer:" the waters of the river are always cool, and there is sufficient pasture for the flocks and herds of the whole tribe.

An Arab whom I met in the tent of one of the Hyta-Bashis, pretended that he was well acquainted with the ruins called *Verhan-Shehr**, of which I had so frequently heard from the natives of Mardin and the Shammar. He described them as being on a hill three days distant from our encampment, and to consist of columns, buildings, and sculptured stones like those of Palmyra. The Turkish Government at one time wished to turn the ancient edifices into barracks, and to place a garrison in the place to keep the Arabs in check.

In the evening Mohammed Emin left us. Suleiman Agha had already invested him with a robe of honor, and had prevailed upon him to join with Ferhan in taking measures for the recovery of the plundered treasure. The scarlet cloak and civil treatment had conciliated the Jebour chief, and when he parted with the Turkish commander in my tent there was an unusual display of mutual compliments and pledges of eternal friendship. Mijwell looked on with indignant contempt, swearing between his teeth that all Jebours were but degenerate, ploughing Arabs, and cursing the whole order of *temminahs*.†

We were detained at Umjerjeh several days by the severe illness of Mr. Hormuzd Rassam. I took the opportunity to visit the tents of the Milli, whose chief, Mousa Agha, had invited us to a feast. On our way thither we passed several encampments of Chichi, Sherrabeen, and Harb, the men and women running out and pressing us to stop and eat bread. The spacious tent of the chief was divided by partitions of reeds tastefully interwoven with colored wool. The coolest part of the salamlik had been prepared for our reception, and was spread with fine carpets and

tion of the Zergan, a small stream coming from Ghours, in the mountains to the west of Mardin; El Tawileh, a large mound fourteen or fifteen miles from Mijdel, and just visible; Om Kaifah, Tal Jahash, and Gutinah. On the river bank opposite to Mijdel, are several groups of mounds called Dibbs. Near Ras-al-Ain is a mound, whether natural or artificial I could not ascertain, called El Chibeseh.

* *I.e.* The ancient ruined city, a name very generally given by the Turks to ruins.

† The form of salutation used by the Turks, consisting of raising the hand from the breast, or sometimes from the ground, to the forehead.

silken cushions. The men of the tribe, amongst whom were many tall and handsome youths, were dressed in clean and becoming garments. They assembled in great numbers, but left the top of the tent entirely to us, seating themselves, or standing at the sides and bottom, which was wide enough to admit twenty-four men crouched together in a row. The chief and his brothers, followed by their servants bearing trays loaded with cups, presented the coffee to their guests.

After some conversation we went to the harem, and were received by his mother, a venerable lady, with long silvery locks and a dignified countenance and demeanor. Her dress was of the purest white and scrupulously clean. Altogether she was almost the only comely old woman I had seen amongst Eastern tribes. The wives and daughters of the chiefs, with a crowd of women, were collected in the tent. Amongst them were many distinguished by their handsome features. They had not the rich olive complexion or graceful carriage of the Bedouin girls, nor their piercing eyes or long black eyelashes. Their beauty was more European, some having even light hair and blue eyes. It was evident, at a glance, that they were of a different race from the wandering tribes of the Desert.

The principal ladies led us into the private compartment, divided by colored screens from the rest of the tent. It was furnished with more than usual luxury. The cushions were of the choicest silk, and the carpets (in the manufacture of which the Milli excel) of the best fabric. Sweetmeats and coffee had been prepared for us, and the women did not object to partake of them at the same time. Mousa Agha's mother described the various marriage ceremonies of the tribe. Our account of similar matters in Europe excited great amusement amongst the ladies. The Milli girls are highly prized by the Kurds. Twenty purses, nearly 100*l.*, we were boastingly told, had been given for one of unusual attractions. The chief pointed out one of his own wives who had cost him that sum. Other members of the same establishment had deserved a less extravagant investiture of money. The prettiest girls were called before us, and the old lady appraised each, amidst the loud laughter of their companions, who no doubt rejoiced to see their friends valued at their true worth. They were all tatooed on the arms, and on other parts of the body, but less so than the Bedouin ladies. The operation is performed by Arab women, who wander from tent to tent for the purpose. Several were present, and wished to give us an immediate proof of their skill upon ourselves.

We declined however. It is usually done at the age of six or seven: the punctures are made by a needle, and the blue color is produced by a mixture of gunpowder and indigo rubbed into the wounds. The process is tedious and painful, as the designs are frequently most elaborate, covering the whole body. The Kurdish ladies do not, like the Mussulman women of the town, conceal their features with a veil; nor do they object to mingle, or even eat, with the men. During my stay at Umjerjeh I invited the harem of the Chichi chief, and their friends, to a feast in my tent — an invitation they accepted with every sign of satisfaction.

The Milli were formerly one of the wealthiest Kurdish tribes. Early in this century, when the hereditary chiefs in different parts of the empire were still almost independent of the Porte, this clan held the whole plain country between the hills of Mardin and the Khabour, exacting a regular *baj*, or black-mail, from caravans and travellers passing through their territories. This was a fruitful source of revenue when an extensive commerce was carried on between Aleppo and Baghdad, and the Aghas were frequently, on account of their wealth and power, raised to the rank of pashas by the Sultan. The last was Daoud Pasha, a chief well-known in Mesopotamia. Like other Kurdish tribes, the Milli had been brought under the immediate control of the local governors, and were now included within the pashalic of Diarbekir. They still possessed all the riches that nomades can well possess, when they were wantonly plundered, and almost reduced to want, by the Turkish troops three years ago. Although the Porte openly condemned the outrage, and had promised compensation, no step whatever had been taken to restore the stolen property, the greater part of which had passed into the government treasury.

We had an excellent dinner in the salamlik, varied by many savoury dishes and delicacies sent from the harem: such as truffles, dressed in different ways, several preparations of milk and cream; honey, curds, &c. After we had retired, the other guests were called to the feast by relays. The chief, however, always remained seated before the dishes, eating a little with all, and leaving his brother to summon those who were invited; such being the custom amongst these Kurds.

Mijwell, during our visit, had been seated in a corner, his eyes wandering from the tent and its furniture to the horses and mares picketed without, and to the flocks pasturing around. He cast, every now and then, significant glances towards me, which said plainly enough, " All this ought to belong to the Bedouins. These people and their property were made for *ghazous*." As we rode away I accused him of evil intentions. "Billah, ya Bej!" said he, "there is, indeed, enough to make a man's heart grow white with envy ; but I have now eaten his bread under your shadow, and should even his stick, wherewith he drives his camel, fall into my hand, I would send it to him." He entertained me, as we returned home, with the domestic affairs of his family. Rathaiyah had offered herself in marriage to Suttum, and not he to her ; a common proceeding, it would appear, among the Bedouins. Suttum had consented, because he thought it politic to be thus allied with the Abde, one of the most powerful branches of the Shammar, generally at war with the rest of the tribe. But his new wife, besides having sent away her rival, had already offended his family by her pride and haughtiness. Mijwell rather looked upon his brother with pity, as a henpecked husband. He himself, although already married to one wife, and betrothed to Maizi, whom he would soon be able to claim, was projecting a third marriage. His heart had been stolen by an unseen damsel, whose beauties and virtues had been the theme of some wandering Arab rhymers, and she was of the Fedhan Aneyza, the mortal enemies of the Shammar. Her father was the sheikh of the tribe, and his tents were on the other side of the Euphrates. The difficulties and dangers of the courtship served only to excite still more the ardent mind of the Bedouin. His romantic imagination had pictured a perfection of loveliness ; his whole thoughts were now occupied in devising the means of possessing this treasure.* He had already apprised the girl of his love by a trusty messenger, one of her own tribe, living with the Shammar. His confidant had extolled the graces, prowess, and wealth of the young Sheikh, with all the eloquence of a Bedouin poet, and had elicited a favorable reply. More than one interchange of sentiments had, by such means, since passed between them. The damsel had, at last, promised him her hand, if he could claim her in her own tent. Mijwell had now planned a scheme which he was eager to put into execution. Waiting until the Fedhan were so encamped that he could approach them without being previously seen, he would mount his deloul, and leading his best mare, ride to the tent of the girl's father. Meat would, of course, be laid before him, and having eaten he would be the guest, and under the protection of the

* Burckhardt remarks that "Bedouins are, perhaps, the only people of the East that can be entitled true lovers." (Notes on Bedouins, p. 155.)

Sheikh. On the following morning he would present his mare, describing her race and qualities, to his host, and ask his daughter; offering, at the same time, to add any other gift that might be thought worthy of her. The father, who would probably not be ignorant of what had passed between the lovers, would at once consent to the union, and give back the mare to his future son-in-law. The marriage would shortly afterwards be solemnised, and an alliance would thus be formed between the two tribes. Such was Mijwell's plan, and it was one not unfrequently adopted by Bedouins under similar circumstances.

A Bedouin will never ask money or value in kind for his daughter, as fathers do amongst the sedentary tribes and in towns, where girls are literally sold to their husbands, but he will consult her wishes, and she may, as she thinks fit, accept or reject a suitor, so long as he be not her cousin. Presents are frequently made by the lover to the damsel herself before marriage, but rarely to the parents. Although the Bedouin chiefs have sometimes taken wives from the towns on the borders of the Desert, such as Mosul, Baghdad, or Aleppo, it is very rare to find townspeople, or Arabs of the cultivating tribes, married to Bedouin women. I have, however, known instances.

The laws of Dakheel, another very remarkable branch of Bedouin legislation, in force amongst the Shammar, are nearly the same as those of the Aneyza and Hedjaz Arabs, of which Burckhardt has given so full and interesting an account. I have little, therefore, to add upon the subject, but its importance demands a few words. No customs are more religiously respected by the true Arab than those regulating the mutual relations of the protected and protector. A violation of Dakheel (as this law is called) would be considered a disgrace not only upon the individual but upon his family, and even upon his tribe, which never could be wiped out. No greater insult can be offered to a man, or to his clan, than to say that he has broken the Dakheel. A disregard of this sacred obligation is the first symptom of degeneracy in an Arab tribe ; and when once it exists, the treachery and vices of the Turk rapidly succeed to the honesty and fidelity of the true Arab character. The relations between the Dakheel and the Dakhal (or the protector and protected) arise from a variety of circumstances, the principal of which are, eating a man's salt and bread, and claiming his protection by doing certain acts, or repeating a certain formula of words. Amongst the Shammar, if a man can seize the end of a string or thread the other end of which is held by his enemy, he immediately becomes his Dakheel.* If he touch the canvas of a tent, or can even throw his mace towards it, he is the Dakheel of its owner. If he can spit upon a man, or touch any article belonging to him with his teeth, he is Dakhal, unless of course, in case of theft, it be the person who caught him. A woman can protect any number of persons, or even of tents.† If a horseman ride into a tent, he and his horse are Dakhal. A stranger who has eaten with a Shammar, can give Dakheel to his enemy; for instance, I could protect an Aneyza, though there is blood between his tribe and the Shammar. According to Mijwell, any person, by previously calling out " Nuffa" (I renounce), may reject an application for Dakheel.

The Shammar never plunder a caravan within sight of their encampment, for as long as a stranger can see their tents they consider him their Dakheel. If a man who has eaten bread and slept in a tent, steal his host's horse, he is dishonored, and his tribe also, unless they send back the stolen animal. Should the horse die, the thief himself should be delivered up, to be treated as the owner of the stolen property thinks fit. If two enemies meet and exchange the " Salam aleikum " even by mistake, there is peace between them, and they will not fight. It is disgraceful to rob a woman of her clothes ; and if a female be found amongst a party of plundered Arabs, even the enemy of her tribe will give her a horse to ride back to her tents. If a man be pursued by an enemy, or even be on the ground, he can save his life by calling out " Dakheel," unless there be blood between them. It would be considered cowardly and unworthy of a Shammar to deprive an

---

* For the very singular customs as to the confinement and liberation of a *haramy*, or robber, and of the relation between a *rabat* and his *rabiet*, or the captor and the captive, see Burckhardt's Notes on the Bedouins, p. 89. I can bear witness to the truth and accuracy of his account, having during my early wanderings amongst the Bedouins witnessed nearly everything he describes. The English reader can have no correct idea of the habits and manners of the wandering tribes of the Desert, habits and manners probably dating from the remotest antiquity, and consequently of the highest interest, without reading the truthful descriptions of this admirable traveller.

† In the winter of the year of my residence in Babylonia, after an engagement near Baghdad, between the Boraij and the Turkish regular troops, in which the latter were defeated, a flying soldier was caught within sight of an encampment. His captors were going to put him to death, when he stretched his hand towards the nearest tent, claiming the Dakheel of its owner, who chanced to be Sahiman, Mijwell's eldest brother. The Sheikh was absent from home, but his beautiful wife Noura answered to the appeal, and seizing a tent-pole beat off his pursuers, and saved his life. This conduct was much applauded by the Bedouins.

enemy of his camel or horse where he could neither reach water or
an encampment. When Bedouins meet persons in the midst of
the Desert, they will frequently take them within a certain distance
of tents, and, first pointing out their site, then deprive them of
their property.

An Arab who has given his protection to another, whether
formally, or by an act which confers the privilege of Dakheel, is
bound to protect his Dakhal under all circumstances, even to the
risk of his own property and life. I could relate many instances of the greatest sacrifices having been made by individuals,
and even of whole tribes having been involved in war with powerful enemies by whom they have been almost utterly destroyed, in
defence of this most sacred obligation. Even the Turkish rulers
respect a law to which they may one day owe their safety, and
more than one haughty Pasha of Baghdad has found refuge and
protection in the tent of a poor Arab Sheikh, whom, during the
days of his prosperity, he had subjected to every injury and wrong,
and yet who would then defy the government itself, and risk his
very life, rather than surrender his guest. The essence of Arab
virtue is a respect for the laws of hospitality, of which the Dakheel
in all its various forms is but a part.

Amongst the Bedouins who watched our camels was one Saoud,
a poet of renown amongst the tribes. With the exception of a
few ballads that he had formerly composed in honor of Sofuk, and
other celebrated Shammar Sheikhs, he chiefly recited extemporary
stanzas on passing events, or on persons who were present. He
would sit in my tent of an evening, and sing his verses in a wild,
though plaintive, strain, to the great delight of the assembled
guests, and particularly of Mijwell, who, like a true Bedouin, was
easily affected by poetry, especially with such as might touch his
own passion for the unknown lady. He would sway his body to
and fro, keeping time with the measure, sobbing aloud as the poet
sang the death of his companions in war, breaking out into loud
laughter when the burden of the ditty was a satire upon his friends,
making extraordinary noises and grimaces to show his feelings,
more like a drunken man than a sober Bedouin. But when the
bard improvised an amatory ditty, the young chief's excitement was
almost beyond control. The other Bedouins were scarcely less
moved by these rude measures, which have the same kind of effect
on the wild tribes of the Persian mountains. Such verses,
chaunted by their self-taught poets, or by the girls of their encampment, will drive warriors to the combat, fearless of death, or

prove an ample reward on their return from the dangers of the *ghazou* or the fight. The excitement they produce exceeds that of the grape. He who would understand the influence of the Homeric ballads in the heroic ages, should witness the effect which similar compositions have upon the wild nomades of the East. Amongst the Kurds and Lours I have not met with bards who chanted extemporary verses. Episodes from the great historical epics of Persia, and odes from their favorite poets, are recited during war or in the tents of their chiefs. But the art of improvising seems innate in the Bedouin. Although his metre and mode of recitation are rude to European ears, his rich and sonorous language lends itself to this species of poetry, whilst his exuberant imagination furnishes him with endless beautiful and appropriate allegories. The wars between the tribes, the *ghazou*, and their struggles with the Turks, are inexhaustible themes for verse, and in an Arab tent there is little else to afford excitement or amusement. The Bedouins have no books; even a Koran is seldom seen amongst them: it is equally rare to find a wandering Arab who can read. They have no written literature, and their traditional history consists of little more than the tales of a few storytellers who wander from encampment to encampment, and earn their bread by chanting verses to the monotonous tones of a one-stringed fiddle made of a gourd covered with sheep-skin.

The extemporary odes which Saoud sung before us were chiefly in praise of those present, or a good-natured satire upon some of our party.

The day of our departure now drew nigh, and Suleiman Agha, to do us honor, invited us to a general review of the irregular troops under his command. The horsemen of the Milli and Chichi Kurds, and of the Arab tribes who encamped with them, joined the Turkish cavalry, and added to the interest and beauty of the display. The Hyta-Bashis were, as usual, resplendent in silk and gold. There were some high-bred horses in the field; but the men, on the whole, were badly mounted, and the irregular cavalry is daily degenerating throughout the empire. The Turkish Government have unwisely neglected a branch of their national armies to which they owed most of their great victories, and at one time their superiority over all their neighbours. The abolition of the Spahiliks, and other military tenures, has, of course, contributed much to this result, and has led to the deterioration of that excellent breed of horses which once distinguished the Ottoman light cavalry. No effort is now made by the government to keep up the race, and the scanty pay of the irregular troops is not sufficient to enable them to obtain even second-rate animals.   Everything has been sacrificed to the regular army, undoubtedly an essential element of national defence; but in a future war the Turks will probably find reason to regret that they have altogether sacrificed to it the ancient irregular horse.

The Kurds, although encumbered by their long flowing garments and huge turbans, are not bad horsemen.   Mijwell, however, as he scanned the motley crowd with his eagle eye, included them all in one expression of ineffable contempt.

The Tent of the Milli Chief.

Volcanic Cone of Koukab

CHAP. XV.

DEPARTURE FROM THE KHABOUR. — ARAB SAGACITY. — THE HOL. — THE LAKE
OF KHATOUNIYAH. — RETURN OF SUTTUM. — ENCAMPMENT OF THE SHAMMAR.
— ARAB HORSES — THEIR BREEDS — THEIR VALUE — THEIR SPEED. — SHEIKH
FERHAN. — YEZIDI VILLAGES. — FALCONS. — AN ALARM. — ABOU MARIA. —
ESKI MOSUL. — ARRIVAL AT MOSUL. — RETURN OF SUTTUM TO THE DESERT.

MR. HORMUZD RASSAM having sufficiently recovered from his dangerous illness to be able to ride a deloul, and no remains, except pottery and bricks, having been discovered in the mounds of Umjerjeh, we left the encampment of Suleiman Agha on the 29th of April, on our return to Mosul. We crossed the Jerujer near its junction with the Khabour, where two mounds, named Al Hasieha and Abou-Bekr, rise on the left bank of the river.

We again visited the remarkable volcanic cone of Koukab. As we drew near to it, Mijwell detected, in the loose soil, the footprints of two men, which he immediately recognised to be those of Shammar thieves returning from the Kurdish encampments. The sagacity of the Bedouin in determining from such marks, whether of man or beast, and, from similar indications, the tribe, time of passing, and business, of those who may have left them, with many other particulars, is well known.   In this respect he resembles the American Indian, though the circumstances differ under which the two are called upon to exercise this peculiar faculty.   The one seeks or avoids his enemy in vast plains, which, for three-fourths of the year, are without any vegetation ; the other tracks his prey through thick woods and high grass.   This quickness of perception is the result of continual observation and of caution encouraged from earliest youth.   When the warriors of a tribe are engaged in distant forays or in war, their tents and flocks are frequently left to the care of a mere child.   He must receive strangers, amongst whom may be those having claims of blood upon his family, and must guard against marauders, who may be lurking about the encampment.   Every unknown sign and mark must be examined and accounted for.   If he should see the track of a horseman he must ask himself why one so near the dwellings did not stop to eat bread or drink water ? was he a spy; one of a party meditating an attack ? or a traveller, who did not know the site of the tents ?   When did he pass ?   From whence did he come?   Whilst the child in a civilised country is still under the care of its nurse, the Bedouin boy is compelled to exercise his highest faculties, and on his prudence and sagacity may sometimes depend the safety of his tribe.

The expert Bedouin can draw conclusions from the footprints and dung of animals that would excite the astonishment of an European. He will tell whether the camel was loaded or unloaded, whether recently fed or suffering from hunger, whether fatigued or fresh, the time when it passed by, whether the owner was a man of the desert or of the town, whether a friend or foe, and sometimes even the name of his tribe.   I have frequently been cautioned by my Bedouin companions, not to dismount from my dromedary, that my footsteps might not be recognised as those of a stranger ; and my deloul has even been led by my guide to prevent those who might cross our path detecting that it was ridden by one not thoroughly accustomed to the management of the animal.   It would be easy to explain the means, simple enough indeed, by which the Arab of the Desert arrives at these results.   In each case there is a train of logical deduction, merely requiring common acuteness and great experience.

We encamped for the night near the mound of Thenenir, and resumed our journey on the following morning.   Bidding farewell

to the pleasant banks of the Khabour, we struck into the Desert
in the direction of the Sinjar. Extensive beds of gypsum, or
alabaster, such as was used in the Assyrian edifices, formed for
some miles the surface of the plain. Its salt and nitrous exuda-
tions destroy vegetation, unless there be sufficient soil about it to
nourish the roots of herbs; generally, only the cracks and fissures
in the strata are marked by lines of grass and flowers crossing the
plain like the meshes of a many-colored net.

We soon approached a dense mass of reeds and rank herbage,
covering a swamp called the Hol, which extends from the Lake of
Khatouniyah to within a short distance of the Khabour. This
jungle is the hiding-place of many kinds of wild beasts : lions
lurk in it, and in the thick cover the Bedouins find their cubs.
As we drew near to the first spring that feeds the marsh, about
eight miles from Thenenir, we saw a leopard stealing from the
high grass. When pursued, the animal turned and entered the
thickets before the horseman could approach it.

When we reached the head spring of the Hol, the Jebours fired
the jungle, and the flames soon spread far and wide. Long after
we had left the marsh we could hear the crackling of the burning
reeds, and until nightfall the sky was darkened by thick volumes
of smoke.

During our journey an Arab joined us, riding on a deloul, with
his wife. His two children were crammed into a pair of saddle
bags, a black head peeping out of either side. He had quarrelled
with his kinsmen, and was moving with his family and little property to another tribe.

After a six hours' ride we found ourselves upon the margin of a
small lake, whose quiet surface reflected the deep blue of the
cloudless sky. To the south of it rose a line of low undulating
hills, and to the east the furrowed mountain of the Sinjar. On all
other sides was the Desert, in which this solitary sheet of water
lay like a mirage. In the midst of the lake was a peninsula,
joined to the mainland by a narrow causeway, and beyond it a
small island. On the former were the ruins of a town, whose falling walls and towers were doubled in the clear waters. It would
be difficult to imagine a scene more calm, more fair, or more un-
looked for in the midst of a wilderness. It was like fairyland.

The small town of Khatouniyah was, until recently, inhabited
by a tribe of Arabs. A feud, arising out of the rival pretensions
of two chiefs, sprang up amongst them. The factions fought, many
persons were killed, and the place was consequently deserted, one

F. Cooper.

John Murray, Albemarle Street, 1852.

N. Chevalier, lith.

Lake & Island of Thatrungah.

party joining the Tai Arabs near Nisibin, the other the Yezidis of Keraniyah. We traced the remains of cultivation, and the dry water-courses, which once irrigated plots of rice and melon beds. The lake may be about six miles in circumference. From its abundant supply of water, and its central position between the Sinjar and the Khabour, Khatouniyah must at one time have been a place of some importance.

The few remains that exist do not belong to an earlier period than the Arab. The small town occupies the whole of the peninsula, and is surrounded by a wall, rising from the water's edge, with a gate opening on the narrow causeway. The houses were of stone, and the rooms vaulted. In the deserted streets were still standing the ruins of a small bazar, a mosque, and a bath.

The water of the lake, although brackish, like nearly all the springs in this part of the Desert, is not only drinkable, but, according to the Bedouins, exceedingly wholesome for man and beast. It abounds in fish, some of which are said to be of very considerable size. As we approached the Bairakdar, seeing something struggling in a shallow rode to it, and captured a kind of barbel, weighing above twenty pounds. Waterfowl and waders, of various kinds, congregate on the shores. The stately crane and the graceful egret, with its snow-white plumage and feathery crest, stand lazily on its margin; and thousands of ducks and teal eddy on its surface round the unwieldy pelican.

Our tents were pitched on the very water's edge. At sunset a few clouds which lingered in the western sky were touched with the golden rays of the setting sun. The glowing tints of the heavens, and the clear blue shadows of the Sinjar hills, mirrored in the motionless lake, imparted a calm to the scene which well matched with the solitude around.

We had scarcely resumed our march in the morning when we spied Suttum and Khoraif coming towards us, and urging their fleet mares to the top of their speed. A Jebour, leaving our encampment at Umjerjeh, when Hormuzd was dangerously ill, had spread a report * in the Desert, that he was actually dead.

---

* The manner in which reports are spread and exaggerated in the Desert is frequently highly amusing. In all encampments there are idle vagabonds who live by carrying news from tribe to tribe, thereby earning a dinner and spending their leisure hours. As soon as a stranger arrives, and relates anything of interest to the Arabs, some such fellow will mount his ready-saddled deloul, and make the best of his way to retail the news in a neighbouring tent, from

To give additional authenticity to his tale he had minutely described the process by which my companion's body had been first
salted, and then sent to Frankistan in a box, on a camel. Suttum,
as we met, showed the most lively signs of grief; but when he saw
the dead man himself restored to life, his joy and his embraces
knew no bounds.

We rode over a low undulating country, at the foot of the Sinjar
hills, every dell and ravine being a bed of flowers. About five
miles from Khatouniyah we passed a small reedy stream, called
Suffeyra, on which the Boraij (Suttum's tribe) had been encamped
on the previous day. They had now moved further into the plain,
and we stopped at their watering-place, a brackish rivulet called
Sayhel, their tents being about three miles distant from us in the
Desert. We pitched on a rising ground immediately above the
stream. Beneath us was the golden plain, swarming with moving
objects. The Khorusseh, and all the tribes under Ferhan, had
now congregated to the north of the Sinjar previous to their summer migration to the pastures of the Khabour. Their mares,
camels, and sheep came to Sayhel for water, and during the whole
day there was one endless line of animals passing to and fro before
our encampment. I sat watching them from my tent. As each
mare and horse stopped to drink at the troubled stream, Suttum
named its owner and its breed, and described its exploits. The
mares were generally followed by two or three colts, who are
suffered, even in their third year, to run loose after their dams, and
to gambol unrestrained over the plain. It is to their perfect freedom whilst young that the horses of the Desert owe their speed and
the suppleness of their limbs.

It may not be out of place to add a few remarks on the subject
of Arab horses. The Bedouins, as it is well known, divide their
thorough-breds into five races, descended, as some declare, from
the five favourite mares of the Prophet. The names, however, of
these breeds vary amongst different tribes. According to Suttum,
who was better acquainted with the history and traditions of the
Bedouins than almost any Arab I ever met, they are all derived

whence it is carried, in the same way, to others. It is extraordinary how rapidly
a report spreads in this manner over a very great distance. Sofuk sent to in-
form the British resident at Baghdad, of the siege and fall of Acre, many days
before the special messenger dispatched to announce that event reached the city;
and I have frequently rejected intelligence received from Bedouins, on account
of the apparent impossibility of its coming to me through such a source, which
has afterwards proved to be true.

from one original stock, the Koheyleh, which, in course of time, was divided, after the names of celebrated mares, into the following five branches:— Obeyan Sherakh, Hedba Zayhi, Manekia Hedrehji, Shouaymah Sablah, and Margoub.* These form the *Kamse*, or the five breeds, from which alone entire horses are chosen to propagate the race. From the *Kamse* have sprung a number of families no less noble, perhaps, than the original five ; but the Shammar receive their stallions with suspicion, or reject them altogether. Among the best known are the Wathna Khersan, so called from the mares being said to be worth their weight in gold ; (noble horses of this breed are found amongst the Arab tribes inhabiting the districts to the east of the Euphrates, the Beni Lam, Al Kamees, and Al Kithere ;) Khalawi, thus named from a wonderful feat of speed performed by a celebrated mare in Southern Mesopotamia ; Jaiaythani†, and Julfa. The only esteemed race in the Desert which, according to Suttum, cannot be traced to the *Kamse*, is the Saklawi, although considered by the Shammar and by the Bedouins of the Gebel Shammar, as one of the noblest, if not the noblest, of all. It is divided into three branches, the most valued being the Saklawi Jedran, which is said to be now almost extinct. The agents of Abbas Pasha, the Viceroy of Egypt, sent into all parts of the Desert to purchase the best horses, have especially sought for mares of this breed. The prices given for them would appear enormous even to the English reader. A Sheikh of the great tribe of the Al Dhofyr was offered and refused for a mare no less than 1200*l.*, the negociation being carried on through Faras, Sheikh of the Montefik, who received handsome presents for the trouble he had taken in the matter. As much as a thousand pounds is said to have been given to Sheikhs of the Aneyza for well-known mares. So that, had the Pasha's challenge been accepted, the best blood in Arabia would have been matched against the English racer. During my residence in the Desert I saw several horses which were purchased for the Viceroy.

To understand how a man, who has perhaps not even bread

---

* According to Burckhardt, the five are, Taueyse, Manekia, Koheyleh, Saklawi, and Julfa. He probably received these names from the Arabs of the Hedjaz, who are less acquainted with the breeds of horses than the Shammar or Aneyza Bedouins. (Notes on Arabs, p. 116., but at p. 253. he observes, that the Nedjd Arabs do not reckon the Manekia and Julfa in the Kamse.)

† A well-known horse, named Merjian, long in my possession, and originally purchased from the Arabs by my friend Mr. Ross, was of this breed.

to feed himself and his children, can withstand the temptation of such large sums, it must be remembered that, besides the affection proverbially felt by the Bedouin for his mare, which might, perhaps, not be proof against such a test, he is entirely dependent upon her for his happiness, his glory, and, indeed, his very existence. An Arab possessing a horse unrivalled in speed and endurance, and it would only be for such that prices like those I have mentioned would be offered, is entirely his own master, and can defy the world. Once on its back, no one can catch him. He may rob, plunder, fight, and go to and fro as he lists. He believes in the word of his Prophet, "that noble and fierce breeds of horses are true riches." Without his mare, money would be of no value to him. It would either become the prey of some one more powerful and better mounted than himself, would be spent in festivities, or be distributed amongst his kinsmen. He could only keep his gold by burying it in some secret place, and of what use would it then be to one who is never two days in the same spot, and who wanders over a space of three or four hundred miles in the course of a few months? No man has a keener sense of the joys of liberty, and a heartier hatred of restraint, than the true Bedouin. Give him the Desert, his mare, and his spear, and he will not envy the wealth and power of the greatest of the earth. He plunders and robs for the mere pleasure and excitement which danger and glory afford. All he takes he divides amongst his friends, and he gladly risks his life to get that which is spent in an hour. An Arab will beg for a whole day for a shirt or a kerchief, and, five minutes after he has obtained it, he will give it to the first person who may happen to admire it.

A mare is generally the property of two or more persons, who have a share in her progeny, regulated by custom, and differing according to the tribe. All the offspring of five celebrated mares belong by usage to the head of the sub-tribe of the Ahl-Mohammed, and whenever horses descended from them are captured by the Shammar from the Aneyza or other tribes, they may be claimed by him. They are merely brought to Ferhan, the present chief, as a matter of form, and he returns them to their captors. Sofuk (his father), however, would frequently insist upon his right, and bestow valuable mares thus obtained upon his immediate retainers. The five breeds are Saklawi Jedran, Emlayah, Margoub, Hedba Enzaii, and Hamdaniyah.

The largest number of horses, as well as those of the most esteemed breeds, are still to be found, as in the time of Burckhardt, amongst the tribes who inhabit Mesopotamia and the great plains watered by the Euphrates and Tigris. These rich pastures, nourished by the rains of winter and spring, the climate, and — according to the Arabs — the brackish water of the springs rising in the gypsum, seem especially favorable to the rearing of horses. The best probably belong to the Shammar and Aneyza tribes, a rivalry existing between the two, and fame giving the superiority sometimes to one, sometimes to the other. The mares of the Aneyza have the reputation of being the largest and most powerful, but as the two tribes are always at war, plundering and robbing one another almost daily, their horses are continually changing owners.

The present Sheikh of the Gebel Shammar, Ibn Reshid, has, I am informed, a very choice stud of mares of the finest breeds, and their reputation has spread far and wide over the Desert. The Nawab of Oude, the Ekbal-ed-Doulah, a good judge of horses, who had visited many of the tribes, and had made the pilgrimage to the holy cities by the little frequented route through the interior of Nedjd, assured me that the finest horses he had ever seen were in the possession of the Shereef of Mecca. The Indian market is chiefly supplied by the Montefik tribes inhabiting the banks of the lower Euphrates; but the purity of their stock has been neglected in consequence of the great demand, and a Montefik horse is not valued by the true Bedouin. Horse-dealers, generally of the mixed Arab tribe of Agayl, pay periodical visits to the Shammar and Aneyza to purchase colts for exportation to India. They buy horses of high caste, which frequently sell for large sums at Bombay. The dealers pay, in the Desert, from 30l. to 150l. for colts of two, three, and four years. The Agayles attach less importance to blood than the Bedouins, and provided the horse has points which seem suited to the Indian market, they rarely ask his pedigree. The Arabs hence believe that Europeans know nothing of blood, which with them is the first consideration.

The horses thus purchased are sent to Bombay by native vessels at a very considerable risk, whole cargoes being lost or thrown overboard during storms every year. The trade is consequently very precarious, and less flourishing now than it used to be. With the exception of one or two great dealers at Baghdad and Busrah, most of those who have been engaged in it have been ruined.

The Arab horse is more remarkable for its exquisite symmetry and beautiful proportions, united with wonderful powers of en-

durance, than for extraordinary speed.   I doubt whether any
Arab of the best blood has ever been brought to England.   The
difficulty of obtaining them is so great, that they are scarcely ever
seen beyond the limits of the Desert.

Their color is generally white, light or dark grey, light chesnut,
and bay, with white or black feet.   Black is exceedingly rare, and
I never remember to have seen dun, sorrel, or dapple.   I refer, of
course, to the true-bred Arab, and not to the Turcoman or to
Kurdish and Turkish races, which are a cross between the Arab
and Persian.

Their average height is from 14 hands to 14¾, rarely reaching
15; I have only seen one mare that exceeded it.   Notwithstanding the smallness of their stature they often possess great
strength and courage.   I was credibly informed that a celebrated mare of the Manekia breed, now dead, carried two men
in chain armour beyond the reach of their Aneyza pursuers.
But their most remarkable and valuable quality is the power of
performing long and arduous marches upon the smallest possible
allowance of food and water.   It is only the mare of the wealthy
Bedouin that gets even a regular feed of about twelve handfuls of
barley, or of rice in the husk, once in twenty-four hours.   During
the spring alone, when the pastures are green, the horses of the
Arabs are sleek and beautiful in appearance.   At other times they
eat nothing but the withered herbs and scanty hay gathered from
the parched soil, and are lean and unsightly.   They are never
placed under cover during the intense heat of an Arabian summer,
nor protected from the biting cold of the Desert winds during
winter.   The saddle is rarely taken from their backs, nor are they
ever cleaned or groomed.   Thus apparently neglected, they are
but skin and bone, and the townsman marvels at seeing an animal,
which he would scarcely take the trouble to ride home, valued
almost beyond price.   Although docile as a lamb, and requiring no
other guide than the halter, when the Arab mare hears the war-cry
of the tribe, and sees the quivering spear of her rider, her eyes
glitter with fire, her blood-red nostrils open wide, her neck is nobly
arched, and her tail and mane are raised and spread out to the
wind.   The Bedouin proverb says, that a high-bred mare when at
full speed should hide her rider between her neck and her tail.

The Shammar Bedouins give their horses, particularly when
young, large quantities of camels' milk.   I have heard of mares
eating raw flesh, and dates are frequently mixed with their food

by the tribes living near the mouth of the Euphrates. The
Shammar and Aneyza shoe their horses if possible, and wandering farriers regularly visit their tents. If an Arab cannot
afford to shoe his mare entirely, he will shoe her fore-feet. The
Chaab (or Kiab) do not usually shoe their horses. The shoes, like
those used in all parts of the East, consist of a thin iron plate
covering the whole foot, except a small hole in the centre. They
are held by six nails, are clumsily made, and usually more clumsily
put on. The Arab horse has but two ordinary paces, a quick
and easy walk, sometimes averaging between four and five miles
an hour, and a half running canter. The Bedouin rarely puts his
mare to full speed unless pursued or pursuing. In racing, the
Arabs, and indeed Easterns in general, have no idea that the
weight carried by the rider makes any difference.

I have frequently pointed out to the Turkish authorities the
fitness of the rich plains watered by the Euphrates and Tigris for
a government stud. It would be difficult, in the present state of
things, to induce the Bedouins to place themselves under the
restraint necessary to such an undertaking; but there are many
half-sedentary tribes, who are well acquainted with the management of horses, and know the best pastures of the Desert. If
properly protected and supported they could defy the Bedouins,
and maintain permanent stations in any part of Mesopotamia. A
noble race of horses, now rapidly becoming extinct, for the breed
of true Arabs is, I believe, daily deteriorating *, and their number
decreasing, might prove a source of strength and wealth to the
empire.

In the evening, as I was seated before my tent, I observed a
large party of horsemen and riders on delouls approaching our
encampment. They stopped at the entrance of the large pavilion
reserved for guests, and picketing their mares, and turning loose
their dromedaries adorned with gay trappings, seated themselves
on the carpets. The chiefs were our old friends, Mohammed
Emin and Ferhan, the great Shammar Sheikh. We cordially
embraced after the Bedouin fashion. I had not seen Ferhan since

---

* Burckhardt states that the number of horses in Arabia did not in his time
exceed 50,000. It has probably considerably decreased since. The defeat of
the Wahabys, the conquest of Arabia, and the occupation of Syria by the
Egyptians, have contributed greatly both to the diminution and deterioration of
the race. I have had no means of ascertaining, even proximately, the number
of horses belonging to such tribes as the Shammar and Aneyza.

the treacherous murder of his father by Nejib Pasha of Baghdad*, to which he alluded with touching expressions of grief, bewailing his own incompetency to fill Sofuk's place, and to govern the divided tribe. He was now on his way with the Jebour Sheikh to recover, if possible, the government treasure, plundered by the Hamoud, for which, as head of the Shammar, he was held responsible by the Porte.

After they had eaten of the feast we were able to prepare for them, they departed about sunset for the tents of the Jebours. I embraced Mohammed Emin for the last time, and saw him no more during my residence in Assyria.

The scene at the watering-place at Sayhel was so changing and varied, that I had little cause to regret a delay of two days on the spot. Long before dawn the sheep and camels gathered round the spring, and it was night before the last shepherd had driven away his flocks. My tents, moreover, were filled with Bedouins from various tribes, who supplied me with information, and entertained me with traditions and tales of the Desert.

On the 4th of May we made a short day's journey of five hours to a beautiful stream issuing from the Sinjar hill, beneath the village of Khersa or Chersa. A Bedouin of the Boraij tribe accompanied us riding on a swift white dromedary of a true Nedjd breed. This animal was scarcely taller than a large English horse. It had been captured by its present owner with another of the same race, and several ordinary camels, during a three months' *ghazou*, or plundering expedition, which he had undertaken with the warriors of his tribe into the interior of Arabia.

Leaving the plain, which was speckled as far as the eye could reach with the flocks and tents of the Bedouins, we skirted the very foot of the Sinjar. Khersa had been deserted by its inhabitants, who had rebuilt their village higher up on the side of the hill.

Since the loss of Hattab, Suttum had never ceased pining for a falcon worthy to take his place. He had been counting the hours to his visit to this part of the Sinjar, known only to yield to the borders of the Persian Gulf in producing the finest and bravest hawks for the chase. The Yezidis carefully preserve their nests as hereditary property, in which certain families have a vested interest. The young birds, with the exception of one left to prevent the parents deserting the place, are taken when half-fledged.

* Nineveh and its Remains, vol. i. p. 113.

They are then sold, generally to the Bedouins, for comparatively large prices, from five to twenty gazees (1*l.* to 4*l.*) being given according to the reputation of the nest, whose peculiar qualities are a matter of notoriety amongst true sportsmen. Three birds only, in each brood, are thought worthy of being trained. The first hatched is the most esteemed, and is called " Nadir;" the second ranks next, and is known as the " Azeez." A hunting-hawk of the Sinjar species brought up by hand is called " Charkh." It strikes its quarry on the ground, and not in the air, and is principally flown at gazelles, bustards, and hares. The young are sold by weight. Suttum sat, scales in hand, examining the unfledged birds with the eye of a connoisseur, and weighing them with scrupulous care. All that were brought to him were, however, rejected, the Sheikh protesting that the Infidels were cheating him, and had sold all the nadirs and azeezs to more fortunate Bedouins.

Next day we made but little progress, encamping near a spring under the village of Aldina, whose chief, Murad, had now returned from his captivity. Grateful for my intercession in his behalf, he brought us sheep and other provisions, and met us with his people as we entered the valley. The Mutesellim was in his village collecting the revenues, but the inhabitants of Nogray had refused to contribute the share assigned to them, or to receive the governor. He begged me to visit the rebellious Yezidis, and the whole day was spent in devising schemes for a general peace. At length the chiefs consented to accompany me to Aldina, and, after some reduction in the Salian, to pay the taxes.

During the negociations, Suttum, surrounded by clamorous Yezidis, was sitting in the shade, examining and weighing unfledged hawks. At length three were deemed worthy of his notice : one being pretty well advanced in days was sent to his tent for education, under the charge of the rider of the Nedjd deloul. The others, being yet in a weak state, were restored to the nest, to be claimed on his return from Mosul. The largest bird, being a very promising specimen, cost five gazees or 1*l.* ; the others, three gazees and a half, as the times were hard, and the tax-gatherers urgent for ready money.

We rode on the following day for about an hour along the foot of the Sinjar hill, which suddenly subsides into a low undulating country. The narrow valleys and ravines were blood-red with gigantic poppies. The Bedouins adorned the camels and horses with the scarlet flowers, and twisted them into their own head-dresses and long garments. Even the Tiyari dressed themselves up in the gaudy trappings of nature, and as we journeyed chanting an Arab war-song, we resembled the return of a festive procession from some sacrifice of old. During our weary marches under a burning sun, it required some such episodes to keep up the drooping spirits of the men, who toiled on foot by our sides. Poetry and flowers are the wine and spirits of the Arab; a couplet is equal to a bottle, and a rose to a dram, without the evil effects of either. Would that in more civilised climes the sources of excitement were equally harmless!

The large artificial mound of Tel Shour rose in the plain to the right of us. About nine miles from our last encamping place we crossed a stream of sweet water named Aththenir, and stopped soon after for the day in the bosom of the hills, near some reedy ponds, called Fukka, formed by several springs. As this was a well-known place of rendezvous for the Bedouins when out on the ghazou, Suttum displayed more than usual caution in choosing the place for our tents, ascending with Khoraith a neighbouring peak to survey the country and scan the plain below.

In the afternoon the camels had wandered from the encampment in search of grass, and we were reposing in the shade of our tents, when we were roused by the cry that a large body of men were to be seen in the distance. The Bedouins immediately sought to drive back their beasts. Suttum unplatting his long hair, and shaking it in hideous disorder over his head and face, and baring his arms to the shoulder, leapt with his quivering spear into the saddle. Having first placed the camp in the best posture of defence I was able, I rode out with him to reconnoitre. But our alarm was soon quieted. The supposed enemy proved to be a party of poor Yezidis, who, taking advantage of our caravan, were going to Mosul to seek employment during the summer.

In the evening Suttum inveighed bitterly against a habit of some travellers of continually taking notes before strangers. I endeavoured to explain the object and to remove his fears. " It is all very well," said the Sheikh, " and I can understand, and am willing to believe, all you tell me. But supposing the Turks, or any body else, should hereafter come against us, there are many foolish and suspicious men in the tribe, and I have enemies, who would say that I had brought them, for I have shown you everything. You know what would be the consequences to me of such a report. As for you, you are in this place to-day, and 100 days' journey off to-morrow, but I am always here. There is not a plot of grass or a spring that that man (alluding to one of our party) does not write down." Suttum's complaints were not unreasonable, and travellers cannot be too cautious in this respect, when amongst independent tribes, for even if they do not bring difficulties upon themselves, they may do so upon others.

We had a seven hours' ride on the delouls, leaving the caravan to follow, to the large ruin of Abou Maria *, passing through Tel Afer. The Jehesh were encamped about two miles from the place. My workmen had excavated for some time in these remarkable mounds, and had discovered chambers and several enormous slabs of Mosul marble, but no remains whatever of sculpture. They had, however, dug out several entire bricks bearing the name of the founder of the north-west palace at Nimroud, but unaccompanied by that of any town or temple. The ruins are of considerable extent, and might, if fully explored, yield some valuable relics.

A short ride of three hours brought us to Eski (old) Mosul, on the banks of the Tigris. According to tradition this is the original site of the city. There are mounds, and the remains of walls, which are probably Assyrian. Upon them are traces of buildings of a far more recent period. My workmen had opened several trenches and tunnels in the principal ruin, and at a subsequent period Awad, with a party of Jehesh, renewed the excavations in it, but no relics throwing any light upon its history were discovered.

Mosul was still nine caravan hours distant, and we encamped the next night at Hamaydat, where many of our friends came out to meet us. On the 10th of May we were again within the walls of the town, our desert trip having been accomplished without any mishap or accident whatever.

Suttum left us two days after for his tents, fearing lest he should be too late to join the warriors of the Khorusseh, who had planned a grand *ghazou* into Nedjd. They were to be away for thirty days, and expected to bring back a great spoil of mares, dromedaries, and camels. As for three days they would meet with no wells, they could only ride their delouls, each animal carrying a spearman and a musketeer, with their skins of water and a scanty stock of provisions. They generally contrive to return from these expeditions with considerable booty. Suttum urged me to accompany them; but I had long renounced such evil habits, and other

---

* I have elsewhere described the ruins and springs of Abou Maria. (Nineveh and its Remains, vol. i. p. 312.)

occupations kept me in Mosul. Finding that I was not to be persuaded, and that the time was at length come for us to part, he
embraced me, crammed the presents we had made to himself and
his wives into his saddle-bags, and, mounting his deloul, rode off
with Mijwell towards the Desert.

Arab Camels